Current Topics in Microbiology and Immunology

216

Editors

Springer
Berlin
Heidelberg
New York
Barcelona
Budapest
Hong Kong
London
Milan
Paris
Santa Clara
Singapore
Tokyo

Pathology of Septic Shock

Edited by E.T. Rietschel and H. Wagner

With 34 Figures

Springer

Professor Dr. E.T. Rietschel

Forschungszentrum Borstel
Zentrum für Medizin und Biowissenschaften
Parkallee 22
D-23845 Borstel
Germany

Professor Dr. Hermann Wagner

Technische Universität München
Institut für Medizinische
Mikrobiologie, Immunologie und Hygiene
Trogerstr. 9
D-81675 München
Germany

Cover illustration: Symbolic space filling models of endotoxin and superantigen (SEB) representing two types of toxins often associated with the sepsis syndrome [Swaminathan et al. (1992) Nature 359: 801; Kastowsky et al. (1992) J Bacteriol 174: 4798]

Cover design: Design & Production, Heidelberg

ISSN 0070–217X
ISBN 3–540–61026–X Springer-Verlag Berlin Heidelberg New York

Library of Congress Catalog Card Number 15 - 12910
Printed in Germany

Typesetting: Scientific Publishing Services (P) Ltd, Madras
SPIN: 10495142 27/3020/SPS – 5 4 3 2 1 0 – Printed on acid-free paper

Preface

Severe sepsis and septic shock are the most serious complications of bacterial infections. Both gram-positive and gram-negative bacteria can trigger these extreme inflammatory responses and, by so doing, cause substantial morbidity and mortality. In the United States alone, over 400 000 patients suffer from septicaemia each year, and approximately 100 000 of these patients die despite optimal intensive care and modern antimicrobial therapy.

These dramatic figures have prompted intensive research to define the bacterial and host factors involved in the septic response. Scientists from many disciplines, including chemistry, physics, biology, medical microbiology, immunology, and pharmacology, have worked closely with clinicians to achieve rapid and profound progress. To translate this newly acquired knowledge into clinical practice, clinical trials have also been performed to evaluate numerous new therapeutic drugs. The disappointing results from these trials have underscored a major lesson, namely, that sepsis constitutes an extremely complex syndrome and that basic and clinical research must be greatly intensified in order to illuminate its molecular mechanisms.

At this stage, the editors of the present volume of *Current Topics in Microbiology and Immunology* considered it would be rewarding to compile a volume summarizing our present basic and clinical knowledge on sepsis. Our particular gratitude extends to those international experts who have followed our invitation and elaborated on particular areas of the basic and clinical aspects of this field.

Their contributions provide a true picture of "the state of the art" and should be a useful reference for design of new sepsis-related research projects. We also thank Doris Walker, Springer-Verlag, for her advice, professional help, and excellent guidance in editing this volume.

Bacterial infections and sepsis continue to represent major threats to human health. We live in an era in which bacterial antimicrobial resistance is increasing and immunosuppressed patients are becoming more common, so that sepsis will have even more opportunities to injure and kill. Sepsis kills quickly and often dramatically. Civilized society should not allow this to happen.

Borstel/München

ERNST T. RIETSCHEL
HERMANN WAGNER

List of Contents

J.L. Vincent
Definition and Pathogenesis of Septic Shock 1

P. Brandtzaeg
Significance and Pathogenesis of Septic Shock 15

E.T. Rietschel, H. Brade, O. Holst, L. Brade, S. Müller-Loennies, U. Mamat, U. Zähringer, F. Beckmann, U. Seydel, K. Brandenburg, A.J. Ulmer, T. Mattern, H. Heine, J. Schletter, H. Loppnow, U. Schönbeck, H.-D. Flad, S. Hauschildt, U.F. Schade, F. Di Padova, S. Kusumoto, and R.R. Schumann
Bacterial Endotoxin: Chemical Constitution, Biological Recognition, Host Response, and Immunological Detoxification 39

K. Heeg, T. Miethke, and H. Wagner
Superantigen-Mediated Lethal Shock: The Functional State of Ligand-Reactive T Cells 83

S. Bhakdi, I. Walev, D. Jonas, M. Palmer, U. Weller, N. Suttorp, F. Grimminger, and W. Seeger
Pathogenesis of Sepsis Syndrome: Possible Relevance of Pore-Forming Bacterial Toxins .. 101

D. Pajkrt and S.J.H. van Deventer
The Cellular Response in Sepsis 119

C.A. Dinarello
Cytokines as Mediators in the Pathogenesis of Septic Shock 133

P. Brouckaert and W. Fiers
Tumor Necrosis Factor and the Systemic Inflammatory Response Syndrome 167

T.J. Evans and J. Cohen
Mediators: Nitric Oxide and Other Toxic
Oxygen Species 189

L.G. Thijs, A.B.J. Groeneveld, and C.E. Hack
Multiple Organ Failure in Septic Shock 209

S. Bahrami, H. Redl, Y.-M. Yao, and G. Schlag
Involvement of Bacteria/Endotoxin Translocation
in the Development of Multiple Organ Failure 239

E. Faist
The Mechanisms of Host Defense
Dysfunction Following Shock and Trauma 259

M. Pollack and C.A. Ohl
Endotoxin-Based Molecular Strategies
for the Prevention and Treatment
of Gram-Negative Sepsis and Septic Shock 275

D. Heumann and M.P. Glauser
Anticytokine Strategies for the Treatment
of Septic Shock: Relevance of Animal Models 299

Subject Index 313

List of Contributors

(Their addresses can be found at the beginning of their respective chapters.)

BAHRAMI, S. 239
BECKMANN, F. 39
BHAKDI, S. 101
BRADE, H. 39
BRADE, L. 39
BRANDENBURG, K. 39
BRANDTZAEG, P. 15
BROUCKAERT, P. 167
COHEN, J. 189
DI PADOVA, F. 39
DINARELLO, C.A. 119
EVANS, T.J. 189
FAIST, E. 259
FIERS, W. 167
FLAD, H.-D. 39
GLAUSER, M.P. 299
GRIMMINGER, F. 101
GROENEVELD, A.B.J. 209
HACK, C.E. 209
HAUSCHILDT, S. 39
HEEG, K. 83
HEINE, H. 39
HEUMANN, D. 299
HOLST, O. 39
JONAS, D. 101
KUSUMOTO, S. 39
LOPPNOW, H. 39
MAMAT, U. 39
MATTERN, T. 39
MIETHKE, T. 83
MÜLLER-LOENNIES, S. 39
OHL, C.A. 275
PAJKRT, D. 119
PALMER, M. 101
POLLACK, M. 275
REDL, H. 239
RIETSCHEL, E.T. 39
SCHADE, U.F. 39
SCHLAG, G. 239
SCHLETTER, J. 39
SCHÖNBECK, U. 39
SCHUMANN, R.R. 39
SEEGER, W. 101
SEYDEL, U. 39
SUTTORP, N. 101
THIJS, L.G. 209
ULMER, A.J. 39
VAN DEVENTER, S.J.H. 119
VINCENT, J.L. 1
WAGNER, H. 83
WALEV, I. 101
WELLER, U. 101
YAO, Y.-M. 239
ZÄHRINGER, U. 39

Definition and Pathogenesis of Septic Shock

J.L. Vincent

1 Definition, Characteristics, and Epidemiology . . . 1
1.1 Incidence . . . 3
1.2 Microbiology . . . 3
1.3 Source . . . 4
2 Clinical Interpretation . . . 4
2.1 Signs of Sepsis . . . 5
2.1.1 Fever . . . 5
2.1.2 White Blood Cell Count . . . 5
2.1.3 Tachypnea . . . 5
2.1.4 Tachycardia . . . 5
2.1.5 Reduced Vascular Tone . . . 5
2.1.6 Cytokines . . . 6
2.2 Signs of Shock . . . 6
2.2.1 Hemodynamic Alterations . . . 6
2.2.2 Myocardial Depression . . . 7
2.2.3 Altered Vasculature . . . 7
2.2.4 Altered Organ Perfusion . . . 7
2.2.5 Blood Lactate . . . 8
3 Organ Dysfunction/Failure . . . 9
3.1 Lung . . . 9
3.2 Coagulation . . . 10
3.3 Liver . . . 10
3.4 Cardiovascular . . . 10
3.5 Central Nervous System . . . 10
3.6 Kidney . . . 11
4 Prognosis . . . 11
References . . . 11

1 Definition, Characteristics, and Epidemiology

Septic shock is the most severe manifestation of infection and appears to be increasingly common, especially in the intensive care unit (ICU). Its incidence is difficult to determine exactly, however, because of the many different definitions of sepsis and its associated complications and because of the varying

Department of Intensive Care, Erasme University Hospital, Route de Lennik 808, Free University of Brussels, 1070 Brussels, Belgium incidences reported for the different populations studied. The following features are commonly seen as defining sepsis:

- Fever (temperature > 38 °C) or sometimes hypothermia (temperature <36 °C)
- Tachycardia (>90 bpm)
- Respiratory alkalosis ($PaCO_2$ < 35 mmHg), tachypnea (>20 breaths per minute)
- Low systemic vascular resistance, increased cardiac output
- Hyperleukocytosis (>12 000 cells/mm^3, with "left shift") or sometimes leukopenia (< 4000 cells/mm^3)
- Isolated thrombocytopenia or disseminated intravascular coagulation
- Increased oxygen consumption, elevated cellular metabolism
- Increased insulin requirements
- Signs of inflammation: increased sedimentation rate, elevated C-reactive protein and fibrinogen levels
- Elevated cytokine levels [tumor necrosis factor (TNF), interleukin (IL) –1, –6, –8 etc.]
- Ophthalmic and/or cutaneous manifestations
- Renal failure, adult respiratory distress syndrome (ARDS), mental obtundation, and other organ dysfunction

As in other forms of acute circulatory failure, septic shock is best described as an imbalance between oxygen demand and oxygen delivery (Fig. 1). A complex mediator network has been implicated in the three major alterations characterizing septic shock: impaired oxygen extraction, increased oxygen needs, and altered myocardial contractility. These abnormalities are described in greater detail below. Other factors than infection may augment the mediator response, such as pain, anxiety, tissue trauma, and severe heart failure (Marchant et al. 1995).

Mortality in the early stages of septic shock is generally associated with profound cardiovascular prolapse (Ruokonen et al. 1993). Despite seemingly adequate resuscitation, at least globally, many septic shock patients die at a later stage due to multiple organ failure (MOF). Ongoing oxygen defects due to incomplete resuscitation at the microvascular level are largely responsible. Splanchnic ischemia may be particularly important since the splanchnic circula-

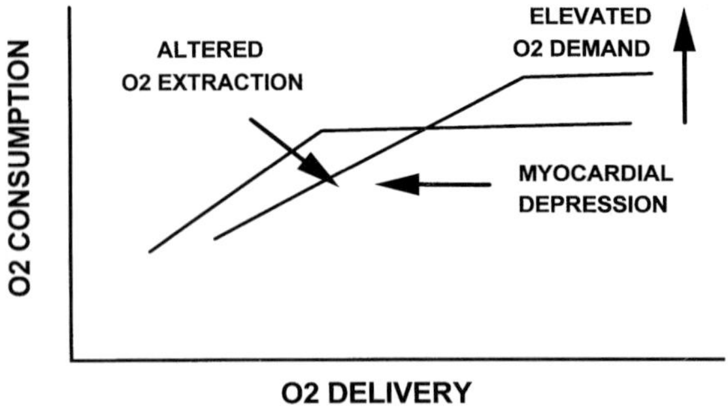

Fig. 1. Oxygen delivery and oxygen consumption in septic shock

tion is particularly at risk and the gut is an immune-rich organ. It has been proposed that splanchnic ischemia results in a breach of the gut mucosal barrier, with secondary translocation of bacteria and their products. A vicious circle of inflammation, ischemia, translocation, and heightened inflammation may be significant in the development of MOF following septic shock. The importance of translocation in humans is still under debate. Nevertheless, even in the absence of translocation the gut is capable of releasing a large quantity of inflammatory mediators, especially in the presence of gut ischemia.

1.1 Incidence

Reports of septicemia more than doubled between 1979 and 1987, from 74 to 176 cases per 100 000 persons (Anonymous 1990). Although such figures depend on the definition used, a reasonable estimate is that sepsis develops in about 1% of hospitalized patients but in 20%–30% of ICU patients.

Despite our improved understanding of the mechanisms involved and the effectiveness of various treatments the incidence of septic shock continues to increase. This is largely because the ICU population contains an increasing proportion of elderly, more compromised, and immunosuppressed patients who have increased risk of infection. Furthermore, advances in modern therapies continue to push the limits of the human system and employ more invasive procedures. These factors may explain why, as some reports suggest, the severity of septic shock has increased, and why mortality from septic shock remains between 40% and 60% in most studies despite the therapeutic advances which have been made (Friedman et al. 1995).

1.2 Microbiology

Traditionally the onset of sepsis was attributed to the destructive effects of the invading micro-organism, but recent improvements in our comprehension of the immunological sequence involved have implicated a complex network of host-synthesized mediators. It should be emphasized that a similar mediator response can be triggered by other events than an infection (Marchant et al. 1995). Indeed it is well documented that infection can be documented in fewer than half of the patients who "look septic" in the posttrauma or injury period. Similarly, the presence of positive blood cultures is not required to establish the diagnosis of septic shock.

Older studies suggested that gram-negative organisms are more commonly associated with septic shock than are gram-positive; however, more recent studies have shown that gram-positive organisms can be just as common. This probably reflects a general change in microbiology related to ongoing antibiotic usage, with a progressive emergence of gram-positive organisms.

While both gram-positive and gram-negative organisms can participate in septic shock, there is no real difference between their influences on hemodynamic presentation or prognosis. A clinically relevant canine model of septic shock (NATANSON et al. 1989a) showed that animals with measurable endotoxemia (*Escherichia coli)* exhibited similar cardiovascular responses to those animals without measurable endotoxemia (S*taphylococcus aureus).* The presence or absence of positive blood cultures seems to have little influence on the course or prognosis of sepsis.

As with gram-negative organisms, gram-positive organisms have been used to invoke septic shock in animals (NATANSON et al. 1989a). Endotoxin has also been used in many experimental models where it has provoked biological effects similar to those observed during septic shock (fever, IL-1 and TNF elevation, complement activation; NATANSON et al. 1989b; ZHANG et al. 1995). Endotoxin has also been used to reproduce the effects of sepsis in human studies involving volunteers. Endotoxin administration resulted in a hyperdynamic cardiovascular condition with dilation and ventricular depression similar to those observed in clinical septic shock (SUFFREDINI et al. 1989). The significance of endotoxemia in patients with septic shock is disputed. About 50% of septic patients may have endotoxin levels in their blood. Endotoxemia may be transient, so that its prevalence depends on the frequency of blood sampling (DANNER et al. 1991).

1.3 Source

The source of septic shock has probably also undergone an evolution. In the 1950s and 1960s the most common sources were often due to invasive procedures. Nowadays urinary tract infections are less common and are also considered more benign. Computed tomography of the abdomen helps to diagnose abdominal complications early. In fact, pulmonary infections are now more common. This may at least in part be associated with the prolonged use of mechanical ventilation.

2 Clinical Interpretation

Acute circulatory failure frequently presents as arterial hypotension. Systolic arterial pressure is typically less than 90 mmHg. To avoid the inclusion of patients with transient hypovolemia the definition includes hypotension which is not immediately responsive to intravenous fluids therapy so that inotropic support is generally required. Signs of abnormal tissue perfusion, such as reduced diuresis (urine output less than 20 ml/h), altered mentation, and increased blood lactate levels, are also present.

2.1 Signs of Sepsis

2.1.1 Fever

Fever is not always present, and hypothermia may even be present in some cases. The degree of fever is not clearly related to prognosis, probably because a high fever may reflect both the severity of the infection and the degree of host response. On the other hand, hypothermia is associated with a poor prognosis (CLEMMER et al. 1992).

2.1.2 White Blood Cell Count

Increased white blood cell count is a feature of inflammation which may result from the stress reaction which is not very specific: it is also observed in other forms of circulatory shock and in other conditions such as trauma or heart failure. The degree of hyperleukocytosis is not well correlated with prognosis. However, some patients may present with acute leukopenia, which reflects a greater peripheral margination and activation of leukocytes. Similarly, an acute fall in the white blood cell count is observed following endotoxin administration in animals and in volunteers. The presence of leukopenia is associated with a worse prognosis.

2.1.3 Tachypnea

Although the mechanism is not entirely understood, sepsis is often associated with increased respiratory rate which frequently leads to hypocapnia (respiratory alkalosis; CLOWES 1974). The sign is neither sensitive nor specific to this condition. Changes in respiratory rate do not reflect well the severity of sepsis. Moreover, many patients with severe sepsis are treated with mechanical ventilation so that the respiratory rate may be at least partly controlled by the ICU physician.

2.1.4 Tachycardia

Elevated heart rate is not specific. It can also be observed in many other conditions, particularly in hypovolemia, anemia, heart failure, and stress. It may represent a mechanism to increase cardiac output or to compensate for a fall in stroke volume.

2.1.5 Reduced Vascular Tone

The loss of vascular tone and distributive blood flow alterations is a hallmark of severe sepsis and septic shock. This has been related to the release of many mediators, but the release of nitric oxide (NO) plays an important role.

2.1.6 Cytokines

Blood cytokine levels are usually elevated in septic shock, and this is related to prognosis (Casey et al. 1993; Pinsky et al. 1993). Interleukin (IL)-6 levels in particular are correlated with the severity of septic shock (Meduri et al. 1995). However, cytokine levels are no more reliable than lactate levels in evaluating the course of septic shock, and they are more expensive to measure (Marécaux et al. 1995). Blood cytokine measurements may help to identify those patients most likely to benefit from new forms of immunotherapy. For example, a study is underway to test whether high IL-6 levels help to identify patients who could benefit from anti-TNF antibodies.

2.2 Signs of Shock

2.2.1 Hemodynamic Alterations

Older studies, namely those from the 1950s, 1960s, and early 1970s, referred to two distinct hemodynamic regimes: the hyperdynamic or hyperkinetic pattern ("warm shock") and the hypodynamic or hypokinetic pattern ("cold shock"). Investigators at that time attempted to correlate the type of pattern to the type of micro-organisms involved (gram-positive or gram-negative).

The hyperdynamic pattern is the typical picture of septic shock, where it is usually present with tachycardia, elevated or normal cardiac output, and decreased systemic vascular resistance (Parker and Parrillo 1983; Parker et al. 1984; Shoemaker 1971; Abraham et al. 1983; Groeneveld et al. 1986; Vincent et al. 1992). A hypodynamic pattern, characterized by low cardiac output, is related primarily to hypovolemia and/or myocardial failure. This pattern can be observed either early in the course of shock, that is, before fluid administration, or very late following severe myocardial failure in dying patients. In a retrospective study of 2469 consecutive ICU patients, Ruokonen et al. (1991) recorded a 73% hospital mortality rate in septic shock. Refractory hypotension was a major early cause of death and MOF was the primary factor later. Cardiac failure was the direct cause of death in only one patient.

These terms are somewhat outdated now, for two main reasons. One is that more aggressive fluid resuscitation and larger use of dobutamine as an inotropic agent has reduced the incidence of the hypokinetic pattern (Vincent et al. 1990; Metrangolo et al. 1995). The second is that regardless of the cardiac output value septic shock is seen as an imbalance between oxygen demand and oxygen supply, so that a cardiac index which is normal or high may still be insufficient to meet an elevated oxygen demand. Hence, one should consider a spectrum of presentations rather than only two extreme patterns. Also the hemodynamic pattern is not clearly related to the type of micro-organism present.

Obtaining an optimal level of cardiac output is an essential factor in the resuscitation of septic shock. The most important question is not whether cardiac output is "high" or "low," but whether it is adequate or inadequate in relation to oxygen requirements and altered oxygen extraction capabilities (Fig. 1).

2.2.2 Myocardial Depression

Advances have been made in understanding the causes of myocardial depression. Altered cellular processes are involved in altering myocardial contractility under the influence of cytokines, including NO (Brady et al. 1992). Importantly, myocardial depression is no longer seen as a terminal event. It is now known that it can be present early, even when cardiac output is normal or high. In such cases the high cardiac output is a result of tachycardia. It is also possible that stroke volume is maintained by ventricular dilation leading to a reduced ventricular ejection fraction. Patients with septic shock have decreased left ventricular ejection fraction and right ventricular ejection fraction. The ventricles dilate to maintain stroke volume and compensate for depressed contractility (by the Frank-Starling mechanism). Interestingly, nonsurvivors are less likely to have decreased left ventricular ejection fraction, probably because they have more severe peripheral alterations, resulting in a reduced left ventricular afterload (Parker et al. 1984). However, nonsurvivors are likely to have lower right ventricular ejection fraction because pulmonary hypertension is an important feature of severe sepsis (Vincent et al. 1992). These observations have important therapeutic implications since inotropic support with dobutamine may be indicated even when cardiac output is not reduced (Vincent et al. 1990; Metrangolo et al. 1995).

2.2.3 Altered Vasculature

Vasodilating mediators can lead to the downregulation of α-receptor activity. O_2 extraction capabilities are reduced by maldistribution of blood flow. Loss of vasomotor regulation, microvascular obstruction, peripheral arteriovenous shunting, and interstitial edema formation can all contribute to the maldistribution of blood flow, with altered oxygen extraction capabilities. In the presence of a very high blood flow the microcirculatory transit time of the red blood cells may also become too short to allow oxygen extraction by the cells.

2.2.4 Altered Organ Perfusion

Shock-related disturbances in tissue perfusion are manifested by signs of altered organ function. In the most severe cases skin perfusion may be altered, as reflected by a mottled skin. However, in the typical hyperdynamic states, the skin perfusion may be normal. Reduced urine output is often an early occurrence. The insertion of a Foley catheter is vital in all patients with shock to facilitate frequent (hourly) measurement of urine output. Reduced renal perfusion is typified by decreased urine sodium and increased urine osmolarity, but once shock is established, isosthenuria typically results from altered tubular function. Sepsis can also result in altered mentation, generally characterized by obtundation, disorientation, and confusion, which is observed in 70% of patients with (BONE 1991). This has been related by several investigators to alterations in cerebral blood flow (BOWTON et al. 1989; PARKER and EMERSON 1977; EKSTROM-JODAL et al. 1982; MILLER et al. 1987), but cytokines are probably also involved.

2.2.5 Blood Lactate

Increased lactate levels are usually indicative of a relative imbalance between the oxygen requirement of the cells and the oxygen supply to them. Anemia and hypoxemia, for example, do not readily result in hyperlactatemia because a fall in hemoglobin level or hemoglobin saturation can be compensated by an increase in cardiac output. On the other hand, hyperlactatemia can be induced experimentally by progressively reducing oxygen delivery until a critical level is reached (ZHANG and VINCENT 1993).

Although increased blood lactate levels are traditionally attributed to tissue hypoxia and the anaerobic mechanism, increased lactate levels in sepsis may be related to other mechanisms. Other derangements that may contribute to elevated lactate levels include altered pyruvate metabolism and increased aerobic glycolysis. In particular, endotoxin can inhibit pyruvate dehydrogenase and thereby increase lactate before tissue hypoxia develops (VARY et al. 1986). PREISER et al. (1990) demonstrated that administration of dichloroacetate, an activator of pyruvate dehydrogenase, can reduce blood lactate but does not have a significant effect on hemodynamic performance. In complex cases it may also be useful to determine pyruvate levels to assess the lactate-to-pyruvate ratio because increased glucose utilization causes pyruvate to increase whereas cellular hypoxia causes the lactate-to-pyruvate ratio to increase. Nevertheless, the clinical applications are quite limited: even patients who are profoundly septic but hemodynamically stable have normal blood lactate levels. Hence increased lactate levels should represent an alarm signal.

Indeed, several studies have shown that blood lactate levels are well correlated with the prognosis in septic shock (BAKKER et al. 1995; VINCENT et al. 1983; Fig. 2). Recently it has been demonstrated that repeated determination of lactate levels provides a more reliable indication of prognosis. In particular, BAKKER et al. (1995) showed that repeated serial lactate measurements gives a reliable indication of not only mortality rate but also incidence and severity of organ failure. Lactate levels are a better clinical indicator than base deficit or bicarbonate level. Recent improvements in measuring instruments have facilitated the measurement of blood lactate levels, and it is to be hoped that such improvements continue so that repeated measurements of blood lactate will be widely available to assess the effects of therapy.

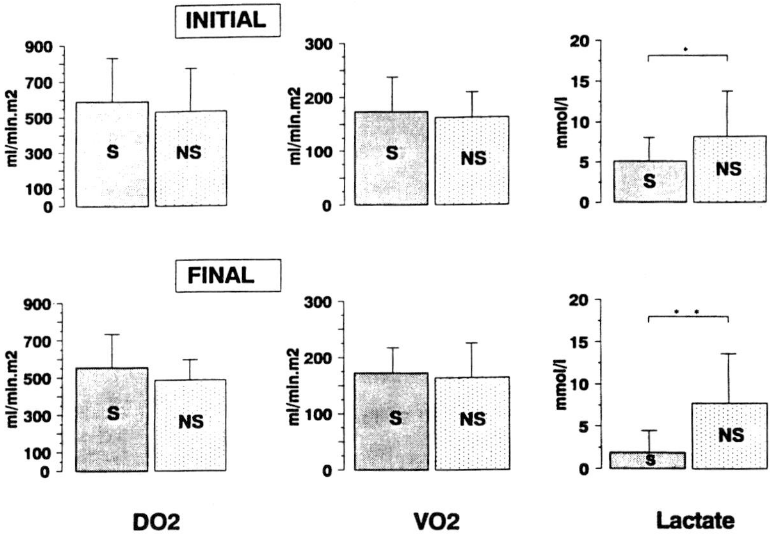

Fig. 2. Values of oxygen delivery (DO_2), oxygen consumption (VO_2), and blood lactate levels in survivors and nonsurvivors from septic shock. In contrast to blood lactate levels, DO_2 and VO_2 are poor prognostic factors in septic shock

3 Organ Dysfunction/Failure

When shock is prolonged, organ dysfunction and eventual MOF are often sustained. If fully developed, MOF is typically associated with very high mortality rates (80%–90%). Organ failure in septic shock has been related to the duration of adrenergic support (Ruokonen et al. 1991).

In assessing the likelihood of MOF-related mortality it is important to consider both the *number of affected systems* and the *degree of dysfunction.* Table 1 presents some important but simple and widely available parameters that can be included in this evaluation. In the early and late stages of sepsis an extremely complex relationship exists between the immune response and tissue hypoxia, in which the interaction of many cytokines such as TNF-α, and IL-1 and other mediators such as NO has been implicated.

3.1 Lung

Pulmonary dysfunction can lead to ARDS and is characterized by bilateral infiltrates on the chest radiograph and by hypoxemia, requiring the use of high FiO_2, and continuous positive airway pressure. ARDS is also frequently associated with reduced lung compliance ("stiff lungs").

Table 1. Simple, widely available parameters for quantifying degree of organ dysfunction in severe sepsis

System	Parameter	Abnormal if
Respiratory	$PaO2/FiO_2$ or respiratory support	<400 mmHg present
Coagulation	Platelet count	<150 000/mm^3
Liver	Bilirubin level	>2mg/dl
Cardiovascular	Requirement for vasopressor agents	Present
Central nervous	Glasgow Coma Score	<15
Renal	Creatinine level or urine output	>1.2 mg/dl or <500 ml/day

3.2 Coagulation

Coagulation defects associated with severe sepsis are common, but aspecific. These range from full-blown disseminated intravascular coagulation to isolated thrombocytopenia. Altered coagulation is most commonly manifested by low platelet count and sometimes by prolonged prothrombin time or activated partial thromboplastin time.

3.3 Liver

Hepatic dysfunction often develops after a few days of septic shock and is associated with a poorer prognosis. It is typically manifested by an increase in serum bilirubin (Franson et al. 1985; Banks et al. 1982; Nolan 1981) and sometimes in liver enzymes. In lipopolysaccharide-stimulated Kupffer cells grown in coculture with hepatocytes Billiar et al. (1989a,b) demonstrated that the release of NO inhibits the function of the hepatocytes, suggesting that NO is involved in the hepatic abnormalities associated with sepsis.

3.4 Cardiovascular

As indicated above, an elevated cardiac index and reduced systemic vascular resistance are hallmarks of sepsis and cannot be considered as signs of cardiovascular dysfunction. Since treatment may vary from one institution to another and even from one physician to another, it would be preferable not to base the degree of organ dysfunction on a therapeutic requirement. Nevertheless, the requirement for vasopressor agents is probably the best factor that one may include.

3.5 Central Nervous System

Severe sepsis is often associated with altered consciousness, without organic substrate. A reduction in the Glasgow Coma Score reflects such alterations.

3.6 Kidney

Renal failure may present with or without oligoanuria. Hemodialysis or continuous hemofiltration may be required in such cases. The role of continuous hemofiltration in the removal of proinflammatory mediators is still controversial (VINCENT and TIELEMANS 1995).

4 Prognosis

The prognosis of septic shock is determined primarily by the underlying status of the patient, including physiological condition, age, type of underlying disease, presence of debilitating factors, such as hematological malignancies, liver cirrhosis, and recent chemotherapy. The prognosis is worse for fungal than for bacterial infections because the former are more common in debilitated patients. Although it has been suggested that infection with *Pseudomonas* is associated with a worse prognosis, the type of bacteria is not very discriminant. As indicated above, the presence of hypothermia or leukopenia is associated with a poorer prognosis.

Even more than the severity of infection, the adequacy of the anti-infectious management is a crucial factor influencing outcome. Appropriate antibiotic therapy and removal of the source (e.g., by surgery or drainage) are of paramount importance. The adequacy of the cardiorespiratory management are other important determinants of survival. Adequate fluid therapy and proper use of vasoactive agents are essential.

Blood levels of many cytokines, adhesion molecules, hormones, and other mediators have been shown to be related to prognosis.

References

Abraham E, Shoemaker WC, Bland RD, Cobo JC (1983) Sequential cardiorespiratory patterns in septic shock. Crit Care Med 11: 799–803

Anonymous (1990) Increase in National Hospital Discharge Survey rates for septicemia – United States, 1979–1987. MMWR 39: 31–34

Bakker J, Gris P, Coffernils M, Kahn RJ, Vincent JL (1995) Serial blood lactate levels can predict the development of multiple organ failure following septic shock. Am J Surg (in press)

Banks JG, Foulis AK, Ledingham IM (1982) Liver function in septic shock. J Clin Pathol 35: 1249–1252

Billiar TR, Curran RD, Stuehr DJ (1989a) Evidence that the activation of Kupffer cells results in the production of L-arginine metabolites that release cell-associated iron and inhibit hepatocyte protein synthesis. Surgery 169: 1467–1472

Billiar TR, Curran RD, Stuehr DJ, West MA, Bentz BG, Simmons RL (1989b) A L-arginine-dependent mechanism mediates Kupffer cell inhibition of hepatocyte protein synthesis in vitro. J Exp Med 169: 1467–1472

Bone RC (1991) Sepsis, the sepsis syndrome, multi-organ failure: a plea for comparable definitions. Ann Intern Med 114: 332–333
Bowton DL, Bertels NH, Prough DS (1989) Cerebral blood flow is reduced in patients with sepsis syndrome. Crit Care Med 17: 399–403
Brady AJ, Poole-Wilson PA, Harding SZ, Warren JB (1992) Nitric oxide production within cardiac myocytes reduces their contractility in endotoxemia. Am J Physiol 32: H1963–H1966
Casey LC, Balk RA, Bone RC (1993) Plasma cytokine and endotoxin levels correlate with survival in patients with the sepsis syndrome [see comments]. Ann Intern Med 119: 771–778
Clemmer TP, Fisher CJ, Bone RC, Slotman GJ, Metz GA, Thomas FO (1992) Hypothermia in the sepsis syndrome and clinical outcome. Crit care Med 20: 1395–1401
Clowes GH (1974) Pulmonary abnormalities in sepsis. Surg Clin North Am 54: 993–1013
Danner RL, Elin RJ, Hosseini JM, Wesley RA, Reilly JM, Parillo JE (1991) Endotoxemia in human septic shock. Chest 99: 169–175
Ekstrom-Jodal B, Haggendal E, Larsson LE (1982) Cerebral blood flow and oxygen uptake in endotoxic shock. An experimental study in dogs. Acta Anaesthesiol Scand 26: 163–170
Franson TR, Hierholzer WJ, LaBrecque DR (1985) Frequency and characteristics of hyperbilirubinemia associated with bacteremia. Rev Infect Dis 7: 1–9
Friedman G, Gomez J, Shahla M et al. (1995) Mortality from septic shock: have we made progress (Abstr)? Am J Respir Crit Care Med 151: A494
Groeneveld AB, Bronsveld W, Thijs LG (1986) Hemodynamic determinants of mortality in human septic shock. Surgery 99: 140–152
Marchant A, Alegre ML, Hakim A, Piérard G, Marécaux G, Friedman G, De Groote D, Kahn RJ, Vincent JL, Goldman M (1995) Clinical and biological significance of interleukin-10 plasma levels in patients with septic shock. J Clin Immunol 15: 266–273
Marécaux G, Pinsky MR, Dupont E, Kahn RJ, Vincent JL (1995) Blood lactate levels are better prognostic indicators than TNF and IL-6 levels in patients with septic shock. Intensive Care Med (in press)
Meduri GU, Headley S, Kohler G, Stentz F, Tolley E, Umberger R, Leeper K (1995) Persistent elevation of inflammatory cytokines predicts a poor outcome in ARDS: plasma IL-1β and IL-6 levels are consistent and efficient predictors of outcome over time. Chest 107: 1062–1073
Metrangolo L, Fiorillo M, Friedman G, Silance PG, Kahn RJ, Novelli GP, Vincent JL (1995) The early hemodynamic course of septic shock. Crit Care Med (in press)
Miller CF, Breslow MJ, Shapiro RM (1987) Role of hypotension in decreasing cerebral blood flow in porcine endotoxemia. Am J Physiol 253: H956–H964
Natanson C, Danner RL, Elin RJ, Hosseini JM, Peart KW, Banks SM, Mac Vittie TJ, Walker RI, Parrillo JE (1989a) Role of endotoxemia in cardiovascular dysfunction and mortality. Escherichia coli and Staphylococcus aureus challenges in a canine model of human septic shock. J Clin Invest 83: 243–251 [published erratum appeared in J Clin Invest (1989) 83(3): 1087]
Natanson C, Eichenholz PW, Danner RL, Eichacker PQ, Hoffman WD, Kuo GC, Banks SM, MacVittie TJ, Parrillo JE (1989b) Endotoxin and tumor necrosis factor challenges in dogs simulate the cardiovascular profile of human septic shock. J Exp Med 169: 823–832
Nolan JP (1981) Endotoxin, reticuloendothelial function, and liver injury. Hepatology 5: 458–465
Parker JL, Emerson TE (1977) Cerebral hemodynamics, vascular reactivity, and metabolism during canine endotoxic shock. Circ Shock 4: 41–53
Parker MM, Parrillo JE (1983) Septic shock: hemodynamics and pathogenesis. JAMA 250: 2324–2327
Parker MM, Shelhamer JH, Bacharach SL, Green MV, Natanson C, Frederick TM, Damske BA, Parrillo JE (1984) Profound but reversible myocardial depression in patients with septic shock. Ann Intern Med 100: 483–490
Pinsky MR, Vincent JL, Deviere J, Alegre M, Schandene L, Kahn RJ, Vandenbussche P, Content J, Dupont E (1993) Serum cytokine levels in human septic shock: relation to multiple-systems organ failure and mortality. Chest 103: 565–575
Preiser JC, Moulart D, Vincent JL (1990) Dichloroacetate administration in the treatment of endotoxin shock. Circ Shock 30: 221–228
Ruokonen E, Takala J, Kari A, Alhava E (1991) Septic shock and multiple organ failure. Crit Care Med 19: 1146–1151
Ruokonen E, Takala J, Kari A, Saxen H, Mertsola J, Hansen EJ (1993) Regional blood flow and oxygen transport in septic shock. Crit Care Med 21: 1296–1303
Shoemaker WC (1971) Cardiorespiratory patterns in complicated and uncomplicated septic shock: physiologic alterations and their therapeutic indications. Ann Surg 174: 119–125

Suffredini AF, Fromm RE, Parker MM, Brenner M, Kovacs JA, Wesley RA, Parrillo JE (1989) The cardiovascular response of normal humans to the administration of endotoxin. N Engl J Med 321: 280–287

Vary TC, Siegel JH, Nakatani T, Sato T, Aoyama H (1986) Effect of sepsis on activity of pyruvate dehydrogenase complex in skeletal muscle and liver. Am J Physiol 250: E634–E640

Vincent JL, Tielemans C (1995) Continuous hemofiltration in severe sepsis: is it beneficial? J Crit Care 10: 27–32

Vincent JL, Dufaye P, Berre J, Leeman M, Degaute JP, Kahn RJ (1983) Serial lactate determinations during circulatory shock. Crit Care Med 11: 449–451

Vincent JL, Roman A, Kahn RJ (1990) Dobutamine administration in septic shock: addition to a standard protocol. Crit Care Med 18: 689–693

Vincent JL, Gris P, Coffernils M, Leon M, Pinsky MR, Reuse C, Kahn RJ (1992) Myocardial depression characterizes the fatal course of septic shock. Surgery 111: 660–667

Zhang H, Vincent JL (1993) Oxygen extraction is altered by endotoxin during tamponade-induced stagnant hypoxia in the dog. Circ Shock 40: 168–176

Zhang H, Spapen H, Manikis P, Rogiers P, Metz G, Buurman WA, Vincent JL (1995) Tirilazad mesylate (U74006F) inhibits the effets of endotoxin in dogs. Am J Physiol 268: H1847–H1855

Significance and Pathogenesis of Septic Shock

P. Brandtzaeg

1 Definition . . . 16
2 Septic Shock and the Multiple Organ Dysfunction Syndrome . . . 16
3 Experimental Shock Research Versus Human Septic Shock . . . 17
4 The Microbial-Host Interaction . . . 17
5 The Nature of Microbial Components That Induce Intravascular Inflammation . . . 17
6 LPS in Plasma and the Development of Septic Shock . . . 18
7 Gram-Positive Cell Wall Products and Septic Shock . . . 19
8 Exotoxins Implicated in Septic Shock . . . 19
9 Bacteria Associated with Septic Shock . . . 20
10 Pathophysiological Effects of Endotoxin in Man . . . 20
11 Circulating LPS and Human Septic Shock . . . 21
12 Pathogenetic Mechanisms Related to the Intravascular Inflammatory Response . . . 23
13 The Proinflammatory Cytokine Response . . . 24
14 The Anti-inflammatory Cytokine Response . . . 25
15 Activation of the Kallikrein Kinin System in Septic Shock . . . 26
16 Activation of the Complement System in Septic Shock . . . 26
17 Activation of the Coagulation System in Septic Shock . . . 28
18 Activation and Inhibition of the Fibrinolytic System in Septic Shock . . . 29
19 Phospholipase A_2 in Septic Shock . . . 30
20 Endorphins . . . 30
21 Vasoactive Intestinal Peptide and Septic Shock . . . 30
22 The Myocardial Response in Septic Shock . . . 31
23 Conclusion . . . 31
References . . . 32

Department of Pediatrics and Clinical Chemistry, Ullevål University Hospital, 0407 Oslo, Norway

1 Definition

Septic shock is a state of inadequate tissue perfusion induced by microbial products and characterized by low blood pressure and biochemical signs of oxygen deficit. The reduced oxygen and nutrient transport to vital organs is caused by a generalized intravascular inflammatory response resulting in vasodilation. Septic shock evolves through different stages and is initially characterized by a hyperdynamic circulation, i.e., high cardiac output and after volume loading low peripheral resistance (PARILLO 1993). Most patients dying of septic shock remain in this circulatory state to the end (PARILLO 1993). In patients with overwhelming bacteremia caused by *Neisseria meningitidis, Streptococcus pneumoniae, Streptococcus pyogenes, Haemophilus influenzae*, and certain other bacteria the initial hyperdynamic circulation is gradually transformed into a hypodynamic circulation due to reduced cardiac performance. These patients die in a state of irreversible vasodilation combined with myocardial failure often with a terminal arrhythmia (MERCIER et al. 1988; BRANDTZAEG 1995).

2 Septic Shock and the Multiple Organ Dysfunction Syndrome

The altered circulation is one of several dysfunctional organ systems induced by the intravascular inflammatory response. The kidneys, lungs, liver, and the vascular system including circulating blood cells may all participate to a variable extent in the multiple organ dysfunction syndrome (MODS). Circulatory failure in connection with severe intravascular bacterial infections have been known for a long time. The increasing interest in septic shock as a clinical entity over the past 40 years has paralleled the increasing number of patients developing septic shock in modern intensive care units (ICU; BONE 1993). This cohort of patients is heterogeneous. The majority of patients developing septic shock today are characterized by a debilitated general condition related to an underlying disease, a recent operation or treatment with cytotoxic drugs, and treatment or surveillance with intravenous devices. The goals of the intensified research into the basic pathophysiological mechanisms of septic shock are to prevent or ameliorate the intravascular inflammation and restore adequate tissue perfusion thereby reducing the persistent high mortality of approximately 50% (BONE 1993; PARILLO 1993).

3 Experimental Shock Research Versus Human Septic Shock

The scientific literature related to possible pathogenetic mechanisms of septic shock is overwhelming. This review addresses major well-established pathogenetic mechanisms and includes a restricted number of results obtained in primate septic shock studies. Human responses to intravenously injected lipopolysaccharides (LPS) are referred to inasmuch as they reflect phenomena observed in septic shock patients. These experimental approaches reflect disease mechanisms typically observed in acute bacteremic patients developing septic shock. Many believe that the majority of shock patients are "sensitized" by underlying diseases to shock provoking mechanisms. It is commonly believed that smaller changes in the various homeostatic systems regulating tissue perfusion may have more devastating effects in diseased than in previously healthy persons. However, we lack good models to study the effect of "sensitization by disease" in primates or other experimental animals closer to man than rodents.

4 The Microbial-Host Interaction

Septic shock develops when specific microbial components gain access to the circulation and are recognized by the immune system, which generates an exaggerated mediator response (Parillo et al. 1990; Young et al. 1991; Glauser et al. 1991, 1994; Bone 1991; Jafari and McCracken 1992; Natanson et al. 1994; Lynn and Cohen 1995). The interplay of bacterial and host factors determines the final outcome. Animal experiments indicate that the response depends on the nature and amount of bacterial components introduced and the time span of the microbial insult, for example, bolus injection versus a long-lasting infusion with the same dose.

5 The Nature of Microbial Components That Induce Intravascular Inflammation

Cell wall constituents such as LPS in gram-negative bacteria and peptidoglycan and teichoic acid in gram-positive bacteria or proteins liberated during growth (exotoxins) are the principal microbial components which induce an intravascular inflammatory reaction leading to septic shock (Rietschel et al. 1991; Chetty and Schwab 1987; Zumla 1992). They share a powerful stimulatory property when recognized by the immune system.

6 LPS in Plasma and the Development of Septic Shock

It is generally believed that LPS must be detached from the outer membrane to exert full biological effect. The toxicity of LPS is associated with lipid A interacting either with a cell receptor initiating signal transduction or host fluid phase cascade systems (Morrison and Ryan 1987; Morrison 1990). LPS is released through complement-mediated bacterial lysis or autolysis. Insertion of the aggregated terminal components 5b-$(9)_n$ (membrane attack complex) of the complement system into the outer membrane destabilize the outer leaflet. In vitro studies indicate that LPS is released through the action of complement although the membrane attack complex does not necessarily disrupt the outer membrane and kill the bacterium (Frank and Fries 1988). Several proteins secreted by neutrophils such as bactericidal permeability increasing protein and defensins may also destabilize the outer membrane (Spitznagel 1990).

Many bacteria produce surplus outer membrane material, termed outer membrane vesicles (OMV), during log phase growth. These phenomena have been studied mostly in vitro, and the relevance for the in vivo situation is unclear. Pathogenic clones of *N. meningitidis* appear to produce large amounts of OMV both in vitro and in vivo (Devoe and Gilchrist 1973; Andersen et al. 1987; Brandtzaeg et al. 1992).

The LPS-releasing effect of antibiotics has been much debated and extensively studied in model systems (Hurley 1992). However, it has never been convincingly documented that initiation of antibiotic treatment precipitates or aggravates septic shock in man (Hurley 1992). On the contrary, inadequate antibiotic treatment is a risk factor associated with mortal outcome in patients with severe bacterial infections. In fulminant meningococcal infections with high pretreatment levels of LPS the infusion of benzylpenicillin and chloramphenicol, which are both bactericidal to *N. meningitidis,* immediately reduces the circulating levels of LPS and various inflammatory mediators (Engebretsen et al. 1986; Brandtzaeg et al. 1989a,c, 1990; Waage et al. 1989).

We know very little about the physicochemical state of LPS molecules released into the circulation. Presumably they do not circulate as single molecules but as complexes due the hydrophobic lipid A region (Brandtzaeg et al. 1992). Many proteins form complexes with lipid A, including high-density lipoproteins, low-density lipoproteins, very low density lipoproteins, albumin, transferrin, and lactoferrin. Collectively these proteins appear to form a "buffer system" that partly reduces the biological effects of lipid A in LPS-responsive cells including monocytes, macrophages, and neutrophils (Flegel et al. 1989; Morrison 1990).

LPS-binding protein (LBP) – a recently discovered acute-phase reactant – forms complexes with lipid A. The LPS-binding protein–lipid A complex associates with CD14, an glycophosphatidyl inositol anchored surface protein on monocytes and macrophages, which greatly enhances the lipid A induced signal transduction to the cell nucleus (Wright et al. 1990). The basic level of LPS-binding protein plasma (approximately 20 μg/ml) appears to be sufficient for generating a maximum response of human monocytes after a LPS challenge. The signal pathway from the cell surface to the nucleus is presently largely unknown but involves G-proteins, various protein kinases, and nuclear factor kB signal transduction pathway.

7 Gram-Positive Cell Wall Products and Septic Shock

Peptidoglycan and teichoic acid represent the major inflammatory inducing principles in the cell wall of gram-positive bacteria. Peptidoglycan is a heteropolymer consisting of 1–4 linked *N*-acetylglucosamine and *N*-acetylmyramic acid cross-linked with short peptides (CHETTY and SCHWAB 1985). The glycan backbone structure is constant whereas the short peptide chains cross-linking the sugars vary. Peptidoglycan is generally regarded as less biologically potent on weight volume base than *Escherichia coli* LPS although this may differ between different species (DZIARSKI 1993). The molecular mechanisms involved in cell activation by peptidoglycan and teichoic acid are largely unknown. Recently a 70–73-kDa surface proteins which is located on various immune cells has been identified as a peptidoglycan receptor (DZIARSKI 1993). This serves a triple function in being a receptor for peptidoglycan, lipid A, and pertussis toxin. The signaling pathway involves a G protein signal transduction system.

We still lack methods sensitive enough to quantify peptidoglycan and teichoic acid in plasma collected from patients with gram-positive bacteremia. Consequently no systematic study in patients has been conducted linking the presence or level of gram-positive cell wall products to activation of various mediator systems. Based on the experience with overwhelming gram-negative bacteremia, high levels of cell wall products are to be expected in plasma collected from patients developing lethal *S. pneumoniae* bacteremia, a condition often linked to lack of splenic function. The spleen plays a particular important role in removing intravascular bacteria when the host lacks specific opsonins (BENACERRAF et al. 1959).

8 Exotoxins Implicated in Septic Shock

Exotoxins are a group of proteins with pleiotropic effects on cell metabolism. They are secreted primarily during bacterial log phase growth. The most intensively studied exotoxins which have been implicated in septic shock are those produced by *S. aureus* (toxic shock syndrome toxin and various enterotoxins) and *S. pyogenes* (streptococcal pyrogenic exotoxins). These toxins are small proteins (20–30 kDa) revealing a significant amino acid homology. They function as superantigens (Zumla 1992). The staphylococcal toxic shock syndrome is usually a pure toxin-mediated disease since cultures of blood usually are negative. The exotoxins are produced outside the vascular system (in the vagina in cases of toxic shock syndrome related to tampon use) gaining access to the circulation. In the streptococcal toxic shock syndrome toxin-producing bacteria are usually present in both the blood and various tissues.

9 Bacteria Associated with Septic Shock

The spectrum of bacteria isolated from cultures of blood or infected extravascular foci in critically ill patients changes over time. Early and intensified treatment with broad-spectrum antibiotics have had a major impact on the isolation rate of live bacteria in blood cultures. New technique employing long-term intravascular devices to treat and monitor patients have contributed to the isolation of increasing numbers of gram-positive skin-related bacteria. Gram-negative bacteria, particularly *E. coli* and other gram-negative rods, however, are still the principal pathogens isolated from blood culture in septic shock patients (Glauser et al. 1994).

10 Pathophysiological Effects of Endotoxin in Man

LPS are the most intensively studied bacterial component related to the development of septic shock (Morrison and Ryan 1987). If 2–4 ng/kg [20–40 endotoxin units (EU)/kg] *E. coli* LPS is injected intravenously in healthy volunteers, profound effects occur within 30–60 min which may last for many days (Martich et al. 1993; Burrell 1994). One hour after the injection the test persons perceive symptoms associated with general malaise such as nausea, headache, myalgia, chills, and rigors which reach a maximum within 2 h (Elin et al. 1981; van Deventer et al. 1990; van Zee et al. 1995). The temperature rises to 38.5°–40 °C. The heart rate increases and the circulation becomes hyperdynamic, with elevated cardiac output, reduced peripheral vascular resistance, and dilation of the left ventricle with reduced performance; these alterations may last for several days (Suffredini et al. 1989b). The pulmonary function is altered, with a fall in PaO_2 and $PaCO_2$ and widening of the alveolar arterial gradient after fluid loading (Suffredini et al. 1992). Broncheoalveolair lavage indicates increased capillary permeability for small molecules but not for albumin (mw 69 kDa).

The coagulation system is activated by upregulation of tissue factor presumably on circulation monocytes. Concomitantly the fibrinolytic system is activated by release of tissue plasminogen activator from endothelial cells. Within 2–3 h the fibrinolysis is gradually inhibited by increasing levels of plasminogen activator inhibitor 1 (SUFFREDINI et al. 1989a; VAN DEVENTER et al. 1990). The net effect is a procoagulant state. The plasma contact system comprising coagulation factors XII and XI, prekallikerin, and high molecular weight kininogen is activated within 2 h. One consequence is the generation of bradykinin, a nonapeptide which causes profound vasodilation and increased transcapillary flux (DELA CADENA et al. 1993).

Cytokines including tumor necrosis factor-α (TNF-α), interleukins (IL) 1, 6, and 8, IL-1 receptor antagonist, and various soluble receptors neutralizing these cytokines are released in a coordinated manner (MICHIE et al. 1988; FONG et al. 1989, 1990; ZABEL et al. 1989; VAN DEVENTER et al. 1990, 1993; CANNON et al. 1990; SPINAS et al. 1990, 1992; MARTICH et al. 1991; GRANOWITZ et al. 1991, 1993; FISCHER et al. 1992; VAN ZEE et al. 1992, 1995; KUHNS et al. 1995).

Circulating neutrophils reveal a dose-dependent change in kinetics ranging from leukocytosis at low doses (0.5–1 ng/kg) to significant leukopenia at a higher dose (2–4 ng/kg) challenge (ELIN et al. 1981; VAN DEVENTER et al. 1990; KUHNS et al. 1995). The difference in kinetics is related to upregulation of endothelial and leukocyte adhesion molecules leading to a close neutrophil–endothelium interaction (KUHNS et al. 1995).

High doses of LPS have been injected into humans, with pronounced adverse effects. A 14-year-old boy developed a life-threatening septic shock and MODS after receiving 100 ml packed erythroyctes heavily contaminated with *Pseudomonas fluorescens* LPS. Plasma from the lot contained 40 000 EU/ml (FOREMAN et al. 1991). A laboratory worker injected 1 mg purified *Salmonella minnesota* LPS intravenously in a suicide attempt. The dose was 3750 times higher (15 000 ng/kg) than the one used in most human experiments. The patient developed persistent hypotension, pulmonary edema, coagulopathy, renal and hepatic impairment and needed treatment with fluid and epinephrine for 50 h. She survived (DA SILVA et al. 1993).

These observations clearly demonstrate that small doses of LPS injected intravenously induce sepsislike symptoms and activate important inflammatory mediators whereas large doses of LPS precipitate life-threatening circulatory collapse and multiple organ failure.

11 Circulating LPS and Human Septic Shock

The introduction of the *Limulus* amebocyte lysate assay into clinical medicine in 1970 made it possible to perform quantitative studies of LPS in plasma (LEVIN et al. 1970, 1972). LEVIN et al. were the first to document that clinical severity is associated with detectable LPS in plasma. Endotoxinemia (LPS ≥ 5 ng/l; 0.05 EU/ml) in platelet-rich plasma is an early marker of a septic response in febrile patients (VAN DEVENTER et al. 1988). MCCARTNEY et al. (1983) detected LPS in plasma of all 31 patients admitted to a large ICU with septic shock of whom 21 eventually died. In patients with overwhelming gram-negative bacteremia caused by *N. meningitidis* the levels of LPS are associated with disease severity. Plasma levels of 800 ng/l (8 EU/ml or higher predict the development of persistent septic shock, impaired renal function, adult respiratory distress syndrome, massive coagulopathy, and ultimately a fatal outcome in half of the patients (BRANDTZAEG et al. 1989a, 1995). Systemic meningococcal disease is so far the only gram-negative infection in which circulating levels of LPS have been related quantitatively to activation of various mediator systems involved in the pathogenesis of septic shock and reveal a dose-dependent association with mortality (BRANDTZAEG and KIERULF 1992; BRANDTZAEG 1995; BRANDTZAEG et al. 1995). In few patients with overwhelming bacteremia caused by *H. influenzae* levels of 500–2000 ng/l (5–20 EU/ml) have been detected (BRAZILIAN PURPURIC FEVER STUDY GROUP 1987; Brandtzaeg, unpublished results).

Low grade endotoxinemia is often observed among neutropenic patients undergoing cytotoxic therapy for hematological malignancies (SHANDS and MCKINNEY 1989; BEHRE et al. 1992). Twenty-one severely neutropenic patients undergoing treatment for leukemia or non-Hodgkin lymphoma were suspected of developing gram-negative sepsis. Of these, 17 (81%) had detectable LPS in plasma (detection limit 14 ng/l, or 0.17 EU/ml) on the first day of antimicrobial treatment, of whom 10 died. None of the four patients without detectable LPS developed septic shock, and all survived. Endotoxinemia was quantitatively associated with fever. Nonsurvivors had significantly higher maximum concentration of LPS than did the survivors. Of 11 patients with endotoxinemia and septic shock 7 (64%) died (BEHRE et al. 1992). SHANDS and MCKINNEY (1989) found no association between endotoxinemia and fever among a similar group of patients.

In 100 patients developing septic shock in an ICU 43% had detectable endotoxinemia during the shock period (DANNER et al. 1991). The development of multiple organ dysfunction occurred ten times more frequently among patients with detectable endotoxin than among those without. However, the majority of patients developing septic shock had no discernable endotoxinemia. LPS were also detected in patients with gram-positive infections (DANNER et al. 1991). GUIDET et al. (1994) studied 93 patients in a medical intensive care study fulfilling criteria of the sepsis syndrome. The 28-day mortality was 53%. Circulating LPS was detected in 47% of the patients. Endotoxinemia was associated with the development of septic shock and organ failure. Among 46 patients with confirmed gram-negative bacterial infection LPS was detected in 67%, as opposed to 28% among patients without documented gram-negative infection. Of 19 patients with gram-negative bacteremia on admission 14 (79%) had detectable LPS in the blood. For the whole group, however, detectable LPS did not predict gram-negative infection or outcome (GUIDET et al. 1994).

Summarizing the literature on the relationship between circulating LPS and the development of septic shock, two different patterns appear to exist. In patients developing acute, overwhelming gram-negative bacteremia which rapidly evolves to septic shock by organisms such as *N. meningitidis* and *H. influenzae* significantly increased levels of LPS have been documented. In patients contracting severe bacteremia caused by *Capnocytophage canimorsus* – a gram-negative rod associated with dog bites and previously termed dysgonic fermenter 2 – LPS may also play an important role in the pathophysiology. The clinical picture reveals features similar to those observed in fulminant meningococcemia (HICKLIN et al. 1987). Patients contracting *N. meningitidis* and *H. influenzae* may have plasma levels of LPS up to 600 pg/ml (6 EU/ml) without developing septic shock (BRANDTZAEG et al. 1989a; Brandtzaeg, unpublished results).

The majority of ICU patients developing septic shock reveal the second pattern. LPS is detectable in fewer than 50% of the patients, and quantitative measurements have no prognostic value. Circulating LPS, when detected, are often associated with bacteremia or a focal infection caused by gram-negative bacteria. A significantly larger proportion of patients with detectable LPS develop septic shock than patients with a negative result on the Limulus amebocyte lysate test. The levels of LPS are usually below 100 pg/ml (1 EU/ml), which is several orders of magnitude lower than that been measured in patients with overwhelming bacterial infections.

12 Pathogenetic Mechanisms Related to the Intravascular Inflammatory Response

Septic shock is the result of an inappropriate inflammatory response which reduces the tone of the resistance vessels and impairs myocardial performance. Two basic phenomena are related to the deficit of circulating volume: (a) relaxation of smooth muscles in the vessel wall and (b) increased endothelial permeability. The combination leads ultimately to tissue hypoxia and altered cell metabolism. It has been documented that mediators related to cascade systems in plasma, cytokines, enzymes, and oxygen radicals elicited by various cells adjacent to or in the vascular compartment play a role in septic shock. It should be stressed that our knowledge is still fragmentary, and that each decade has its favorite mediator system considered particular important in the pathophysiology of septic shock. So far, no single mediator or mediator system can explain all features of this clinical entity. Moreover, a very complex interplay appears to exist linking the various systems to each other. Many systems are activated before overt circulatory failure becomes obvious. Finally, the ultimate dysfunction occurs within cells of the vessel wall and myocardium, areas not readily accessible to investigation in patients.

13 The Proinflammatory Cytokine Response

The principal proinflammatory cytokines which have been implicated in human septic shock are TNF-α, IL-1, IL-6, and IL-8 (Waage et al. 1987, 1989; Girardin et al. 1988; Debets et al. 1989; Damas et al. 1989; Calandra et al. 1990, 1991; Offner et al. 1990; Cannon et al. 1990; Pinsky et al. 1993; Halstensen et al. 1993; van Deuren et al. 1994; Casey et al. 1993; Marty et al. 1994). Recently leukemia-inhibiting factor has been added as a putative proinflammatory cytokine of septic shock (Waring et al. 1994). Among the members of the growing cytokine family TNF-α is regarded as a major mediator of shock. This conclusion is based on extensive animal experiments, observations in human septic shock patients, and studies in human volunteers. In the baboon shock model pretreatment with TNF-α neutralizing antibodies prevents otherwise lethal septic shock. (Tracey et al. 1987). In the human endotoxin challenge model 2–4 ng/kg *E. coli* LPS induces a significant rise in plasma TNF-α, which partly induces or augments the production of other cytokines (Michie et al. 1988; Zabel et al. 1989; van Deventer 1990; Fong et al. 1990). Infusion of recombinant TNF-α reproduces pathophysiological alterations associated with human septic shock (van der Poll et al. 1990, 1991a-c, 1992). The most important production site of TNF-α after an intravenous LPS challenge is the splancnic area, possibly Kuffper cells in the liver (Fong et al. 1990). Peripheral mononuclear cells appear to contribute little to the cytokine response in plasma (Monnoz et al. 1991; Granowitz et al. 1993).

Shortly after its discovery TNF-α was identified as an important marker of severity in meningococcal septic shock (Waage et al. 1987, 1989, 1994; Girardin et al. 1988; van Deuren et al. 1994. By employing a bioassay (Wehi 164 clone 13 mouse fibrosarcoma cell line) Waage et al. (1989) documented that TNF-α circulates as a functionally active molecule, i.e., not in complex with soluble TNF-receptors in patients with overwhelming sepsis. In a few patients with exceptionally high levels of meningococcal LPS in plasma bioactive IL-1, which increases the toxicity of TNF-α, was present (Waage et al. 1989). Circulating IL-1β is usually not detected in septic shock patients and is considered a primarily local mediator.

Studies comprising a more mixed population of patients developing septic shock have disclosed that TNF-α is not uniformly present and varies over time (De Groote et al. 1989; Debets et al. 1989; Damas et al. 1989; Calandra et al. 1990; Offner et al. 1990; Pinsky et al. 1993; Casey et al. 1993). Summarizing 10 years of experience, two different patterns of TNF-α release are discernable. In acute overwhelming bacteremia caused either by gram-negative or gram-positive bacteria high levels of bioactive TNF-α are present. Effective antibacterial therapy stops further bacterial proliferation and reduces the levels of circulating proinflammatory cytokines, including TNF-α, within few hours.

In patients developing septic shock from other causes than overwhelming bacteremia TNF-α is not universally present in plasma and may fluctuate over time. The persistence of TNF-α – more than its level – indicates a grave prognosis. Whether TNF-α is present in a bioactive form has not been well documented in the latter group of patients since most clinical studies have been performed with immunassays rather than bioassays. In many surgical patients TNF-α is biologically inactive, presumably after complex formation with soluble TNF-receptors (WAAGE and AASEN 1992).

14 The Anti-inflammatory Cytokine Response

The biological effects of proinflammatory cytokines are balanced by several anti-inflammatory cytokines, notably IL-10, IL-4, IL-13, and transforming growth factor-β. IL-10 has been detected in septic shock patients induced by gram-negative and gram-positive bacteria and in children with fulminant meningococcal septicemia (MARCHANT et al. 1994; DERKX et al. 1995).

Functional studies employing septic shock plasma containing high levels of native meningococcal LPS and proinflammatory cytokines (TNF-α, IL-6, IL-8) were unable to activate human monocytes collected from a blood donor (BRANDTZAEG et al. 1996a). Removing IL-10 from the samples restored the proinflammatory effects of plasma on human monocytes. These results suggest that IL-10 is a functionally important principle counteracting the effect of LPS and proinflammatory cytokines in overwhelming bacteremia. IL-4 and transforming growth factor-β are not detected in plasma from meningococcal septic shock patients.

The lack of monocyte activation by LPS-containing shock plasma explains the dissociation between the levels of cytokines in plasma and concomitant lack of cytokine production in peripheral blood mononuclear cells, which has described in patients with the sepsis syndrome (MONNOZ et al. 1991). IL-10 may partly explain the development of tolerance after LPS stimulation (CAVAILLON et al. 1994). The proinflammatory cytokines are apparently counteracted by two different principles: (a) soluble receptors and receptor antagonists and (b) anti-inflammatory cytokines. Preliminary results suggest that the levels of IL-10 are 10–100 times higher and are elicited at an earlier stage than TNF-α (BRANDTZAEG et al. 1996a).

15 Activation of the Kallikrein Kinin System in Septic Shock

The contact activation system of plasma comprises the coagulation, the fibrinolytic, and the kallikrein kinin systems. Coagulation factor XII (Hageman factor) circulates in complex with prekallikerin and high molecular weight kininogens and may react with negatively charged surfaces and bacterial cell wall products such as lipid A or peptidoglycans, initiating a cascade reaction. One effect is the generation of bradykinin, a nonapeptide which induces an increased transcapillary permeability, reduces arterial resistance, and causes hypotension in experimental animals (Cochrane 1985; Proud and Kaplan 1988). In a classic primate shock study Nies et al. (1968) showed that endotoxin extracts from *E. coli* and *Pseudomonas pseudomallei* induce a prolonged hypotension with reduced peripheral vascular resistance and decreased cardiac output which occurs concomitantly with increased levels of bradykinin and reduced levels of kininogen. The hypotensive effect of live *E. coli* in the baboon shock model can be significantly ameliorated, converting a lethal septic shock to nonlethal hypotension by a monoclonal antibody reacting with factor XII (Pixley et al. 1993). Infusion of a low dose of *E. coli* LPS (4 ng/kg) in human volunteers activates the kallikrein kinin system (Dela Cadena et al. 1993). In meningcoccal septic shock low levels of prekallikrein and kallikrein inhibitor have been documented suggesting a significant and long-lasting activation (Brandtzaeg and Kierulf 1992). Persistently low levels of factor XII and high molecular weight kininogens are associated with unfavorable outcome in patients with systemic inflammatory response syndrome (Pixley et al. 1995). The hypotensive effect of bradykinin is partly explained by its ability to induce nitric oxide in the endothelial cells.

16 Activation of the Complement System in Septic Shock

The complement system is a key mediator system in defending the host against intruding micro-organisms. Uncontrolled activation leading to massive generation of anaphylatoxins (C3a, C4a, and most importantly C5a) may, however, have a detrimental effect on the host. Low doses, (2–4 ng/kg) of *E. coli* LPS injected intravenously into human volunteers do not activate the complement system, indicating that the threshold is higher than for other cascade systems (Moore et al. 1987, van Deventer et al. 1990). The importance of the complement system in the development of septic shock has been clearly documented in the baboon septic shock model. Infusion of increasing amounts of live *E.coli* leads to a dose-dependent activation of the complement system (de Boer et al. 1993). Neutralization of C5a by polyclonal antibodies converts an otherwise lethal septic shock to a sublethal septic state (HANGEN et al. 1989).

The complement system is activated in septic shock patients. Initially the observations were based on a significant decrease in the levels of complement factors (McCABE 1973; WHALEY et al. 1979; SPRUNG et al. 1986). Complement activation has recently been monitored more closely by quantitation of activation products of key complement factors that expose neoepitopes (HACK et al. 1989; BRANDTZAEG et al. 1989b; LIN et al. 1993). By this technique the degree of complement activation and the relative contribution of the two different activation pathways have been monitored in three different groups of patients developing septic shock. In meningococcal patients the level of complement activation was closely associated with the plasma level of LPS and was significantly higher in patients with septic shock than in other clinical categories (BRANDTZAEG et al. 1989b). A follow-up study of the same patients demonstrated activation of both the alternative and the classical pathways. The massive activation in shock patients was, however, caused predominantly by alternative pathway activation (BRANDTZAEG et al. 1996a).

HACK et al. (1989) documented formation of C1 and C1-esterase inhibitor complexes as well as increased levels of C4 activation products, indicating classical pathway activation in a group of ICU patients developing septic shock. In a similar population LIN et al. (1993) concluded that the alternative pathway activation is quantitatively the most important. Septic shock is thus associated with significant complement activation, and both pathways may contribute to the activation. The alternative pathway presumably contributes the most to C3 and C5b-9 activation.

Vitronectin and clusterin, two plasma proteins inhibiting the aggregation and insertion of C5b-9 in membranes, are reduced in shock patients, suggesting consumption during activation of the complement cascade (HØGÅSEN et al. 1994). The levels are negatively correlated to levels of the terminal complement complex (vitronectin–C5b-9) and vitronectin–thrombin-antithrombin complexes. Formation of these complexes, which are rapidly removed from the circulation, may contribute to the low levels of complement inhibitors (HØGÅSEN et al. 1994). Low levels of inhibitors of the coagulation system, such as protein C and antithrombin III, have previously been demonstrated in the same patients (BRANDTZAEG et al. 1989d).

Serial studies of patients with profound meningococcal septic shock have disclosed that the complement system is increasingly activated during the first 12–24 h of treatment (BRANDTZAEG et al. 1989b). This pattern differs markedly from the rapid downregulation of bioactive cytokines (TNF-α, IL-1, IL-6, IL-8), vasoactive intestinal peptide, and markers of coagulation and fibrinolysis observed in the same patients (BRANDTZAEG et al. 1989c, 1990; WAAGE et al. 1989, 1994; HALSTENSEN et al. 1993). Intervention strategies to ameliorate complement activation in septic shock, particularly in patients with overwhelming bacteremia, should be tested in the future.

17 Activation of the Coagulation System in Septic Shock

The coagulation system is activated either via the extrinsic (tissue factor) or the intrinsic (via factor XII) pathways. The interaction of factor VIIa with tissue factor represents the physiological activation pathway of coagulation leading to formation of the prothrombinase complex (activated factors X and V on platelets lipid membranes; FURIE and FURIE 1992; LEVI et al. 1993). Prothrombin is converted to thrombin, splitting fibrinogen to monomeric fibrin and several small peptides.

When a low doses of *E. coli* LPS (2–4 ng/kg) were injected as a bolus to human volunteers, LPS was detectable in plasma for approximately 15–30 min, with a maximum of 13 pg/ml (VAN DEVENTER et al. 1990). Activation of the coagulation system was indicated by increased levels of the prothrombin activation peptides F_1 and F_2 and thrombin-antithrombin and lasted for 4–6 h. Activation of the plasma contact system via factor XII (the Hagemann factor–prekallikrein–high molecular weight kininogen complex) was not detected during this experiment.

In the baboon septic shock model activation of the coagulation system is associated with the development of shock. By inhibiting the tissue factor–factor VII–factor X–factor V–prothrombin pathway at various levels, the baboons survived otherwise lethal septic shock (TAYLOR et al. 1987, 1988, 1991; CREASEY et al. 1993). Baboon experiments which block the activation of Factor XII and thus the plasma contact system ameliorate the hypotensive response but not the activation of the coagulation system (PIXLEY et al. 1993). The mechanisms through which activation of the coagulation system influence the vascular tone is not well understood. Activation of the coagulation system enhances other inflammatory mediator systems. The use of tissue factor pathway inhibitor that inactivates the tissue factor–factor VII–factor X complex, effectively downregulates the IL-6 response in baboons challenged with *E. coli* (CREASEY et al. 1993). Principles which inhibit activation of the coagulation system protect various organs from formation of microthrombi and consequently hypoxia and impaired function.

Markers of disseminated intravascular fibrin formation and increased fibrinolysis are readily observed in patients with severe septic shock. Reduced levels of fibrinogen, platelets, various coagulation factors, protein C and antithrombin III concomitantly with increased levels of fibrinopeptide A, thrombin-antithrombin complexes, D dimers, and other fibrin split products all indicate increased coagulation and fibrinolysis. The extrinsic pathway is thought to be activated in septic shock either by upregulation of tissue factor on circulating monocytes or at the luminal side of the endothelial cells. In the baboon *E. coli.* shock model monocytes in the peripheral blood reveal procoagulant activity, i.e., upregulation of tissue factor, whereas endothelial cells do not (DRAKE et al. 1993). Circulating monocytes may represent the key cell which induces the coagulopathic response in septic shock in man. They develop a transitory procoagulant state by a upregulation of tissue factor and downregulation of the fibrinolytic potential after LPS stimulation which later is reversed (Kierulf and Andersen 1990). Increased levels of tissue factor have been detected in monocytes isolated from patients with fulminant meningococcemia (Østerud and Flaegstad 1983). Whether the endothelium really upregulates tissue factor in septic shock has so far never been compellingly demonstrated in man.

The procoagulant state induced by bacteremia appears to be a phylogenetical old pattern of reaction. It is observed in the horseshoe crab (*Limulus polyphemus*), a "living fossil." While this may be an appropriate response to wall off intruding micro-organisms locally in tissues, the response is highly inappropriate when it causes generalized intravascular fibrin formation, plugging of the microvasculature, and bleeding.

One should, however, bear in mind that the baboon experiments which compellingly document the link between coagulation per se and the general inflammatory response may reflect only the situation in overwhelming human bacteremia and not that in the majority of patients.

18 Activation and Inhibition of the Fibrinolytic System in Septic Shock

Activation of the fibrinolytic system is closely coordinated with activation of the coagulation system. When human volunteers were challenged with 2–4 ng/kg *E. coli* LPS intravenously, increased fibrinolysis was observed during the first 1–3 h (Suffredini et al. 1989a; van Deventer et al. 1990). Tissue plasminogen activator released from the endothelium converts plasminogen to plasmin. Later the fibrinolysis is increasingly inhibited through the release of plasminogen activator inhibitor 1 from the endothelial cells which complexes with and inactivates tissue plasminogen activator (plasminogen activator inhibitor). The net effect of the marked inhibition of the fibrinolysis is a shift in the balance of coagulation and fibrinolysis to a procoagulant state. The same has been observed in massive endotoxinemia caused by *N. meningitidis* (Engebretsen et al. 1886; Brandtzaeg et al. 1990). The pronounced inhibition of fibrinolysis may partly explain thrombotic organ complications in kidneys, adrenal, liver, lungs, and spleen. Postmortem examinations of baboons in septic shock and patients dying of fulminant meningococcemia reveal diffuse formation of thrombi in these organs, which apparently are dissolved at a later stage (Brandtzaeg 1995). The reduced organ perfusion contributes to the elevated markers of tissue hypoxia and cell injury in plasma (lactic acid, creatinine, uric acid) and lowered pH (Creasey et al. 1993).

19 Phospholipase A_2 in Septic Shock

Phospholipase A_2 induces a profound hypotensive response if injected intravenously in animals (PRUZANSKI and VADAS 1991). TNF-α and IL-1 enhance the production and release of phospholipase A_2. Infusion of *Salmonella abortus equi* LPS (4 ng/kg) in cancer patients as part of an experimental treatment protocol induced a 20-fold increase in circulating levels of phospholipase A_2 (PRUZANSKI et al. 1992). In 12 patients with septic shock the levels of PLA_2 were 14- to 260-fold higher than baseline (VADAS et al. 1988). The source of phospholipase A_2 in these patients is unknown. The levels did not correlate with the levels of pancreatic enzymes, suggesting another mechanism than that observed during acute pancreatitis. The role of phospholipase A_2 in human septic shock is still unclear, but it may play an important role in the generation of arachidonic acid metabolites and platelet activating factor.

20 Endorphins

β-Endorphin and lipotropin are elevated in the primate *E. coli* LPS septic shock model (GURLL et al. 1988). Blockade with the opiate receptor antagonist naloxone converts a lethal septic shock to sublethal hypotension in the same model. β-Endorphin is increased in plasma in human volunteers challanged with LPS (CASALE et al. 1990). Treatment of septic shock in patients with naloxone has improved certain circulatory parameters but not survival (NAPOLITANO and CHERNOW 1988).

21 Vasoactive Intestinal Peptide and Septic Shock

Vasoactive intestinal peptide derived from the autonomous peripheral nervous system is a potent vasodilator which is released into the circulation during canine and porcine septic shock (BONE et al. 1980; FREUND et al. 1981; REVHAUG et al. 1988).

The release occurs primarily in the splancnic area, and higher VIP levels are detected in the splancnic than the general circulation. The local release of vasoactive intestinal peptide may represent an effort to increase the perfusion of the gut in septic shock. It is not released during experimental porcine cardiogenic shock (REVHAUG et al. 1988). Significantly increased levels of vasoactive intestinal peptide have been detected in fulminant meningococcal septic shock (BRANDTZAEG et al. 1989c).

22 The Myocardial Response in Septic Shock

Contrary to previous belief, cardiac performance is basically altered in septic shock (PARRILLO 1993). Both ventricles are often dilated and reveal a decreased ejection fraction. The changes may represent a compensatory mechanism since it is more readily observed in surviving than nonsurviving patients. The changes are normalized within 7–10 days. In the human *E. coli* LPS challenge model the same changes are observed (SUFFREDINI et al. 1989b). The reason for the reduced myocardial contractility is presently unknown. A set of explanations have been proposed over the years, including altered myocyte ATP metabolism, reduced coronary flow, cardiodepressant activity of proinflammatory cytokines (TNF-α), unidentified humoral principles collectively called myocardial depressant factors, and recently nitric oxide in the myocardium (PARILLO 1993). Recent human studies suggest that the coronary flow is not significantly reduced in human septic shock and therefore cannot explain the myocardial dysfunction (CUNNION et al. 1986). A local dysregulation of the myocardial circulation may, however, exist (DHAINAUT et al. 1987).

23 Conclusion

Septic shock is a multimediator disease. We have identified a large number homeostatic systems that directly or indirectly may contribute to the vasodilation. Septic shock caused by acute overwhelming bacteremia represents one clinical entity. Most animal models developed to study septic shock reflect this clinical situation. Studies of patients with fulminant meningococcal septic shock have been particularly valuable in identifying disease mechanisms associated with this clinical entity.

We know less about the pathogenetic mechanisms associated with the development of septic shock in the majority of our patients with underlying diseases. The association with plasma LPS or other bacterial cell wall components is less consistent. Other principles may induce the inflammatory response. We lack the means to evaluate the "degree of sensitization" for bacterial components of a particular patient. New animal models should be developed which reflect this situation.

In the future we may discover that the disease mechanisms leading to hypotension, hypoperfusion, and organ dysfunction differ significantly among the different patients groups and thus require individual treatment approaches. We will undoubtedly discover new mediators that contribute to the circulatory failure and learn more about the relative contribution of each system at the different stages of septic shock. Research so far suggests that septic shock involves all component of inflammation. It is not likely that we ever will discover one "master" mediator which we easily can manipulate to the benefit of our patients.

References

Andersen BM, Solberg O, Bryn K, Fröholm LO, Gaustad P, Høiby EA, Kristiansen BE, Bøvre K (1987) Endotoxin liberation from *Neisseria meningitidis* isolated from carriers and clinical cases. Scand J Infect Dis 19: 409–419

Behre G, Schedel I, Nentwig B, Wörmann B, Essink M, Hiddemann W (1992) Endotoxin concentration in neutropenic patients with suspected gram-negative sepsis: correlation with clinical outcome and determination of antiendotoxin core antibodies during therapy with polyclonal immunoglobulin M-enriched immunoglobulins. Antimicrob Agents Chemother 36: 2139–2146

Benacerraf B, Seberstyen MM, Schlossman S (1959) A quantitative study of the kinetics of blood clearance of p^{32}-labelled *Escherichia coli* and staphylococci by the reticuloendothelial system. J Exp Med 110: 27–48

Bone R, Said S, Hiller C, Wilson F (1980) Vasoactive intestinal polypeptide in endotoxin shock. Am Rev Respir Dis 121: 113

Bone RC (1991) The pathogenesis of sepsis. Ann Intern Med 115: 457–469

Bone RC (1993) Gram-negative sepsis: a dilemma of modern medicine. Clin Microbiol Rev 6: 57–68

Brandtzaeg P (1995) Pathogenesis of meningococcal infections. In: Cartwrit K (ed) Meningococcal disease. Wiley, Chichester, pp 71–114

Brandtzaeg P, Kierulf P (1992) Endotoxin and meningococcemia. Intravascular inflammation induced by native endotoxin in man. In: Ryan JL, Morrison DC (eds) Immunopharmacology and pathophysiology. CRC Press, Boca Raton, pp 327–346 (Bacterial endotoxic eipopolysaccharides, vol 2)

Brandtzaeg P, Kierulf P, Gaustad P, Skulberg A, Bruun JN, Halvorsen S, Sørensen E (1989a) Plasma endotoxin as a predictor of multiple organ failure and death in systemic meningococcal disease. J Infect Dis 159: 195–204

Brandtzaeg P, Mollnes TE, Kierulf P (1989b) Complement activation and endotoxin levels in systemic meningococcal disease. J Infect Dis 160: 58–65

Brandtzaeg P, Øktedalen O, Kierulf P, Opstad PK (1989c) Elevated VIP and plasma endotoxin in human gram-negative septic shock. Regul Pept 24: 37–44

Brandtzaeg P, Sandset M, Joø GB, Øvstebø R, Kierulf P, Abildgaard U (1989d) The quantitative association of plasma endotoxin, antithrombin, protein C, extrinsic coagulation pathway inhibitor and fibrinopeptide A in systemic meningococcal disease. Thromb Res 55: 459–470

Brandtzaeg P, Joø GB, Brusletto B, Kierulf P (1990) Plasminogen activator inhibitor 1 and 2, alpha-2-antiplasmin, plasminogen, and endotoxin levels in systemic meningococcal disease. Thromb Res 57: 271–278

Brandtzaeg P, Bryn K, Kierulf P, Øvstebø R, Namork E, Aase B, Jantzen E (1992) Meningococcal endotoxin in lethal septic shock plasma studied by gas chromatography, mass-spectrometry, ultracentrifugation, and electron microscopy. J Clin Invest 89: 816–823

Brandtzaeg P, Øvstebø R, Kierulf P (1995) Bacteremia and compartmentalization of LPS in meningococcal disease. In: Levin J, Alving CR, Munford R, Redl H (eds) Lipopolysaccharides from genes to therapy. Wiley, New York pp 219–233

Brandtzaeg P, Osnes L, Øvstebø R, Joø GB, Westvik ÅB, Kierulf P (1996a) Net inflammatory capacity of human septic shock plasma evaluated by a monocyte based target cell assay. J Exp Med (in press)

Brandtzaeg P, Høgåsen K, Kierulf P, Mollnes TE (1996b) The excessive complement activation in fulminant meningococcal septicemia is predominantly caused by alternative pathway activation. J Infect Dis 173: 647–655

Brazilian Purpuric Fever study Group (1987) Epidemic purpura fulminans associated with antecedent purulent conjunctivitis. Lancet 2: 757–763

Burrell R (1994) Human responses to bacterial endotoxin. Circ Shock 43: 137–153

Calandra T, Baumgartner JD, Grau GE, Wu MM, Lambert PH, Schellekens J, Verhoef J, Glauser MP, and the Swiss-Dutch J5 Immunoglobulin Study Group (1990) Prognostic values of tumor necrosis factor/cachectin, interleukin-1, interferon-alpha, and interferon-gamma in the serum of patients with septic shock. J Infect Dis 161: 982–987

Calandra T, Gerain J, Heumann D, Baumgartner JD, Glauser MP, and the Swiss-Dutch J5 Immunoglobulin Study Group (1991) High circulating levels of interleukin-6 in patients with septic shock: evolution during sepsis, prognostic value, and interplay with other cytokines. Am J Med 91: 23–29

Cannon JC, Tompkins RG, Gelfand JA, Michie HR, Stanford GG, van der Meer JWM, Enders S, Lonnemann G, Corsetti J, Chernow B, Wilmore DW (1990) Circulating interleukin-1 and tumor necrosis factor in septic shock and experimental endotoxin fever. J Infect Dis 161: 79–84

Casale TB, Ballas ZK, Kaliner MA, Keahey TM (1990) The effect of intravenous endotoxin on various host-effector molecules. J Allergy Clin Immunol 85: 45–51

Casey LC, Balk RA, Bone RC (1993) Plasma cytokine and endotoxin levels correlate with survival in patients with the sepsis syndrome. Ann Intern Med 119: 771–778

Cavaillon JM, Pitton C, Fitting C (1994) Endotoxin tolerance is not a LPS-specific phenomenon: partial mimicry with IL-1, IL-10 and TGFβ. J Endotoxin Res 1: 21–29

Chetty C, Schwab JH (1987) Endotoxin-like products of grampositive bacteria. In: Rietschel ET (ed) Chemistry of endotoxin. Elsevier Amstrdam, pp 376–410 (Handbook of endotoxin, vol 1)

Cochrane CG (1985) The contact activation system in septic shock In: Hinshaw LB (ed) Pathophysiology of endotoxin. Elsevier, Amsterdam, pp 286–298 (Handbook of endotoxin, vol 2)

Creasey AA, Chang ACK, Feigen L, Wün TC, Taylor FB, Hinshaw LB (1993) Tissue factor pathway inhibitor reduces mortality from *Escherichia coli* septic shock. J Clin Invest 91: 2850–2860

Cunnion RE, Schaer GL, Parker MM, Natanson C, Parillo JE (1986) The coronary circulation in human septic shock. Circulation 73: 637–644

Da Silva AMT, Kaulbach HC, Chuidian FC, Lambert DR, Suffredini AF, Danner RL (1993) Shock and multiple organ dysfunction after self administration of Salmonella endotoxin. N Engl J Med 328: 1457–1460

Damas P, Reuter A, Gysen P, Demonty J, Lamy M, Franchimont P (1989) Tumor necrosis factor and interleukin-1 serum levels during severe sepsis in humans. Crit Care Med 17: 975–978

Danner RL, Elin RL, Husseini JM, Wesley RA, Reilly JM, Parillo JE (1991) Endotoxemia in human septic shock. Chest 99: 169–175

De Boer JP, Creasey AA, Chang A, Roem D, Eerenberg AJM, Hack CE, Taylor FB (1993) Activation of the complement system in baboons challenged with live *Escherichia coli*: correlation with mortality and evidence for a bifasic activation pattern. Infect Immun 61: 4293–4301

De Groote MA, Martin MA, Densen P, Pfaller MA, Wenzel RP (1989) Plasma tumor necrosis factor levels in patients with presumed sepsis. JAMA 262: 249–251

Debets JMH, Kampmeijer R, van der Linden MPMH, Buurman WA, van der Linden CJ (1989) Plasma tumor necrosis factor and mortality in critically ill septic patients. Crit Care Med 17: 489–494

DeLa Cadena RA, Suffredini AF, Page JD, Pixley RA, Kaufman N, Parrillo JE, Colman RW (1993) Activation of the kallikrein-kinin system after endotoxin administration to normal human volunteers. Blood 81: 3313–3317.

Derkx B, Marchant A, Goldman M, Bijlmer R, van Deventer S (1995) High levels of interleukin-10 during the initial phase of fulminant meningococcal septic shock. J Infect Dis 171: 229–232

DeVoe IW, Gilchrist JE (1973) Release of endotoxin in the form of cell wall blebs during in vitro growth of *Neisseria meningitidis*. J Exp Med 138: 1156–1167

Dhainaut JF, Huyghebaert MF, Monsallier JF et al. (1987) Coronary hemodynamics and myocardial metabolism of lactate, free fatty acid, glucose, and ketones in patients with septic shock. Circulation 75: 533–541

Drake TA, Cheng J, Chang A, Taylor FB (1993) Expression of tissue factor, thrombomodulin, and E-selectin in baboons with lethal *Escherichia coli* sepsis. Am J Pathol 142: 1458–1470

Dziarski R (1993) The 70 kDa peptidoglycan/lipopolysaccharide receptor. In: Levin J, Alving CR, Munford RS, Stütz PL (eds) Bacterial endotoxin: recognition and effector mechanisms. Elsevier, Amsterdam, pp 151–165

Elin RJ, Wolff SM, McAdam KPWJ, Chedid L, Audibert F, Bernard C, Oberling F (1981) Properties of reference *Escherichia coli* endotoxin and its phthalylated derivative in humans. J Infect Dis 144: 329–336

Engebretsen LF, Kierulf P, Brandtzaeg P (1986) Extreme plasminogen activator inhibitor and endotoxin values in patients with meningococcal disease. Tromb Res 42: 713–716

Fischer E, van Zee KJ, Marano MA, Rock CS, Kenny JS, Poutsiaka DD, Dinarello CA, Lowry SF, Moldawer LL (1992) Interleukin-1-receptor antagonist circulates in experimental inflammation and in human disease. Blood 79: 2196–2200

Flegel WA, Wolpl A, Mannel DN, Northoff N (1989) Inhibition of endotoxin-induced activation of human monocytes by human lipoproteins. Infect Immun 57: 2237–2245

Fong Y, Moldawer LL, Marano M, Wei H, Tatter SB, Clarick RH, Santhanam U, Sherris D, May LT, Sehgal PB, Lowry SF (1989) Endotoxemia elicits increased circulating β_2-INF/IL-6 in man. J Immunol 142: 2321–2324

Fong Y, Marano MA, Moldawer LL, Calvano SE, Kenny JS, Allison AC, Cerami A, Shires GT, Lowry SF (1990) The acute splancnic and peripheral tissue metabolic response to endotoxin in humans. J Clin Invest 85: 1896–1904

Foreman NK, Wng WC, Cullen EJ, Stidham GL, Pearson TA, Shenep JL (1991) Endotoxic shock after transfusion of contaminated red blood cells in a child with sickle cell disease. Pediatr Infect Dis J 10: 624–626

Frank MM, Fries LF (1988) The role of complement in defence against bacterial disease. Clin Immunol Allergy 2: 335–361

Freund H, Ebeid AM, Fischer JE (1981) An increase in vasoactive intestinal peptide levels in canine endotoxin shock. Surg Gynecol Obstet 152: 604–606

Furie B, Furie BC (1992) Molecular and cellular biology of blood coagulation. N Engl J Med 326: 800–806

Girardin E, Gray GE, Dayer JM, Roux-Lombard P, the J5 Study Group, Lambert PH (1988) Tumor necrosis factor and interleukin-1 in the serum of children with severe infectious purpura. N Engl J Med 319: 397–400

Glauser MP, Zanetti G, Baumgartner JD, Cohen J (1991) Septic shock: pathogenesis. Lancet 338: 732–739

Glauser MP, Heumann D, Baumgartner JD, Cohen J (1994) Pathogenesis and potential strategies for prevention and treatment of septic shock: an update. Clin Infect Dis 18: S205–S216

Granowitz EV, Santos AA, Poutsiaka DD, Cannon JG, Wilmore DW, Wolff SM, Dinarello CA (1991) Production of interleukin-1-receptor antagonist during experimental endotoxinaemia. Lancet 228: 1423–1424

Granowitz EV, Porat R, Mier JW, Orencole SF, Kaplanski G, Lynch EA, Ye K, Vannier E, Wolff SM, Dinarello CA (1993) Intravenous endotoxin suppresses the cytokine response of peripheral blood mononuclear cells of healthy humans J Immunol 151: 1637–1645

Guidet B, Barakett V, Vassal T, Petit JC, Offenstadt G (1994) Endotoxemia and bacteremia in patients with sepsis syndrome in the intensive care unit. Chest 106: 1194–1201

Gurll HJ, Reynolds DG, Holaday JW (1988) Evidence for a role of endorphins in the cardiovascular pathophysiology of primate shock. Crit Care Med 16: 521–530

Hack CE, Nuijens JH, Felt-Bersma RJF, Schreuder WO, Eerenberg-Belmer AJM, Paardekooper J, Bronsveld W, Thijs LG (1989) Elevated plasma levels of anaphylatoxins C3a and C4a are associated with a fatal outcome in sepsis. JAMA 86: 20–26

Halstensen A, Ceska M, Brandtzaeg P, Redl H, Naess A, Waage A (1993) Interleukin-8 in serum and cerebrospinal fluid from patients with meningococcal disease 167: 471–475

Hangen DH, Stevens JH, Satoh PS, Hall EW, O'Hanley PT, Raffin T (1989) Complement levels in septic primates treated with anti-C5a antibodies. J Surg Res 46: 195–199

Hicklin H, Verghese A, Alvarez S (1987) Dysgonic fermenter 2 septicemia. Rev Infect Dis 9: 884 – 890

Høgåsen K, Mollnes TE, Brandtzaeg P (1994) Low levels of vitronectin and clusterin in acute meningococcal disease are closely associated with formation of the terminal complement complex and vitronectin-thrombin-antithrombin complex. Infect Immun 62: 4874–4880

Hurley JC (1992) Antibiotic–induced release of endotoxin: a reappraisal. Clin Infect Dis 15: 840–854

Jafari HS, McCracken GH (1992) Sepsis and septic shock: a review for clinicians. Pediatr Infect Dis J 11: 739–749

Kierulf P, Andersen ÅBW (1990) Lipopolysaccharide-instructed, cryopreserved, human monocytes sequentially convert plasma fibrinogen to fibrin and lyse the fibrin formed. Pathobiology 59: 197–199

Kuhns DB, Alvord WG, Gallin JI (1995) Increased circulating cytokines, cytokine antagonists, and E-selectin after intravenous administration of endotoxin in humans. J Infect Dis 171: 145–152

Levi M, ten Cate H, van der Poll T, van Deventer SJH (1993) Pathogenesis of disseminated intravascular coagulation in sepsis. JAMA 270: 975–979

Levin J, Poore TE, Zauber NP, Oser RS (1970) Detection of endotoxin in the blood of patients with sepsis due to gram-negative bacteria. N Engl J Med 283: 1313–1316
Levin J, Poore TE, Young NS, Margolis S, Zauber NP, Townes AS, Bell WR (1972) Gram-negative sepsis: detection of endotoxemia with the Limulus test. Ann Intern Med 76: 1–7
Lin RY, Astiz ME, Saxon JC, Saha DC, Rackow EC (1993) Alterations in C3, C4, factor B, and related metabolites in septic shock. Clin Immunol Immunopathol 69: 136–142
Lynn WA, Cohen J (1995) Adjunctive therapy for septic shock: a review of experimental approaches. Clin Infect Dis 20: 143–158
Marchant A, Devière J, Byl B, De Groote D, Vincent JL, Goldman M (1994) Interleukin-10 production during septicaemia. Lancet 343: 707–708
Martich GD, Danner RL, Ceska M, Suffredini AF (1991) Detection of interleukin 8 and tumor necrosis factor in normal humans after intravenous endotoxin: the effect of antiinflammatory agents. J Exp Med 173: 1021–1024
Martich GD, Boujoukos AJ, Suffredini AF (1993) Response of man to endotoxin. Immunobiology 187: 403–416
Marty C, Misset B, Tamion F, Fitting C, Carlet J, Cavaillon JM (1994) Circulating interleukin-8 concentrations in patients with multiple organ failure of septic and non septic origin. Crit Care Med 22: 673–679
McCabe WR (1973) Serum complement levels in bacteremias due to gram-negative organisms. N Engl J Med 288: 21–23
McCartney AC, Banks JG, Clements GB, Sleigh JD, Tehrani M, Ledingham IM (1983) Endotoxaemia in septic shock: clinical and post mortem correlations. Intensive Care Med 9: 117–122
Mercier JC, Beaufils F, Hartmann JF, Azéma D (1988) Hemodynamic patterns of meningococcal shock in children. Crit Care Med 16: 27–33
Michie HR, Manogue KR, Spriggs DR, Revhaug A, O'Dwyer S, Dinarello CA, Cerami A, Wolff SH, Wilmore DW (1988) Detection of circulating tumor necrosis factor after endotoxin administration. N Engl J Med 318: 1481–1486
Monnoz C, Misset, B, Fitting C, Blériot JP, Cavaillon JM (1991) Dissociation between plasma and monocyte-associated cytokines during sepsis. Eur J Immunol 21: 2177–2184
Moore FC, Moss NA, Revhaug A, Wilmore D, Mannick JA, Rodrick ML (1987) A single dose of endotoxin activates neutrophils without activating complement. Surgery 102: 200–205
Morrison DC (1990) Diversity of mammalian macromolecules which bind to bacterial lipopolysaccharides. In: Nowotny A, Spitser JJ, Ziegler EJ (eds) Cellular and molecular aspects of endotoxin reactions. Elseviers, Amsterdam, pp 183–189
Morrison DC, Ryan JL (1987) Endotoxins and disease mechanisms. Annu Rev Med 38: 417–432
Napolitano L, Chernow B (1988) Endorphins in circulatory shock. Crit Care Med 16: 566–567
Natanson C, Hoffman WD, Suffredini AF, Eichacker PQ, Danner RL (1994) Selected treatment strategies for septic shock based on proposed mechanisms of pathogenesis. Ann Intern Med 120: 771–783
Nies AS, Forsyth RP, Williams HE, Melmon KL (1968) Contribution of kinins to endotoxin shock in unanathetizied rhesus monkeys. Circ Res 22: 155–164
Offner F, Vogelaers JPD, Colardyn F, Baele G, Baudrihaye M, Vermeulen A, Leroux-Roels G (1990) Serum tumor necrosis factor levels in patients with infectious disease and septic shock. J Lab Clin Med 116: 100–105
Østerud B, Flaegstad T (1983) Increased tissue tromboplastin activity in monocytes of patients with meningococcal infection: related to unfavourable prognosis. Thromb Haemostas 49: 5–7
Parillo JE (1993) Pathogenetic mechanisms of septic shock. N Engl J Med 328: 1471–1477
Parillo JE, Parker MM, Natanson C, Suffredini AF, Danner RL, Cunnion RE, Ognibene FP (1990) Septic shock in humans. Ann Intern Med 113: 227–242
Petrak RA, Balk RA, Bone RC (1989) Prostaglandins, cyclo-oxygenase inhibitors, and thromboxane synthesis inhibitors in the pathogenesis of multiple organ failure. Crit Care Clin 5: 303–314
Pinsky MR, Vincent JL, Alegre M, Kahn RJ, Dupont E (1993) Serum cytokine levels in human septic shock. Chest 103: 565–575
Pixley RA, DeLa Cadena R, Page JD, Kaufman N, Wyshock EG, Chang A, Taylor FB, Colman RW (1993) The contact system contributes to hypotension but not disseminated intravascular coagulation in lethal bacteremia. J Clin Invest 91: 61–68
Pixley RA, Zellis S, Bankes P, DeLa Cadena PA, Page JD, Scott CF, Kappelmayer J, Wyshock EG, Kelly JJ, Colman RW (1995) Prognostic value of assessing contact system activation and factor V in systemic inflammatory response syndrome. Crit Care Med 23: 41–51

Proud D, Kaplan AP (1988) Kinin formation: mechanisms and role in inflammatory disorders. Annu Rev Immunol 6: 49–83

Pruzanski W, Vadas P (1991) Phospholipase A_2 – a mediator between proximal and distal effectors of inflammation. Immunol Today 12: 143–146

Pruzanski W, Mackensen A, Engelhardt R, Stefanski E, Vadas P (1992) Induction of circulating phospholipase A_2 activity by intravenous infusion of endotoxin in patients with neoplasia. J Immunother 12: 242–246

Revhaug A, Lugren I, Jenssen TG, Giercksky KE, Burhol PG (1988) Vasoactive intestinal peptide in sepsis and shock. Ann N Y Acad Sci 527: 536–545

Rietschel ET, Seydel U, Zähringer U, Schade UF, Brade L, Loppnow H, Feist W, Wang MH, Ulmer AJ, Flad HD, Brandenburg K, Kirikae T, Grimmecke D, Holst O, Brade H (1991) Bacterial endotoxin: molecular relationship between structure and activity. Infect Dis Clin North Am 5: 753–779

Shands JW, McKinney C (1989) Plasma endotoxin: increased levels in neutropenic patients do not correlate with fever. J Infect Dis 159: 777–780

Spinas GA, Bloesch D, Kaufmann MT, Keller U, Dayer JM (1990) Induction of plasma inhibitors of interleukin 1 and TNF-α activity by endotoxin administration to normal humans. Am J Physiol 259: R993–R997

Spinas GA, Keller U, Brockhaus M (1992) Release of soluble receptors for tumor necrosis factor (TNF) in relation to circulating TNF during experimental endotoxinemia. J Clin Invest 90: 533–536

Spitznagel JK (1990) Antibiotic proteins of human neutrophils. J Clin Invest 86: 1381–1386

Sprung CL, Schultz DR, Marcial E, Caralis PV, Gelbard MA, Arnold PI, Long WM (1986) Complement activation in septic shock patients. Crit Care Med 14: 525–528

Suffredini AF, Harpel PC, Parrillo JE (1989a) Promotion and subsequent inhibition of plasminogen activation after administration of intravenous endotoxin to normal subjects. N Engl J Med 320: 1165–1172

Suffredini AF, Fromm RE, Parker MM, Brenner M, Kovacs JA, Wesley RA, Parrillo JE (1989b) The cardiovascular response of humans to the administration of endotoxin. N Engl J Med 321: 280–287

Suffredini AF, Shelhamer JH, Neumann RD, Brenner M, Baltaro RJ, Parrillo JE (1992) Pulmonary and oxygen transport effects on intravenously administered endotoxin in normal humans. Am Rev Respir Dis 145: 1398–1403

Taylor FB, Chang A, Esmon CT, D'Angelo A, Vigano-D'Angelo S, Blick KE (1987) Protein C prevents the coagulopathic and lethal effects of *Escherichia coli* infusion in the baboon. J Clin Invest 79: 918–925

Taylor FB, Emerson TE, Jordan R, Chang AK, Blick KE (1988) Antithrombin-III prevents the lethal effects of *Escherichia coli* infusion in baboons. Circ Shock 26: 227–235

Taylor FB, Chang A, Ruf W, Morrisey JH, Hinshaw L, Catlett R, Blick K, Edgington TS (1991) Lethal *E. coli* septic shock is prevented by blocking tissue factor with monoclonal antibody. Circ Shock 33: 127–135

Tracey KJ, Fong Y, Hesse Dg, Manogue KR, Lee AT, Kuo GC, Lowry SF, Cerami A (1987) Anti-cachectin/TNF monoclonal antibodies prevent septic shock during lethal bacteremia. Nature 330: 662–664

Vadas P, Pruzanski W, Stefanski E, Sternby B, Mustard R, Bohnen J, Fraser I, Farewell V, Bombardier C (1988) Pathogenesis of hypotension in septic shock: correlation of circulating phospholipase A_2 levels with circulatory collapse. Crit Care Med 16: 1–7

Van der Poll T, Büller HR, ten Cate H, Wortel CH, Bauer KA, van Deventer SJH, Hack CE, Sauerwein HP, Rosenberg RD, ten Cate JW (1990) Activation of the coagulation after administration of tumor necrosis factor to nornal subjects. N Engl J Med 322: 1622–1627

Van der Poll T, Romijn JA, Endert E, Borm JJJ, Büller HR, Sauerwein HP (1991a) Tumor necrosis factor mimics the metabolic response to acute infection in healthy humans. Am J Physiol 261: E457–E465

Van der Poll T, Levi M, Büller HR, van Deventer SJH, de Boer JP, Hack CE, ten Cate JW (1991b) Fibrinolytic response to tumor necrosis factor in healthy subjects. J Exp Med 174: 729–732

Van der Poll T, van Deventer SJH, Büller HR, Sturk A, ten Cate JW (1991c) Comparison of the early dynamics of systemic prostacyclin release after administration of tumor necrosis factor and endotoxin to healthy humans. J Infect Dis 164: 599–601

Van der Poll T, van Deventer SJH, Hack CE, Wolbink GJ, Aarden LA, Büller HR, ten Cate JW (1992) Effects on leukocytes following injection of tumor necrosis factor into healthy humans. Blood 79: 693–698

Van Deuren M, van der Ven-Jongekrijg J, Demacker PNM, Bartelink Akm, van Dalen R, Sauerwein RW, Gallati H, Vannice JL, van der Meer JWM (1994) Differential expression of proinflammatory cytokines and their inhibitors during the course of meningococcal infections. J Infect Dis 169: 157–161

Van Deventer SJH, Buller H, ten Cate JW, Sturk A, Pouw W (1988) Endotoxaemia: an early predictor of septicemia in febrile patients. Lancet 1: 605–609

Van Deventer SJH, Büller HR, ten Cate JW, Aarden LA, Hack E, Sturk A (1990) Experimental endotoxinemia in humans: analysis of cytokine release and coagulation, fibrinolytic, and complement pathways. Blood 76: 2520–2526

Van Deventer SJH, Hart M, van der Poll T, Hack CE, Aarden LA (1993) Endotoxin and tumor necrosis factor-α-induced interleukin-8 release in humans. J Infect Dis 167: 461–464

Van Zee KJ, Kohno T, Fischer E, Rock CS, Moldawer LL, Lowry SF (1992) Tumor necrosis factor soluble receptors circulate during experimental and clinical inflammation and can protect against excessive tumor necrosis factor α in vitro and in vivo. Proc Natl Acad Sci USA 89: 4845–4849

Van Zee KJ, Coyle SM, Calvano SE, Oldenburg HSA, Stiles DM, Pribble J, Catalano M, Moldawer LL, Lowry SF (1995) Influence of IL-1 receptor blockade on the human response to endotoxinemia. J Immunol 154: 1499–1507

Waage A, Aasen AO (1992) Different role of cytokin mediators in septic shock related to meningococcal disease and surgery/polytrauma. Immunol Rev 127: 221–230

Waage A, Halstensen A, Espevik T (1987) Association between tumor necrosis factor in serum and fatal outcome in patients with meningococcal disease. Lancet 1: 355–357

Waage A, Brandtzaeg P, Halstensen A, Kierulf P, Espevik T (1989) The complex pattern of cytokines in serum from patients with meningococcal septic shock. J Exp Med 169: 33–38

Waage A, Halstensen A, Espevik T, Brandtzaeg P (1994) Cytokines in meningococcal disease. Baillières Clin Infect Dis 1: 97–108

Waring PM, Waring JL, Metcalf D (1994) Circulating leukemia inhibiting factor levels correlate with disease severity in meningococcemia. J Infect Dis 170: 1224–1228

Whaley K, Khong TY, McCartney AC, Ledingham IM (1979) Complement activation and its control in gram negative endotoxic shock. J Clin Lab Immunol 2: 117–124

Wright SD, Ramos RA, Tobias PS, Ulevitch RJ, Mathison JC (1990) CD14, a receptor for complexes of lipopolysaccharide (LPS) and LPS binding protein. Science 249: 1431–1433

Young LS, Proctor RA, Beutler B, McCabe WR, Sheagren JN (1991) University of California/Davis interdepartmental conference on gram-negative septicemia. Rev Infect Dis 13: 666–687

Zabel P, Wolter DT, Schönharting MM, Schade UF (1989) Oxpentifyllin in endotoxaemia. Lancet 2: 1474–1477

Zumla A (1992) Superantigens, T cells and microbes. Clin Infect Dis 15: 313–320

Bacterial Endotoxin: Chemical Constitution, Biological Recognition, Host Response, and Immunological Detoxification*

E.T. Rietschel[1], H. Brade[1], O. Holst[1], L. Brade[1], S. Müller-Loennies[1], U. Mamat[1], U. Zähringer[1], F. Beckmann[1], U. Seydel[1], K. Brandenburg[1], A.J. Ulmer[1], T. Mattern[1], H. Heine[1], J. Schletter[1], H. Loppnow[1], U. Schönbeck[1], H.-D. Flad[1], S. Hauschildt[2], U.F. Schade[3], F. Di Padova[4], S. Kusumoto[5], and R.R. Schumann[6]

1 Introduction 40

2 Structure of Endotoxin 41

2.1 O-Specific Chain and Core 41
2.2 Lipid A 43
2.3 Chemical Structure of Lipid A 43
2.4 Physical Structure 45

3 Endotoxic Activity of LPS and Lipid A 46

3.1 Humoral LPS Targets 47
3.2 Cellular LPS Targets 47
3.3 Cellular LPS Receptors 49

4 Lipopolysaccharide-Binding Protein 50

5 The CD14 Molecule 51

6 LPS-Induced Signal Transduction 54

7 Regulation of LPS-Induced Cytokine-Production in Macrophages by Lipoxygenase Products 55
7.1 13-Hydroxylinoleic Acid and Macrophage Activation 56

8 Lipid A Determinants Involved in Binding to and Activation of Cells 57

9 Inhibition of Lipid A Binding to Cells and Cellular Activation 57

10 The Pathophysiological Cascade of Events Leading to Gram-Negative Septic Shock 60

11 Anti-Endotoxin Antibodies 61
11.1 Anti-O-Chain Antibodies 62
11.2 Anti-Lipid A Antibodies 62
11.3 Polyclonal Anti-Core Antibodies 64
11.4 Cross-Reactive and Cross-Protective Monoclonal Anti-Core Antibodies 66

12 Final Remarks 68

References 69

---

*Dedicated to our friend and esteemed colleague, Professor Roland Schauer, University of Kiel, Germany, on the occasion of his 60th birthday (April 8, 1996).

[1]Forschungszentrum Borstel, Zentrum für Medizin und Biowissenschaften, Parkallee 22, 23845 Borstel, Germany
[2]Institut für Zoologie und Immunbiologie, Universität Leipzig, Talstr. 33, 04103 Leipzig, Germany
[3]Universitätsklinikum Essen, Zentrum für Chirurgie, Hufelandstr. 55, 45122 Essen, Germany
[4]Sandoz, Preclinical Research, Kohlenstr., 4002 Basel, Switzerland
[5]Department of Chemistry, Faculty of Science, Osaka University, Japan
[6]Max-Delbrück-Centrum für Molekulare Medizin, Abteilung für Onkologie und Angewandte Molekularbiologie, Robert-Rössle-Str. 10, 13125 Berlin, Germany

1 Introduction

The discovery of endotoxin dates from the late nineteenth century when Richard Pfeiffer, then working in Berlin, characterized endotoxins as heat-stable and cell-associated molecules (Westphal et al. 1977), thus distinguishing them from the heat-labile and proteinous exotoxins which are actively secreted by bacteria (Bhakdi et al. 1994). They were first found to be produced by *Vibrio cholerae* bacteria and later by *Salmonella* and *Serratia*. Endotoxins, due to their various potent biological activities soon attracted worldwide scientific interest. Initial chemical analyses of purified endotoxin indicated that it consists essentially of polysaccharide and lipid, and it was therefore termed lipopolysaccharide (LPS). Today the terms endotoxin (Wolff 1904) and lipopolysaccharide (Shear and Turner 1943) are used synonymously for the same molecule.

Endotoxins are known to constitute amphiphilic macromolecules located on the surface of gram-negative bacteria (Raetz 1990; Rietschel and Brade 1992; Rietschel et al. 1994). Released from bacteria in vivo or administered in an isolated form, endotoxins exert both physiological and powerful pathophysiological effects in higher organisms and thus represent important virulence factors of gram-negative bacteria. With the ultimate aim of pharmacologically or immunologically controlling the endotoxic effects observed during severe gram-negative infections, various research groups attempted to characterize the nature of the biologically active region of LPS and to analyze the mode of action of endotoxin in higher organisms. Subsequently it was shown that the lipid component of LPS, termed lipid A, constitutes its toxic and immunostimulatory principle (Galanos et al. 1985; Loppnow et al. 1989), and the chemical structure of lipid A was elucidated (Zähringer et al. 1994). It was further demonstrated that certain humoral host factors interact with lipid A, and that specific cellular receptors recognize lipid A or LPS, particularly if associated with serum components (Schumann et al. 1990; Wright et al. 1990; Raetz et al. 1991; Ulevitch 1993; Morrison et al. 1993). The result of the specific interaction of LPS with cells may be the activation of cells, for example, cellular proliferation as observed with murine B-lymphocytes and, more important in the context of endotoxicity, the formation and secretion of bioactive mediators as produced by murine or human monocytes, macrophages, and vascular cells (Galanos et al. 1992; Loppnow and Libby 1989, 1990). It was also realized that the physiological and pathophysiological host response to endotoxin is mediated by such endogenous mediators which possess intrinsic biological activities (Dinarello et al. 1984; Beutler and Cerami 1989). If formed in small amounts, these mediators are important for the proper functioning of the immune system and its struggle with invading microorganisms. If, however, they are abundantly produced, they cause toxic effects such as pyrogenicity, leukopenia, and ultimately irreversible shock. With the gradual evaluation of the pathway of events: bacteria → endotoxin → mediators → toxic effects, determination of rational approaches designed to interfere with the endotoxin-induced devastating cascade of events became possible.

This chapter summarizes recent advances in our knowledge of the chemical structure, host recognition of endotoxin, and response of the host to this microbial toxin. A brief description is also presented of new immunological strategies to prevent or treat endotoxin-caused systemic inflammation.

2 Structure of Endotoxin

Over the past two decades significant progress has been made in understanding the chemistry (primary structure) and physics (mono- and multimolecular conformation) of the endotoxin molecule. Knowledge of the chemical and physical LPS structure also allowed the establishment of structure-activity relationships, some aspects of which are briefly reviewed in this chapter.

2.1 O-Specific Chain and Core

LPS of various gram-negative bacterial families are made up according to the same architectural principle, i.e., they consist of a polysaccharide portion and the covalently bound lipid A component (Fig. 1). In the classical case of Enterobacteriaceae (e.g., *Salmonella enterica, Escherichia coli*), the polysaccharide portion consists of two regions, the O-specific chain and the core oligosaccharide (LÜDERITZ et al. 1982). The O-specific chain is usually composed of a polymer of repeating oligosaccharide units. A peculiar situation is encountered in *Legionella pneumophila* serogroup 1, where up to 75 residues of a single sugar (5-acetamidino-7-acetamido-8-*O*-acetyl-3,5,7,9-tetradeoxy-D-glycero-L-galacto-

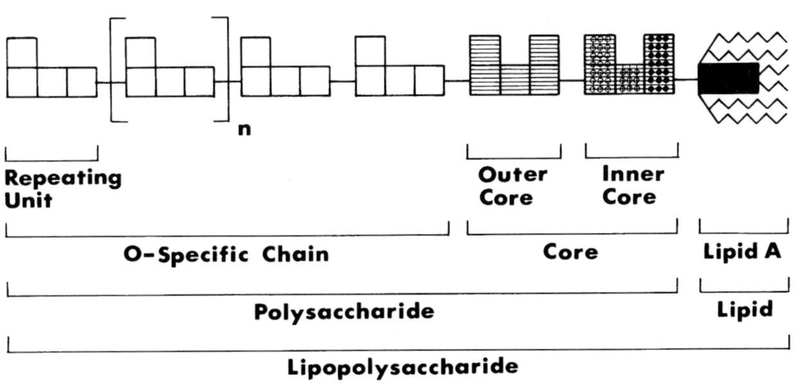

Fig. 1. Schematic structure of a *Salmonella* LPS. Phosphate and other substituents have been omitted from the structure. (From LÜDERITZ et al. 1982)

nonulosonic acid; legionaminic acid) form the O-specific chain, which in this case constitutes a homopolymer (KNIREL et al. 1994; ZÄHRINGER et al. 1995). The O-specific chain is characteristic of and unique for a given LPS and its bacterial origin (serotype). In view of the existence of thousands of serotypes, comparison of different gram-negative bacteria reveals enormous structural variability of the O-specific chain region.

The biosynthesis of the O-specific chain is determined by a cluster of genes termed *rfb*. Mutants harboring a defect in the *rfb* locus or lacking it are able to grow in vitro, indicating that the O-specific chain is dispensable for bacterial viability. In vivo, however, Enterobacteriaceae survive, i.e., resist phagocytosis and killing by bactericidal serum factors such as complement (C'), only if they express an O-specific chain. The O-chain therefore constitutes an important bacterial virulence factor and at the same time a promising target for antibacterial therapy. Recently a 262-bp DNA fragment of *Acetobacter methanolicus* phage Acm1 was found to be sufficient for the suppression of O-specific chain biosynthesis in various enterobacterial strains, including *Escherichia coli*, *Salmonella enterica*, and *Klebsiella pneumoniae* (MAMAT et al. 1991). The results strongly suggest that the Acm1-derived DNA fragment codes for a *trans*-acting regulatory antisense RNA of 97 nucleotides, designated lbi RNA (LPS biosynthesis interfering), being obviously able to interact with conserved target RNA sequences which are linked directly or indirectly to LPS biosynthesis (MAMAT et al. 1995). This antisense RNA-mediated downregulation of O-specific side chain biosynthesis could represent a new approach for the construction of attenuated live vaccine strains applicable to a wide spectrum of gram-negative bacteria.

The core portion consisting of the O-chain proximal outer core and the lipid A proximal inner core is structurally less variable. In the case of *S. enterica* and *E. coli* the outer core exhibits some variability due to different locations and linkages of its constituents (HOLST and BRADE 1992). Nevertheless, common elements are recognized, such as the presence of the pyranosidic hexoses D-glucose (Glc), D-galactose (Gal), and 2-amino-2-deoxy-D-glucose (GlcN). Hexoses are also present in the outer core of *K. pneumoniae* (SÜSSKIND et al. 1995) and *Proteus mirabilis* (RADZIEJEWSKA-LEBRECHT et al. 1990). The inner core of all LPS-expressing gram-negative bacteria contains 2-keto-3-deoxyoctulosonic acid (Kdo) and in *Enterobacteriaceae* and many other families heptose residues of the L-glycero-D-manno configuration which are often phosphorylated. Thus the inner core region of various LPS contains a chemically similar region. In fact a common and typical core segment can be defined that is structurally conserved in all *S. enterica* and *E. coli* serotypes and various other bacteria. The schematic architecture of this core element is shown in Fig. 2. The presence of this region allowed the recent development of a monoclonal antibody (see Sect. 11.4) recognizing an epitope within this region and therefore cross-reacting with all serotypes of *S. enterica*, *E. coli*, and some other genera (DI PADOVA et al. 1993a,b). Shared structures which may constitute epitopes for cross-reactive and possibly cross-protective antibodies are also found in other gram-negative bacterial groups. For example, in LPS of all *Pseudomonas aeruginosa* serotypes,

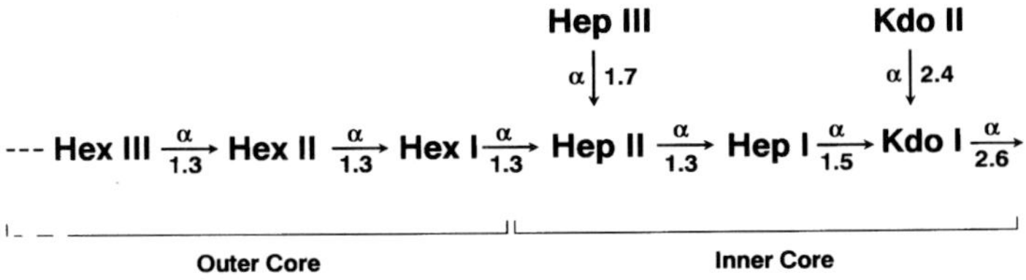

Fig. 2. Common structure present in LPS core types of E.coli and S.enterica. *Hex,* D-Glc, D-Gal, D-GlcNAc; *Hep,* L-glycero-D-manno-heptose; *Kdo,* 2-keto-3-deoxyoctulosonic acid. (From Holst and Brade 1992)

L-glycero-D-manno-heptose, carrying a carbamoyl residue in position 7 was identified as an ubiquitous constituent (Beckmann et al. 1995).

The lipid A component constitutes the chemically most uniform portion of LPS, and its structure is described in more detail below.

Obviously, structural variability of LPS decreases from the surface-exposed O-specific chain over the core region to the polysaccharide-covered and membrane-embedded lipid A component. The reason for this gradient of variability may be the evolutionary pressure exerted by phagocytic cells and antibodies on gram-negative bacteria (Nikaido 1970). It is conceivable that the microbes evaded this pressure by changing their most exposed surface structure, i.e., the O-specific chain, and to a lesser degree the outer core.

2.2 Lipid A

Lipid A represents the covalently bound lipid component of LPS. It can be separated from the polysaccharide region by treatment of LPS with mild acid, which preferentially cleaves the linkage between the Kdo I residue of the inner core and lipid A (Fig. 1). Lipid A has been shown to constitute the endotoxic principle of LPS (Galanos et al. 1985). With the aim of defining those structural and conformational peculiarities which endow it with endotoxic properties, lipid A has been chemically analyzed in detail.

2.3 Chemical Structure of Lipid A

Figure 3 shows the structure of lipid A of four different types of gram-negative bacteria which all express biologically highly active LPS (*E. coli, Haemophilus influenzae, Chromobacterium violaceum,* and *Neisseria meningitidis*; Zähringer et al. 1994). Structurally, these lipid A share a 1,4′-bisphosphorylated β1,6-linked D-GlcN disaccharide (lipid A backbone) which is acylated by four (*R*)-3-hydroxy fatty acids at positions 2, 3, 2′, and 3′. In each case the acyl group at position 2′ of GlcN II carries at its 3-hydroxyl group a further (fifth) fatty acid. The four structures, however, differ in the location of a sixth acyl group (R1 or R2) and the chain length of fatty acids (symbols m, n, and o). As Fig. 3 shows, lipid A of *E. coli* and *H.influenzae* carries this sixth fatty acid at GlcN II and thus possesses an asymmetric distribution of acyl groups over GlcN I and GlcN II (4+2), whereas a symmetric acyl arrangement (3+3) is present in lipid A of *C.violaceum* and *N. meningitidis*. Importantly, the average length of acyl chains is smaller in the latter group (mainly 12 carbon atoms) than in the former (mainly 14 carbon atoms).

These structural examples demonstrate that lipid A of various origins exhibits a similar architecture, but that variations exist concerning the nature and location of acyl groups. More striking examples for structural variability of lipid A are found, for example, in *Rhizobium, Chlamydia,* and *Legionella* where very long chain (C22–C32) acyl groups, also in the form of (W-1)-hydroxy or oxo-derivatives, are present (Hollingsworth and Lill-Elghanian 1989; Nurminen et al. 1985; Brade et al 1986; Moll et al. 1992; Sonesson et al. 1993). In *Rhodobacter sphaeroides* and *R. capsulatus*, part of the 3-hydroxy fatty acids is replaced by 3-oxo fatty acids (Mayer et al. 1990; Takayama and Qureshi 1992); interestingly, these latter lipid A preparations are biologically inactive (Loppnow et al. 1990;

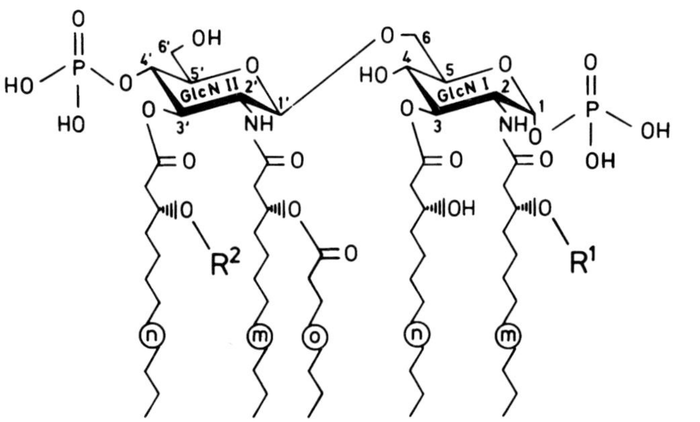

	Nature of		Number of Carbon Atoms		
	R^1	R^2	m	n	o
Escherichia coli	H	14:0	14	14	12
Haemophilus influenzae	H	14:0	14	14	14
Neisseria meningitidis	12:0	H	14	12	12
Chromobacterium violaceum	12:0	H	12	10	12

12:0, dodecanoic acid 14:0, tetradecanoic acid

Fig. 3. Chemical structure of the lipid A component of various gram-negative bacteria. (From Rietschel et al. 1994; Zähringer et al. 1994)

QURESHI et al. 1991; SCHÖNBECK et al. 1994). Further variations of the lipid A structure concern the substitution of phosphate groups and the nature of backbone glycosyl residues (not shown in Fig. 3). Thus GlcN may be replaced by 2,3-diamino-2,3-dideoxy-D-glucose (GlcN3N), as is the case, for example, in *Rhodopseudomonas viridis, Campylobacter jejuni,* and *L.pneumophila* (for literature see ZÄHRINGER et al. 1994). Finally, phosphate groups may be absent from lipid A. This is the case, for example, in *Bacteroides fragilis* lipid A which lacks the nonglycosidic phosphate group (WEINTRAUB et al. 1989). A most remarkable phosphate-free structural variant of lipid A has been identified in *R. leguminosarum* bv. *phaseoli* (BHAT et al. 1994; PRICE et al. 1994). Here during biosynthesis the 4'-phosphate group is enzymatically removed and replaced by an α-linked D-galacturonic acid (GalA) residue. Also, the glycosidic phosphate is cleaved off, followed by oxidation of GlcN I at C1, yielding 2-amino-2-deoxygluconic acid carrying (long-chain) acyl residues in amide and ester linkage. Thus lipid A of *R.leguminosarum* carries a total of two negative charges at the nonreducing and the reducing backbone unit, however, not in the form of phosphate residues, as is the case in *E. coli* but rather as carboxylate groups.

Based on the results of chemical analyses *E. coli* type lipid A has been chemically synthesized (KUSUMOTO 1992). The demonstration of identity of bacterial and synthetic *E. coli* lipid A in all chemical, physicochemical, physical, and biological parameters unequivocally verified the previously deduced and proposed structure to be correct.

2.4 Physical Structure

Lipid A (and LPS) represents amphiphiles and therefore forms aggregates in aqueous medium above the critical micellar concentration. The availability of defined and homogenous synthetic lipid A and partial structures allowed analyses of the nature of such aggregates, i.e., their three-dimensional physical structure. According to present knowledge, biologically active lipid A adopts, at physiological ambient conditions [37 °C, pH 7, high (>90%) water content, presence of Mg^{2+}] exclusively nonlamellar structures (SEYDEL et al. 1993). These nonlamellar structures are either cubic (*S. enterica* sv. *minnesota*) or hexagonal (*R. gelatinosus*). In contrast, biologically nonactive lipid A (*R. capsulatus*) adopt lamellar structures (BRANDENBURG et al. 1993). This suggests that endotoxicity is determined by a defined conformation of lipid A or LPS. The three-dimensional structure of lipid A multimers and the shape of the corresponding monomers of active and nonactive lipid A is shown schematically in Fig. 4. Whether bioactivity of endotoxin requires the presence of a supramolecular structure (in the form of nonlamellar aggregates) is under debate at present (SHNYRA et al. 1993). More likely, however, endotoxicity is expressed by individual endotoxin or lipid A molecules possessing a conformation which, at higher concentration leads to the observed three-dimensional nonlamellar structures (SCHROMM et al. 1995). Recent studies employing disaggregated LPS support the hypothesis that en-

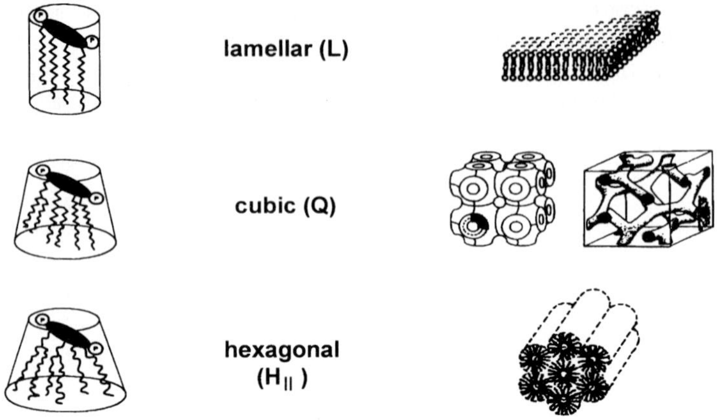

Fig. 4. Physical structure of endotoxically active and nonactive lipid A. Shown are the three-dimensional arrangements and the shape of the corresponding monomer. (From SEYDEL et al. 1993; BRANDENBURG et al. 1993)

dotoxic activity is mediated by single molecules (TAKAYAMA et al. 1994). It further appears that a peculiar spatial shape of lipid A is required for the expression of endotoxicity. We have termed this unique arrangement of lipid A the "endotoxic conformation" (RIETSCHEL et al. 1987).

The concept of a unique lipid A conformation implies that a fitting recognition molecule (receptor) for lipid A should be present either in the circulation or on endotoxin target cells. As the following sections show, the interaction of lipid A with humoral and cellular host factors, leading to a biological response, is in fact highly specific and requires a unique conformation of lipid A on the LPS side and peculiar recognition molecules on the host side.

3 Endotoxic Activity of LPS and Lipid A

The mechanisms involved in the activity of endotoxin are well understood today (MORRISON and RYAN 1987). When gram-negative bacteria multiply or die, endotoxin is released in the form of free LPS or complexed to the outer membrane protein A (FREUDENBERG et al. 1991). Circulating endotoxins constitute a particular class of toxins which induce in host organisms the production of bioactive mediators ultimately responsible for the effects observed during endotoxemia. According to present knowledge, LPS after associating with certain serum factors interacts with receptors expressed by endotoxin target cells such as granulocytes, lymphocytes, vascular cells, and in particular monocytes/macrophages. In response to LPS these cells form and secrete endogenous mediators which are endowed with potent intrinsic bioactivities and which ultimately induce the typical endotoxin effects.

The following sections describe some of the early humoral and cellular LPS targets and some of the consequences of their interaction with lipid A.

3.1 Humoral LPS Targets

Among the humoral factors interacting with LPS figures prominently high-density lipoprotein (HDL) which is able, as is low-density lipoprotein, to attenuate LPS effects (FREUDENBERG et al. 1980; ULEVITCH et al. 1981; MUNFORD et al. 1982; FLEGEL et al. 1989). Other circulating LPS-neutralizing proteins are the bactericidal permeability-increasing protein (BPI; present in serum at a concentration of approximately 1 ng/ml) and the cationic proteins CAP 18, CAP 37, and P15A/P15B (WEISS et al. 1992; ELSBACH and WEISS 1993; MARRA et al. 1990; LEVY et al. 1993; LARRICK et al. 1991; HIRATA et al. 1994). Further serum proteins which associate with LPS include albumin, transferrin, lactoferrin, hemoglobin, lysozyme and a 28-kDa mannose-binding protein. These latter proteins do not modify endotoxin bioactivity to any significant extent (MORRISON 1990).

With regard to the expression of endotoxicity the most important serum protein appears to be the LPS binding protein (LBP), which dramatically augments LPS and lipid A activity, rendering femto- to picogram amounts of LPS bioactive (TOBIAS et al. 1986, 1988). The structure and properties of LBP are described in more detail below. A recently described class of serum proteins, termed septin, is described to possess similar properties as LBP (WRIGHT et al. 1992). Septin could be distinguished from human LBP by means of a monoclonal antibody (WRIGHT 1994). Further studies are, however, indicated to elucidate the peculiarities of septin in comparison to LBP.

3.2 Cellular LPS Targets

The most important target cells of endotoxin are components of the cellular immune system. Thus certain defense cells of almost all species have the ability to recognize minute amounts of LPS, thereby sensing invading micro-organisms. Four cell types can be distinguished which recognize LPS but respond to it in different ways, i.e., by phagocytosis, differentiation or proliferation, and mediator secretion.

1. Polymorphonuclear leukocytes (PMN) take up bacteria and bacterial membrane fragments including LPS, and their phagocytic capacity is greatly enhanced by LPS (SCHADE et al. 1987). As phagocytosis may be regarded as a first and direct measure of the host to eliminate threatening micro-organisms, the phagocytosis-augmenting property of LPS emphasizes its important role in early nonspecific steps of the host defense against microbial invasion. On the other hand, PMN have enzymes which degrade (i.e., de-O-acylate and depho-

sphorylate) LPS and lipid A to nontoxic partial structures (Munford and Hall 1989; Erwin and Munford 1992). Further, PMN contain polycationic proteins which are capable of interacting with LPS. Among these BPI a 55-kDa cationic protein has been shown to be toxic for bacteria and to bind strongly to LPS, thereby efficiently neutralizing its bioactivity (Weiss et al. 1992; Marra et al. 1990, 1992; Gazzano-Santoro et al. 1994; Little et al. 1994; Theofan et al. 1994). Also, the PMN proteins CAP 18 and CAP 37 possess LPS-inactivating properties (Larrick et al. 1991). Finally, PMN may attach to endothelial cells and, by causing damage to the endothelial lining and by penetrating through the vessel wall into the tissue, may contribute substantially to inflammatory reactions.

2. B-lymphocytes of murine origin are stimulated by LPS to proliferation, differentiation and secretion of antibodies (Anderson et al. 1973). This polyclonal activation may also be regarded as an early defense mechanism of the host against pathogenic microorganisms as it yields antibodies of various antibacterial specificities. Also T-lymphocytes (Th1 type) of human origin are, in a monocyte-dependent fashion, activated by LPS to proliferate and to secrete lymphokines (Mattern et al. 1994). Thus LPS may also be involved in the cellular immunity against pathogenic microorganisms. In addition, LPS acts on a subset of murine T-lymphocytes ($CD8^+$ $CD4^-$) which are capable of suppressing the humoral immune response to bacterial polysaccharides such as pneumococcal type III polysaccharide (Baker 1993). Lipid A has been shown to downregulate these T-suppressor cells, thereby augmenting anti-polysaccharide antibody formation (Baker 1993).

3. Monocytes and tissue macrophages are activated by LPS to produce a large variety of bioactive protein mediators which include interleukin (IL)-1, IL-6, IL-8, macrophage migration-inhibitory factor (Bernhagen et al. 1994), and, in particular, tumor necrosis factor α (TNF; Nathan 1987). Many host cells carry receptors for these mediators and are capable of responding to them, for example, by enhanced activity or by apoptosis. If produced in small amounts, these monokines help to eliminate and inactivate invading micro-organisms, such as by causing moderate fever, inducing leukocytosis, attracting defense cells to the infectious focus, activating intracellular microbicidal mechanisms, and initiating an acute-phase response (Echternacher et al. 1990; Havell 1989). If overproduced, however, these hormonelike mediators become a threat to the host organism by causing damage to cells and organs (Tracey et al. 1986; Waage et al. 1987; Vogel and Hogan 1990; Li et al. 1995). Thus multiorgan failure and irreversible shock may be the result of severe gram-negative infection resulting from overwhelming mediator production. On exposure to endotoxin macrophages produce (Nathan 1987), in addition to monokines, reduced oxygen species (superoxide anion, hydrogen peroxide, hydroxyl radical and nitric oxide), bioactive metabolites of arachidonic acid [prostaglandins, thromboxane, and leukotrienes (LT); (Schade et al. 1989a,b; Lüderitz et al. 1989) and of linoleic acid [e.g., (*S*)-13-hydroxylinoleic acid; Schade et al. 1987], and platelet-activating factor (Braquet et al. 1987). Studies using such isolated or corresponding synthetic molecules, biosynthesis inhibitors, antibodies, inactivating enzymes, and

receptor antagonists have confirmed these secondary nonprotein mediators to be intimately involved in the pathophysiology of endotoxicosis.

4. Finally, vascular cells such as endothelial or smooth muscle cells have the capacity upon stimulation by LPS to produce proteinaceous mediators such as the cytokines IL-1, IL-6, and IL-8 (LOPPNOW and LIBBY 1989, 1990, 1992; SCHÖNBECK et al. 1995). These cells also produce a variety of other mediators, including prostacyclin, nitric oxide, platelet-activating factor, interferons, and colony-stimulating factors (MANTOVANI and BUSSOLINO 1991; LIBBY et al. 1991). In addition to mediator release, the expression of adhesion molecules is of great importance in the regulation of inflammatory responses. The expression of adhesion molecules is induced by LPS as well as by IL-1 and TNF (YU et al. 1986; DOHERTY et al. 1989). The activation by endotoxin induces expression of adhesion molecules on cultured vascular endothelial cells as potently as IL-1 (SCHÖNBECK et al. 1994).

3.3 Cellular LPS Receptors

Some 2 decades ago the association of LPS with host cells resulting in their activation was believed to be based on an interdigitation of the lipid A acyl residues into the membrane of cellular targets (KABIR et al. 1978). Thus a rather nonspecific mechanism was thought to cause the initiation of endotoxic events. As a result of the pioneering work of MORRISON, WRIGHT, ULEVITCH, TOBIAS, and others this view has since changed dramatically, and today different types of binding proteins and receptors are known to be involved in the specific recognition of LPS by the host.

Some of these cellular structures, such as the scavenger receptor and the CD11b/CD 18 receptors, mediate binding, uptake, and detoxification of LPS and thus are involved in early steps of host defense (WRIGHT et al. 1989). They do not, however, participate in LPS-induced activation of cells (but see Sect. 5). Also other macrophage/monocyte membrane proteins of 18, 25, 38, and 40 kDa have been implicated in LPS binding. The biological significance of these proteins in LPS-mediated cell activation remains to be elucidated (for literature compare ULEVITCH 1993; MORRISON et al. 1993; KIRIKAE et al. 1991, 1993b).

This is also true for a 73-kDa protein first described in 1988 (LEI and MORRISON 1988a,b). A large body of experimental evidence underlined the importance of the 73-kDa protein in LPS binding (MORRISON 1989; MORRISON et al. 1993). A recent study, however, reported this 73-kDa protein to be identical to cell surface associated albumin (DZIARSKI 1994). This study also showed that cell-bound albumin is not required for cell-activation, thus questioning the significance of the 73-kDa protein in endotoxicity.

The only known structure present on the surface of endotoxin-responsive cells which is unequivocally important for both binding and LPS-induced cellular activation is the CD14 molecule. This glycoprotein, the significance of which was recognized in 1990, is therefore dealt with in some detail below (WRIGHT et al. 1990; LEE et al. 1992; SCHÜTT et al. 1988). Its soluble form, sCD14, is also capable of binding to LPS. It may inhibit LPS activity (SCHÜTT et al. 1992), but interestingly LPS/sCD14 complexes are capable of inducing biological responses in certain CD14-negative cells such as endothelial cells (FREY et al. 1992; PUGIN et al. 1993; ARDITI et al. 1993) and smooth muscle cells (LOPPNOW et al. 1995).

The following sections discuss in more detail the humoral LPS-binding protein LBP and the membrane-associated LPS-binding protein CD14, together with its soluble form.

4 Lipopolysaccharide-Binding Protein

LBP is synthesized in hepatocytes as a glycosylated 58-kDa protein which is constitutively secreted into the blood stream (SCHUMANN et al. 1990; SCHUMANN 1992; TOBIAS and ULEVITCH 1993). The concentration of LBP in normal serum is approximately 14–22 μg/ml with maximum of up to 200 μg/ml in acute-phase serum 24 h after induction (TOBIAS and ULEVITCH 1993; GALLAY et al. 1994a). LBP was first purified from rabbit and human plasma and more recently from murine sources (TOBIAS et al. 1986; SCHUMANN 1992; TOBIAS and ULEVITCH 1993; GALLAY et al. 1993a,b, 1994a). Rabbit and human LBPs have been sequenced, and 69% homology was revealed between the two species (SCHUMANN et al. 1990). The domain responsible for LPS binding is located in the N-terminal region. It is characterized by an accumulation of positively charged amino acids and expresses hydrophobic properties. LBP shares 44% sequence homology with human BPI, the lipid A binding region of which is also located in the N-terminal region (SCHUMANN et al. 1990; LITTLE et al. 1994). A putative LPS-binding domain, characterized by the presence of positively charged amino acids and located in the proximity of a hydrophobic domain is also present in the structurally well-characterized endotoxin-neutralizing protein (ENP) naturally present in the hemolymph of the horseshoe crab *Limulus polyphemus* (HOESS et al. 1993). It is tempting to speculate that ENP, BPI, and LBP adopt a similar conformation exposing cationic and neighboring hydrophobic LPS-binding sites involved in binding of lipid A. However, BPI and ENP do not (as LBP) augment but rather inhibit LPS bioactivity. These different properties of the three LPS-binding proteins appears to depend on structural features of the carboxy terminus.

The only known biological activity of LBP is its capacity to bind to LPS with high affinity (10^{-9} *M*; TOBIAS and ULEVITCH 1993) and to augment LPS bioactivity. The well-known LPS activity enhancing effect of serum in biological systems is at least in part due to the presence and action of LBP. Thus the LPS-induced TNF production and the expression of TNF mRNA is strongly enhanced by LBP (ULEVITCH et al. 1990). Also, LPS-induced responses of PMN can be greatly augmented by the addition of LBP (VOSBECK et al. 1990). On the other hand, neither LPS-induced B-cell activation nor LPS effects caused by large doses of LPS are enhanced by LBP.

The biological activity of LBP is therefore based on its capacity of associating with LPS and thereby augmenting the biological activity of minute amounts of LPS. Most likely LBP associates with LPS or lipid A in molar stoichiometry, and it is possible that LBP singles out LPS monomers from the three-dimensional aggregates described (see Sect. 2.4). LBP may thus be regarded as a biological amplifier which enables the host organism to detect small amounts of LPS that signal the invasion of gram-negative bacteria, i.e., infection. The host organism therefore uses LPS/LBP complexes to activate its defense system to deal appropriately with the invading micro-organism.

The mechanism by which LBP augments LPS activity has been elucidated in great detail in recent years. It is based on the ability of LBP to transfer and deliver LPS molecules to a cellular binding site, i.e., the CD14 protein (HAILMAN et al. 1994). It should be mentioned, however, that according to a recent study LBP catalytically transfers LPS not only to CD14 but also to HDL. Kinetic experiments involving LPS, LBP, and HDL show that initially the augmenting effect of LBP predominates whereas with longer incubation times LBP mediates diffusion of LPS into HDL, thus catalyzing endotoxin neutralization (WURFEL et al. 1994).

5 The CD14 Molecule

The membrane-bound CD14 molecule (mCD14) constitutes a 53-kDa glycoprotein present on the surface of myeloid cells including monocytes and macrophages (WRIGHT et al. 1990; ULEVITCH 1993; ZIEGLER-HEITBROCK and ULEVITCH 1993). It is embedded in the plasma membrane via a glycerophosphatidyl inositol (GPI) anchor (HAZIOT et al. 1988), a property shared by other cell surface molecules involved in cellular responses such as CD24, Thy-1, and CD58/LFA-3 (ROBINSON 1991).

The mCD14 molecule has been implicated as playing a role in cellular adhesion (BEEKHUIZEN et al. 1991, 1993). Its particular significance as a primary LPS receptor was recognized only 5 years ago (WRIGHT et al. 1990). It was then observed that CD14 serves as an LPS-binding site provided LPS has been allowed to complex to LBP (SCHUMANN et al. 1990; WRIGHT et al. 1990; ULEVITCH 1993). LPS binding takes place via the lipid A component and is of high affinity ($K_D=3\times10^{-8}$ *M*) (KIRKLAND et al. 1993) , and the binding domain of CD14 has been defined to reside between amino acids 57 and 64 (JUAN et al. 1995; VIRIYAKOSOL and KIRKLAND 1995). Certain (endotoxin-free) anti-CD14 monoclonal antibodies (mAbs) have been found to mimic LPS effects in monocytes, such as the formation of reduced oxygen species and cytokines (LAUENER et al. 1990). Other anti-CD14 mAbs (3C10, 60b) inhibit LPS/LBP complex induced TNF production (WRIGHT et al. 1990) and protein tyrosine phosphorylation (WEINSTEIN et al. 1991; see also Sect. 6). Further, certain cell types transfected with CD14 become highly reactive to LPS/LBP complexes (LEE et al. 1992).

CD14 is not a transmembrane molecule, but it is linked to the cell surface via a GPI anchor (Haziot et al. 1988). As this type of membrane anchor does not allow direct signal transduction, the mechanism of LPS/LBP-induced and CD14-mediated cell activation is presently not understood. Several observations bearing on this problem are mentioned below.

Monocytes of patients suffering from paroxysomal nocturnal hemoglobinuremia are unable to form a GPI anchor and therefore lack membrane-bound CD14. Nevertheless, such monocytes can be activated by LPS (with or without LBP) to form cytokines (Couturier et al. 1992; Schütt and Schumann 1993). Also, vascular cells lacking CD14 mRNA and CD14-protein respond to LPS (Loppnow et al. 1995). This shows that a CD14-independent pathway must exist involving another, perhaps membrane-associated receptor. One possible model therefore suggests that LPS is guided to cells by LBP, where it first associates with CD14 which subsequently interacts with or presents LPS in monomeric form to a second, so far unknown receptor which can transduce an intracellular signal. This scenario is schematically shown in Fig. 5.

CD11c/CD18 has very recently been demonstrated to function as a transmembrane signaling receptor for endotoxin which, however, appears to be operative in the absence of CD14 (Ingalls and Golenbock 1995). In looking for a CD14-dependent functional LPS receptor we used a ligand blotting assay to investigate the binding of LPS to membrane proteins of the human monocytic cell line Mono-Mac-6. Among membrane proteins an 80-kDa protein was identified which binds LPS or free lipid A only in the presence of serum (Schletter et al. 1995). Subsequent experiments identified the serum factors mediating binding of lipid A to the 80-kDa membrane protein as sCD14 and LBP. Thus the 80-kDa protein fulfills an important prerequisite of a putative signal transducting molecule: it recognizes LPS or lipid A only in the context of LBP/CD14. The 80-kDa membrane protein is also present in membrane preparations of human peripheral blood monocytes and endothelial cells.

Several investigators have reported CD14-independent LPS binding at high (μg/ml range) LPS concentrations (Couturier et al. 1992). In this case LPS may interact directly with the unknown second receptor; this latter pathway, however, requires larger amounts of LPS for cell activation. Furthermore, as with other GPI-linked surface molecules, CD14 is internalized after binding of the ligand. It is possible that this event constitutes an important step in cell activation, perhaps involving a cytosolic LPS receptor.

CD14 also exists in soluble form (sCD14). As such it is present in the circulation, at a concentration of 2–6 μg/ml (or 5×10^{-8} *M*; Frey et al. 1992). In normal serum several types of sCD14 (48-, 53-, 55-kDa) are present which either result from shedding of membrane-bound CD14 or from cellular production of GPI-free CD14 forms (Bažil and Strominger 1991; Durieux et al. 1994). It has recently been demonstrated that sCD14 is capable of interacting directly with LPS (Hailman et al. 1994). In this case LBP seems to play a catalytic role in facilitating the association of LPS with sCD14 without participating in complex formation (Tobias et al. 1995). The sCD14/LPS complex is capable of binding to

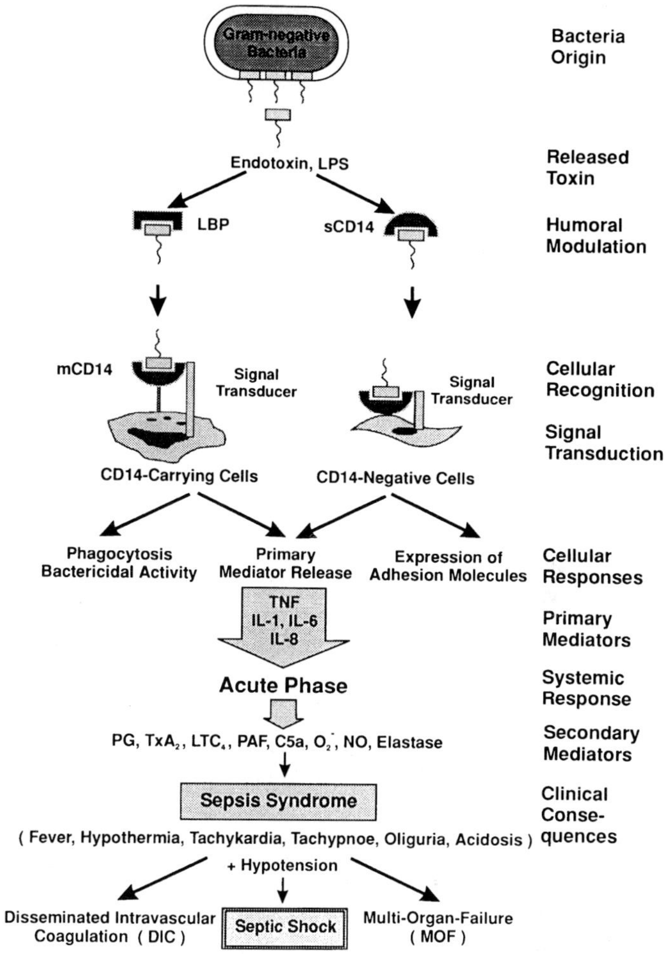

Fig. 5. Intracellular pathways involved in endotoxin-induced signal transduction in monocytes

CD14-negative cells such as endothelial cells and of activating these to produce cytokines (Frey et al. 1992; Pugin et al. 1993; Arditi et al. 1993; Haziot et al. 1993). Endothelial cells apparently possess a receptor for sCD14/LPS complexes, the nature of which awaits further characterization. It is tempting to speculate that the recently discovered 80-kDa protein (Schletter et al. 1995) which recognizes LPS in the presence of sCD14 and LBP represents this cellular receptor molecule.

Recently it has been shown that both untreated and activated monocytes shed CD14-bearing microparticles into plasma. It remains to be elucidated whether microparticle-associated CD14 participates in LPS transport and cell stimulation (Satta et al. 1994).

It should be mentioned that activation of monocytes by certain other bacterial immunomodulators such as peptidoglycan, arabinogalactan, and lipoarabinomannan (Weidemann et al. 1994; Heumann et al. 1994; Pugin et al. 1994) also proceeds via mCD14. The structural requirements of these glycoconjugates for binding to mCD14, cell activation, and cytokine secretion are presently not understood. On the other hand, monocyte activation by glycosphingolipids of *Sphingomonas paucimobilis* which share chemical and physical features with LPS (Kawahara et al. 1991) is not dependent on mCD14 (Krziwon et al. 1995).

6 LPS-Induced Signal Transduction

The signal transduction events that follow binding of LPS to target cells remain to be fully defined. Several studies suggest that treatment of macrophages with LPS results in the activation of phospholipase C as determined by the formation of inositol tris-phosphate ($InsP_3$). The response is modest and occurs 1–20 min after activation (Prpic et al. 1987; Chang et al. 1990). According to Hurme et al. (1992) and Drysdale et al. (1987) this slow turnover of $InsP_3$ is not accompanied by a rise in $[Ca^{2+}]_i$. Thus, it remains unclear whether $InsP_3$ formation constitutes an obligatory cellular event in LPS-triggered activation (Chang et al. 1990). In addition, it has been reported that responses to LPS may involve a pertussis toxin sensitive G protein (Jackway and De Franco 1986; Dziarski 1989) and activation of protein kinase C (PKC) (Liu et al. 1994; Novotney et al. 1991; Bakouche et al. 1992; Hurme and Serkkola 1991). There may be two alternative pathways by which LPS activates PKC (Chen et al. 1986). Activation occurs either via diacylglycerol and $InsP_3$ or via calpain-catalyzed cleavage of PKC, yielding the 40-kDa catalytic domain (PKM). The cleavage of PKC to PKM by calpain frees it from the regulatory constraint and releases PKM into the cytosol, where it can phosphorylate a distinct set of proteins.

LPS has been shown to induce changes in protein phosphorylation in murine macrophages, including phosphorylation of a 65-kDa protein which has been purified and characterized (Shinomiya et al. 1991). An early event triggered in monocytes/macrophages by LPS is tyrosine phosphorylation of a series of 38- to 45-kDa proteins known as mitogen-activated protein (MAP) kinases (Weinstein et al. 1992, 1993; Han et al. 1993; Liu et al. 1994; Lee et al. 1994). The p38 MAP kinase appears to be distinct from MAPK1 (44 kDa) and MAPK2 (40–42 kDa) showing sequence homologies to the products of the *Saccharomyces cerevisiae* osmosensing gene HOG1 (Han et al. 1994). Furthermore, phosphorylation of p38 occurs via the human MAP kinase kinases MKK3 and MKK4, whereas MAPK1 and MAPK2 are activated by MAP kinase kinases MEK1 and MEK2 (Derijard et al. 1994). Phosphorylation of these serine/threonine protein kinases contributes to their increased enzymatic activity. Proteins such as 90-kDa ribosomal S6 protein kinase (Chung et al. 1991; Sturgill et al. 1988) and C-jun transcription factor (Pulverer et al. 1991) have been found to be efficient in vitro substrates of MAP kinases, and this may be indicative of MAP kinase function in vivo.

Recently we have found that endotoxin causes in human monocytes phosphorylation of two cytosolic proteins of 36 and 38 kDa which are distinct from MAP kinases (Heine et al. 1995). Associated with the LPS-induced activation of macrophages is the change in the ADP-ribosylation state of a 33-kDa cytosolic protein (Hauschildt et al. 1994). As with phosphorylation, ADP-ribosylation constitutes a covalent modification by which cells regulate protein functions. Whether this reversible protein modification is related to the functional activation of macrophages is not yet clear. We have, however, demonstrated that suppression of ADP-ribosylation inhibits phosphorylation of the 36- and 38-kDa proteins and at the same time the formation of TNF and IL-6 mRNA as well as protein production (Heine et al. 1995). This indicates that ADP-ribosylation is involved in LPS-induced alteration of the phosphorylation state of the two proteins p36 and p38, and that the latter are involved in processes leading to monocytic cytokine formation.

Treatment of cells with LPS leads to induction of proto-oncogenes such as the tyrosine kinase, proto-oncogenes *hck* and *lyn*, and the transcription factors *myc* and *fos* (Introna et al. 1986; Ziegler et al. 1988; Müller et al. 1993; Stefanova et al. 1994). LPS also activates nuclear factor κB and c-Rel/p50, which may activate genes in vivo by binding the κB elements located in the enhancer region (Cordle et al. 1993).

7 Regulation of LPS-Induced Cytokine-Production in Macrophages by Lipoxygenase products

Several lines of evidence suggest that lipoxygenases are involved in the regulation of endotoxic activities (Fig. 5). For example, lipoxygenase inhibitors protect from endotoxicity in experimental models (Keppler et al. 1987; Tiegs and Wendel 1988) and suppress the formation of TNF as determined in serum of LPS-treated mice (Schade et al. 1989b). In vitro it has been found that inhibitors of lipoxygenase (ETYA, BW 755C) but not inhibitors of cyclo-oxygenase (indomethacin, aspirin) interfere with LPS- or *Staphylococcus aureus*-induced production of IL-1 by human peripheral monocytes (Dinarello et al. 1984). These findings suggested that product of the lipoxygenase pathway promotes the synthesis of IL-1. The compound in question, however, was not characterized. In subsequent studies other investigators analyzed various lipoxygenase products for their capacity to modulate IL-1 induction in macrophages. The results reported by different groups are at variance. For example, LTB_4 and LTD_4 have been described as stimulating the synthesis of IL-1 in human monocytes and mouse macrophages (Chensue and Kunkel 1985; Rola-Plesczcynski and Lemaire

1985). However, this has not been confirmed by others (BRANDWEIN 1986) who treated human monocytes with LTB_4 and failed to determine any activity on IL-1 synthesis. Therefore the question as to the nature of the lipoxygenase product involved in IL-1 synthesis remained open.

7.1 13-Hydroxylinoleic Acid and Macrophage Activation

In our laboratory it was found that lipoxygenase inhibitors also block LPS-induced TNF formation in murine macrophage cultures (SCHADE et al. 1989a,b). These results have been confirmed with other inhibitors and in different macrophage populations (BADGER et al. 1989; MOHRI et al. 1990). In analyses aiming at identifying the presumable product of lipoxygenase activity, LTB_4, LTC_4, LTD_4, or LTE_4 could not be detected in the supernatants of mouse macrophages stimulated with LPS. However, when LPS-activated mouse macrophages were hydrolyzed and extracts of the hydrolysates were analyzed, the lipoxygenase product 13-hydroxyoctadecadienoic acid (13-HODD), which is derived from linoleic acid, was identified (SCHADE et al. 1987). Analysis of the absolute configuration showed that the product is (*S*)-13-HODD, which confirms its enzymatic origin (SCHADE et al. 1987). 13-HODD was found to be enriched in LPS-activated cells. Furthermore, the effects of site-specific lipoxygenase inhibitors in comparison to less specific inhibitors were tested on LPS-induced TNF formation, zymosan-induced synthesis of LTC_4, and synthesis of 13-HODD in mouse peritoneal macrophages (SCHADE et al. 1991). The inhibitors TZI-41127 and VZ 65 reduced both TNF formation and synthesis of LTC_4 and 13-HODD (although with both substances there is a preference for inhibition of LTC_4 vs. 13-HODD). As expected, the 5-lipoxygenase inhibitors MK 886 and CGS 8515 inhibited the synthesis of LTC_4, whereas the formation of 13-HODD was unaffected. Most interestingly, neither 5-lipoxygenase inhibitors had any effect on the formation of TNF.

These results indicate a good quantitative correlation between TNF formation and the inhibition of a 15-lipoxygenase (formation of 13-HODD), but not between TNF formation and inhibition of a 5-lipoxygenase (synthesis of LTC_4). These findings are in accord with the notion that 13-HODD is produced increasingly during LPS activation, a process required for LPS-induced TNF formation (SCHADE et al. 1987, 1989b). In agreement with these results, MK 886 and other specific 5-lipoxygenase inhibitors do not impede IL-1 production in mouse macrophages stimulated with LPS (PARKAR et al. 1990).

The influence of authentic 13-HODD and other lipoxygenase products on LPS mitogenicity has recently been tested in murine spleen cell cultures (ELEKES et al. 1993). Lipoxygenase inhibitors suppressed the proliferation of spleen cells and in parallel TNF synthesis induced by LPS. It was further found that 13-HODD, unlike other lipoxygenase products, is able to counteract the inhibitors and to restore the mitogenic response. TNF was shown to have essential functions in LPS mitogenicity in murine spleen cells (JAKOBS and SCHADE 1994). It is therefore possible that the inhibitory effect of lipoxygenase inhibitors and the augmenting effect of 13-HODD on mitogenicity is due to its action on TNF synthesis.

Summarizing these data, it appears that the linoleic acid metabolite 13-HODD plays an important role in endotoxin-induced TNF production in macrophages. Its position in the LPS-triggered signaling pathway and the molecular basis of 13-HODD action is presently under investigation.

8 Lipid A Determinants Involved in Binding to and Activation of Cells

The initial key event in endotoxin action is the interaction of the lipid A component of LPS with defined humoral and, notably, cellular recognition molecules. It is reasonable to assume that at least two parameters determine the result of the interaction between lipid A and host cells, i.e., between the endotoxic conformation of individual lipid A molecules and the host recognition system resulting in the activation of monocytes and the production of bioactive mediators. These parameters are the capacity of lipid A specifically to *bind* to the receptor and subsequently the ability of lipid A to *activate* the cell by a receptor-mediated mechanism.

Recent studies in our laboratory have shown that the hydrophilic lipid A backbone (bis-phosphorylated D-GlcN disaccharide together with two fatty acids) mediates the specific *binding* of lipid A or LPS to the cellular receptor (Kirikae et al. 1993a). On the other hand, it is the number, nature, and distribution of fatty acids, i.e., the acylation pattern which determines the *activation* capacity and mediator-inducing capacity of lipid A. These investigations also revealed that the binding of lipid A is a necessary but not sufficient event for the activation of cells. It appears that an additional receptor-mediated triggering event is required for which the acylation pattern of lipid A plays a most critical role, and which provides the signal for cellular stimulation (Rietschel et al. 1993).

9 Inhibition of Lipid A Binding to Cells and Cellular Activation

The separation of the "binding" and "activation" events suggested the possibility that endotoxically inactive LPS or partial structures may bind to target cells without activating them, thus representing possible inhibitors of LPS-induced effects.

Indeed, we and later others have demonstrated that tetraacyl lipid A (Fig. 6A), an intermediate in lipid A biosynthesis (precursor Ia) in its chemically synthesized form (compound 406), inhibits, in a dose-dependent manner the monokine production by human monocytes induced by LPS but not by *Staphylococcus epidermidis*, lipoprotein, or bacille Calmette-Guérin (Loppnow et al. 1989; Wang et al. 1991; Flad et al. 1993).These results show that the inhibitory capacity of compound 406 is highly specific for the LPS-cell interaction, and that it is not due to suppression of monokine release. Rather, as has been shown by northern blot analysis, compound 406 suppresses the LPS-induced formation of mRNA for TNF and IL-1, indicating that monokine production is inhibited before or at the level of transcription (Flad et al. 1993). Similar results have been reported for the isostructural natural precursor Ia (also termed lipid IVa) isolated from *S. enterica* sv.typhimurium (Lynn and Golenbock 1992; Kovach et al. 1990) and for nontoxic lipid A species of *R.sphaeroides* (Qureshi et al. 1991) and *R. capsulatus* (Fig. 6B; Loppnow et al. 1990, 1993). Compound 406 also inhibits LPS-but not IL-1-induced expression of the adhesion molecule ICAM-1(Schönbeck et al. 1994) and the (serum-dependent) formation of IL-6 in endothelial cells (Loppnow et al. 1993). These observations are in accordance with the previous demonstration that partially deacylated LPS is capable of inhibiting LPS-induced adherence of granulocytes to vascular endothelium (Pohlmann et al. 1987) and also with the concept that the observed inhibition is LPS specific. Very recently an LPS antagonist, the structure of which is based on that of *R. capsulatus* lipid A, was synthesized (compound E5531). This molecule protects mice from endotoxin-induced lethality and, when administered together with an antibiotic, from lethal *E. coli* induced peritonitis (Christ et al. 1995).

The mechanism of the inhibitory action of lipid A partial structures such as compound 406 has been studied in various cells types (Ulmer et al. 1992; Lynn and Golenbock 1992; Kitchens et al. 1992; Kirikae et al. 1993a; Flad et al. 1993; Heine et al. 1994; Kitchens and Munford 1995). Our and other results suggest that in the case of J774.1 cells and peripheral human monocytes inhibition is based on competitive binding to a specific LBP. This concept is supported by the finding that the degree of inhibition depends on the dose of the competitor, and that inhibition is overcome by higher LPS doses. A true competitive, i.e., antagonistic, mechanism is also suggested by the fact that if the LPS-binding data obtained in the presence or absence of an inhibitor are plotted in a double reciprocal manner (according to Lineweaver and Burk), both regression lines cross the y-axis (Heine et al. 1994). These and other results lead us to conclude that antagonistically active compounds such as precursor Ia and lipid A of *R. sphaeroides* (Kirikae et al. 1994) or compounds E5531 (Christ et al. 1995) bind to the same cellular (or humoral) recognition site as lipid A and LPS, yet lack the capability to functionally engage the receptor. Antagonistic preparations thus block the receptor effector binding site, subsequently denying access of LPS and thereby inhibiting the activation of cells.

The nature of the cellular binding molecule involved remains to be established. There is ample evidence, however, that the CD14 molecule plays a crucial

A

B

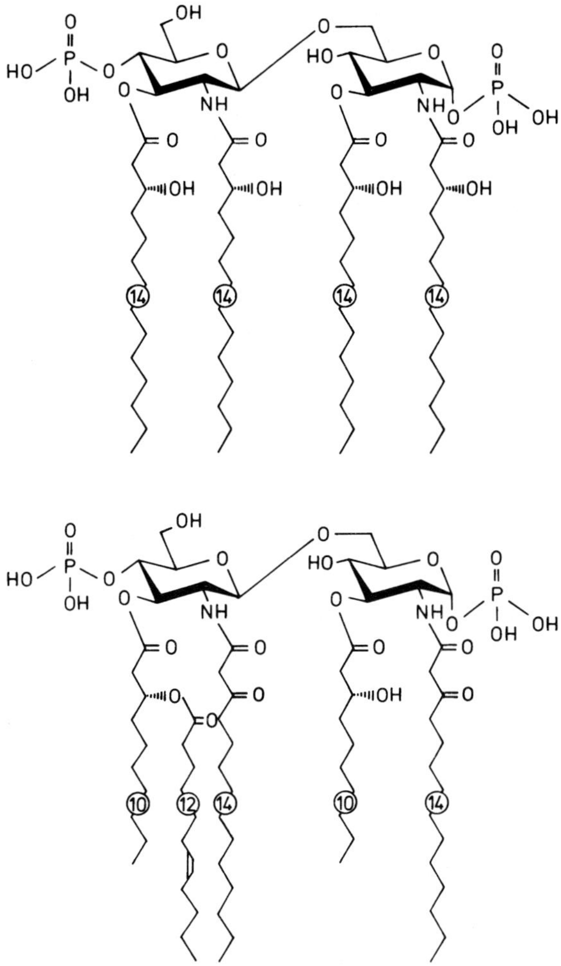

Fig. 6A,B. Chemical structure of synthetic compound 406. **A** Precursor Ia. **B** Nontoxic lipid A of *Rhodobacter capsulatus.* (From MAYER et al. 1990; ZÄHRINGER et al. 1994)

role in the antagonistic inhibitory pathway described. For example, CD14 has been shown to be capable of interacting with compound 406 (HAILMAN et al. 1994), and the antagonistic activity of compound 406 is inhibited by anti-CD14 antibodies. A CD14-dependent but nonantagonistic mechanism has been proposed in a system employing THP-1 cells (KITCHENS et al. 1992). This study showed that compound 406 interacts with CD14 without, as expected, stimulating cells, yet still allowing LPS binding to CD14. It was therefore argued that compound 406 exerts its biosuppressive activity in THP-1 cells in a non-antagonistic fashion. One could speculate that compound 406 binds to a second structure located in close topographical neighborhood to CD14 (such as the signal transducer), thereby inhibiting the LPS-CD14 initiated signal transduction. It is also possible that the association of precursor Ia with CD 14 or an associated molecule initiates a suppressive signal, which actively counteracts the activation signal generated by the interaction of LPS with CD14 (KITCHENS and MUNFORD 1995).

10 The Pathophysiological Cascade of Events Leading to Gram-Negative Septic Shock

Considering the pleiotropic action of endotoxin and the multiplicity of endogenous mediators produced, it is difficult to define the exact mechanisms which are causally involved in the phenomenon of endotoxin-induced shock. Nevertheless, some principal reactions can be described which take place in vivo after injection of endotoxin or its release from bacteria (see Fig. 5). That portion of LPS which is not detoxified by humoral (e.g., BPI or HDL) or cellular (e.g., PMN) host components interacts with LBP, septin, or sCD14 and subsequently activates target cells to produce and release endogenous mediators or to express adhesion molecules. The primary target cells include monocytes/macrophages, endothelial cells, granulocytes, and lymphocytes. Prominent mediators such as TNF, IL-1, IL-6, and IL-8 are produced which are capable of activating susceptible (i.e., cytokine receptor carrying) cells to produce lipid mediators (including platelet-activating factor, LTs, and prostaglandins), reduced oxygen species, nitric oxide, and proteases such as elastase and collagenase (JOCHUM et al. 1992). If this inflammatory reaction cascade is limited in its capacity to produce mediators, and if it is confined to a local area, it may be highly beneficial as it helps to activate the defense system and to destroy invading micro-organisms (MASTROENI et al. 1993). However, overproduction of these mediators and release into the circulation may result in a systematic inflammatory reaction. Low mediator concentrations may also become harmful if the host organism is in a hyperreactive state to LPS. Hyperreactivity to endotoxin may be caused by exotoxins, chronic infection, and growing tumors. Important endogenously formed molecules contributing to sensitization to LPS have been identified as migration-inhibitory factor, interferons β and α and γ, and granulocyte-macrophage colony-stimulating factor (GALANOS and FREUDENBERG 1993; BENDER et al. 1993; BERNHAGEN et al. 1994; CARR et al. 1994; TIEGS et al. 1994).

Endogenous mediators subsequently stimulate receptor-carrying cells and organs, resulting in secondary and clinically relevant reactions. For example, TNF, IL-1, and IL-6 act as endogenous pyrogens on the hypothalamus to cause fever (or hypothermia). The hypophysis releases ACTH, cortisol, and migration-inhibitory factor, and the liver responds to IL-6 with the production of acute-phase proteins (BAUMANN and GAULDIE 1994; STEEL and WHITEHEAD 1994). Some acute-phase proteins dampen the LPS-dependent inflammatory reaction while others, such as LBP, intensify it. Blood clotting factors are consumed, and the complement system is activated, producing other potent cytomodulators. These factors as well as TNF cause leukocytes to adhere to endothelial cells, resulting in leukopenia, blood vessel injury, and capillary leakage. On the other hand, endothelial cells are stimulated to produce nitric oxide, which causes vasodilation followed by hypotension and increased cardiac output. T-cells also take part in this scenario. They release colony-stimulating factor and interferon γ, the former acting on the bone marrow causing leukocytosis and the latter activating macrophages, which in turn overproduce toxic mediators.

These effects clearly resemble the clinical picture of gram-negative septic shock, being characterized by fever, chills, leukopenia followed by leukocytosis, disseminated intravascular coagulation, hemorrhage, hypotension, tachycardia, and metabolic acidosis leading to dysfunction or failure of vital organs such as lung, kidney, liver, heart, gastrointestinal tract, and central nervous system (GLAUSER et al. 1991; PARILLO 1993). Due to this clear association between LPS and sepsis endotoxin is implicated as a main factor involved in the pathological manifestations and consequences of gram-negative bacterial sepsis.

11 Anti-Endotoxin Antibodies

Gram-negative sepsis remains associated with high morbidity and mortality, the mortality rate being in the order of 40%–60%. These figures have not changed over the past two decades despite the development of new antibiotics and the use of improved and highly sophisticated intensive-care measures (BONE et al., 1987, 1989; CENTERS for DISEASE CONTROL 1990; YOUNG and GLAUSER 1991). The fact that in the United States alone an estimated 100 000 deaths annually result from septic shock demands intensified efforts to devise new ways in the fight against the life-threatening sepsis syndrome. Indeed, the failure of antibiotics to reduce gram-negative septic mortality below 40% is probably related in part to their ability to liberate endotoxin from infecting microbes. Experimental studies indicate that the mechanisms underlying gram-negative bacterial lethality can be altered by antibiotics, such that the resulting abrupt and sustained increases in endotoxaemia contribute even more importantly to the lethal outcome (JOHNSTON and GREISMAN 1984).

New therapeutic strategies may focus on the initial stimulus (endotoxin); enhancing factors (LBP, sCD14, septin); receptors of LPS target cells (CD14); LPS-induced signal transduction leading to production of cytokines; TNF, IL-1, and IL-6; cytokine receptors of cytokine target cells; cytokine-induced signal transduction; or on other parameters. Indeed, various approaches have recently been developed for interfering with distinct phases of the septicemic process and supplementing the (obligatory) administration of antibodies. Many of these approaches have proven effective in experimentally induced endotoxemia or sepsis; however, clinically no benefit has been observed so far (NATANSON et al. 1994).

Of particular interest are newly discovered anti-endotoxin preventive or therapeutic agents such as mAbs to endotoxin and other endotoxin-neutralizing agents. Here we deal only with immunological approaches against LPS, i.e., anti-LPS antibodies (for an update concerning other therapeutic approaches see POLLACK 1990; see also M. POLLACK and C.A. OHI, this volume). The targeting of endotoxins by immunoglobulin has several potential advantages over other approaches. These include the blockade of a very early step in the cascade leading to gram-negative sepsis, particular properties of antibodies such as long half-life, low immunogenicity, their capacity to activate complement and to interact with Fc receptors, the possibility to produce mAbs antibodies with defined specificity and of high affinity, and the choice of defined antibody classes and isotypes.

In principle, all segments of the LPS molecule are immunogenic and antigenic. Polyclonal antisera and mAbs have been prepared against the O-specific chain, the core and lipid A, and their properties have been evaluated in vitro and in vivo.

11.1 Anti-O-Chain Antibodies

Polyclonal antibodies (SAXEN et al. 1986) and mAbs (SOLWELL et al. 1984; KIRKLAND and ZIEGLER 1984; BAUMGARTNER et al. 1990) directed against determinants of the O-specific chain are highly protective in endotoxin and infection models. However, in view of the wide structural variability of the O-specific chain of those pathogenic gram-negative bacteria that are significant in clinical settings (*E. coli, Klebsiella, Pseudomonas*) and the restricted usefulness of such antisera or antibodies directed only against the homologous LPS or bacterial serotype, O-specific immunoglobulins are unlikely to gain broad use. They appear, however, to be useful in the form of multispecific, i.e., polyclonal, immunoglobulin preparations (SCHEDEL et al. 1991; DOMINIONI et al. 1991) and in exceptional and defined cases such as the prevention of *Pseudomonas* infection associated with cystic fibrosis (CRYZ et al. 1994).

11.2 Anti-Lipid A Antibodies

Lipid A has been considered a logical target for the generation of protective antibodies for two reasons. (a) Lipid A of various pathogenic gram-negative bacteria is structurally highly conserved (RIETSCHEL et al. 1994; ZÄHRINGER et al. 1994). It can therefore be expected that anti-lipid A antibodies would cross-react with LPS of different gram-negative bacteria. (b) Lipid A constitutes the toxic principle of LPS (GALANOS et al. 1985). Therefore such antibodies could be expected to neutralize bioactivity of lipid A and endotoxin.

The immunogenicity of lipid A remained unknown for a long time because lipid A antibodies are not engendered upon immunization of experimental animals with S- or R-form bacteria or the corresponding LPS. In 1972, however, it was shown that lipid A induces specific antibodies, provided lipid A is administered in a form devoid of the LPS polysaccharide component and if complexed to a suitable carrier (GALANOS et al. 1984). Polyclonal lipid A antisera were shown to be cross-reactive with polysaccharide-free lipid A of various bacterial origin (GALANOS et al. 1984). Such antisera, however, lacked antiendotoxic and anti-infectious properties unless very peculiar in vivo assay conditions were used (RIETSCHEL and GALANOS 1977). When monoclonal anti-lipid A antibodies were available (DUNN et al. 1986; BOGARD et al. 1987), their serological properties were studied and their epitopes characterized using various partial structures of LPS, in particular, synthetic lipid A (BRADE et al. 1988). The most important results of these studies can be summarized as follows: (a) The epitopes recognized by various monoclonal lipid A antibodies reside without exception in the hydrophilic backbone region (phosphorylated disaccharide; KUHN et al. 1992; BRADE et al. 1993). (b) Fatty acids modulate the exposure of these epitopes but are not part of the determinants (BRADE et al. 1987b). (c) Lipid A mAbs cross-react with a large variety of free lipid A (i.e., polysaccharide-deprived lipid A) of distinct bacterial origin (GALANOS et al. 1984; BRADE et al. 1988). (d) Lipid A mAbs, however, do not cross-react with LPS, i.e., with lipid A carrying the polysaccharide component (GALANOS et al. 1984).

This lacking cross-reactivity is not fully understood at the present time. When present as part of LPS, the lipid A epitopes may not be expressed, suggesting that lipid A represents a neoantigen due to the exposure of determinants not present in LPS (such as the primary hydroxyl group of GlcN II which could constitute part of such an epitope). Our recent finding that synthetic lipid A preparations harboring a free primary hydroxyl group at GlcN II do cross-react with lipid A antibodies, whereas derivatives being methylated at this hydroxyl-group do not react, favors this interpretation (BRADE et al. unpublished results). On the other hand, Kdo and other LPS constituents may sterically hinder the binding of antibodies to lipid A.

Because of these properties lipid A antibodies are highly unlikely to express protective activity in endotoxic and septicemic states. Nevertheless, two types of monoclonal antibodies [Centoxin (HA-1A)] and [XMMEN-OE5 (E5)] have been described to be reactive with lipid A in the context of LPS. They have been evaluated intensively in vitro and for protective activity in experimental animals and septic patients in vivo (GREENMAN et al. 1991; ZIEGLER et al. 1991). Both antibodies are of the IgM class and are likely to be derived from CD5-positive B-cells. The lipid A epitopes recognized by either antibody are not identified (FUJIHARA et al. 1993). Cross-reactivity with LPS has been demonstrated only under very favorable conditions, i.e., with very high coating density of LPS in enzyme-linked immunosorbent assay. Furthermore, the reactivity of mAb HA-1A is not restricted to lipid A or LPS, as it also binds to other amphiphilic compounds including gentabioseoctaacetate and chitobiose (TENG et al. 1985). Recently, mAb HA-1A has been identified as a cold agglutinin (BHAT et al. 1993).

Animal studies suggest that mAbs HA-1A and E5 are protective in endotoxemic and bacteremic models (TENG et al. 1985; YOUNG et al 1989). Other investigations, however, have failed to confirm these results (BAUMGARTNER and GLAUSER 1993). In an *E. coli* sepsis model even lethality-enhancing properties of mAb HA-1A were revealed (QUEZADO et al. 1993). Furthermore, the clinical trials performed in septic patients did not show any significant protective effect of either mAb HA-1A (BAUMGARTNER and GLAUSER 1993) or mAb E5 (WENZEL et al. 1991). It thus appears that lipid A antibodies as obtained by immunization with free lipid A or LPS do not fulfill the promise as an universal immunotherapeutic anti-endotoxic drug (BAUMGARTNER and GLAUSER 1993; WARREN et al. 1993).

11.3 Polyclonal Anti-Core Antibodies

Chemical and genetic investigations performed in the 1960s and 1970s showed that LPS of enterobacterial R mutants lacks the O-specific chain but contains (parts of) the core oligosaccharide and lipid A, thus representing a partial structure of S-form LPS (LÜDERITZ et al. 1982), and further that the constituents of the core oligosaccharide are similar or identical in various gram-negative bacterial serotypes, species, genera, and families. These findings led to the view that (a) the LPS core of various bacteria may harbor common determinants, (b) these determinants are immunogenic in R mutants, (c) they are exposed in S-form LPS and wild-type bacteria, (d) antisera against R-mutant bacteria or their LPS would cross-react with S-form LPS and S-form bacteria, and (e) such R-mutant antisera would cross-protect in endotoxemic and infection models.

Two research groups (BRAUDE and DOUGLAS 1972; MCCABE 1972) followed these ideas, one group (BRAUDE and DOUGLAS 1972) working with an Rc mutant of *E. coli* O111 (termed J5) and the other using mutants of the Re chemotype (strain R595) derived from *S. enterica* sv. *minnesota* (MCCABE 1972). The chemical structure of the Re-mutant LPS (ZÄHRINGER et al. 1985) and of the *E. coli* Rc-mutant LPS (MÜLLER-LOENNIES et al. 1994) is shown in Fig. 7. The Re-mutant LPS consists of an $\alpha 2 \rightarrow 4$ linked Kdo disaccharide which is bound to lipid A via the primary hydroxyl group of GlcN II of the lipid A backbone. The J5 (Rc-mutant) structure contains, in addition, a GlcN-carrying heptose trisaccharide and a terminal Glc residue (MÜLLER-LOENNIES et al 1994). Whereas in some studies Rc (J5) and Re (R595) antisera obtained in rabbits exhibited cross-reactivity and cross-protective activity (ZIEGLER et al. 1973; BRUINS et al. 1977), other studies have not confirmed these serological and biological data (GREISMAN et al. 1978; GREISMAN and JOHNSTON 1988; TRAUTMAN and HAHN 1985). The former studies, however, failed to define the protective epitope or the fine specificity of the protective antibodies, thus rendering a comparison between these conflicting studies impossible.

We have recently addressed this problem by analyzing the antibody specificities present in polyclonal Re-mutant antisera using defined bacterial or synthetic partial structures as antigens (ROZALSKI et al. 1989). Surprisingly, the abundant antibody specificities were identified as being reactive with the α(2–4)-Kdo-disaccharide together with parts of the lipid A region. These antibodies bind *specifically* to Re-mutant LPS and do *not* cross-react with other R- or S-form LPS. Accordingly, they cannot be expected to provide cross-protection. In Re antisera trace amounts of antibodies were also found which are specific for a terminal α-pyranosidic Kdo residue. Of the latter specificity it is known (in the form of mAbs such as clone 20), that it cross-reacts with other LPS, i.e., those LPS which expose a lateral Kdo residue (BRADE et al. 1987a). In our hands, however, these Kdo-specific, in vitro cross-reactive antibodies do not exhibit protective activity in endotoxin test systems such as pyrogenicity in rabbits (RIETSCHEL et al., unpublished data). A similar situation was encountered with polyclonal antisera raised against Rd chemotypes. Again, the main specificity was Rd-specific and not cross-reactive with other R-mutant or S-form LPS (SWIERZKO et al. 1993, 1994). It is more than likely that the same situation holds true for the Rc chemotype, i.e., *E. coli* J5.

In view of these data it is difficult to understand how Re and Rc antisera, which must be considered as essentially chemotype-specific, should confer cross-protection. In fact in the six clinical trials performed with J5 antisera only two (ZIEGLER et al. 1982; BAUMGARTNER et al. 1983) reported a significant ther-

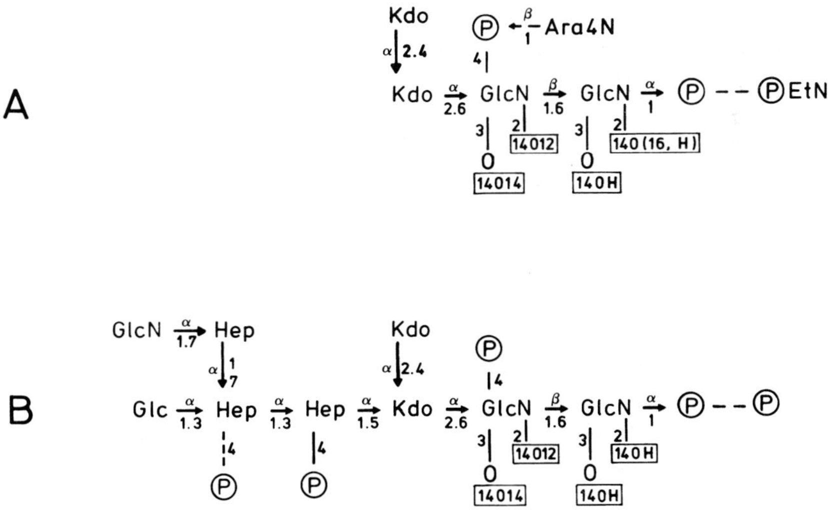

Fig. 7. Chemical structure of the LPS of immunotherapeutically important enterobacterial R mutants., **A** Re mutant of *S.enterica* serovar *minnesota* (strain R595)., **B** Rc mutant of *E.coli* 0111 (strain J5). (From ZÄHRINGER et al. 1985; MÜLLER-LEONNIES et al. 1994)

apeutic efficacy (improved survival). Whether in fact this effect was due to LPS antibodies or to other serum factors was not established. In summary, therefore, serological data render the protective approach of using polyclonal antisera obtained by immunization with R mutants highly questionable, and the vast majority of serological and biological data available do not support it.

11.4 Cross-Reactive and Cross-Protective Monoclonal Anti-Core Antibodies

Only very recently it has been shown that despite the trials and data described above, anti-core antibodies are capable of expressing cross-reactive and cross-protective properties (Di Padova et al. 1993a,b). Using O-chain defective LPS as immunogen and NZB mice, we generated broadly cross-reactive mAbs recognizing the core region of *E. coli*, *S. enterica*, and *Shigella* R- and S-form LPS. These antibodies also exhibit potent anti-endotoxic properties. One of these mAbs (WN1 222-5, IgG2a) was selected for further evaluation in vivo (Di Padova et al. 1993b). It was found to prevent effectively lethality in mice and pyrogenicity in rabbits induced by S-form LPS of different *E. coli* serotypes harboring different core types and *S. enterica* serotypes. The serological properties of mAb WN1 222-5 are quite different from those of an antibody described by Nnalue et al. (1992), as this mAb of IgM class recognizes only *Salmonella* and *E. coli* R2.

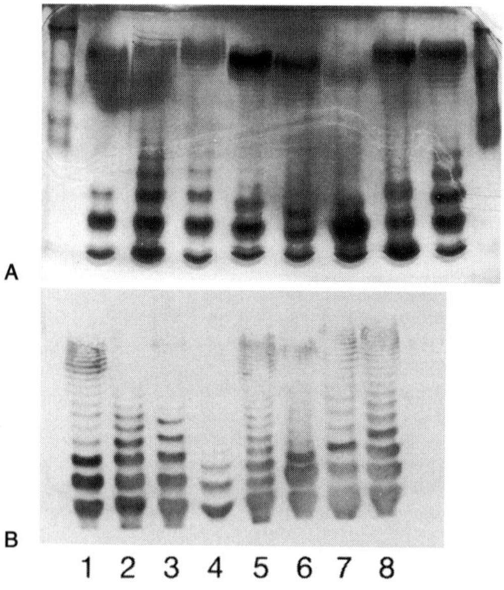

Fig. 8. Gel electrophoresis (**A**) and immunoblot (**B**) with cross-reactive and chimerized anti-LPS monoclonal antibody SDZ 219–800; 1 *E. coli* 0111; 2 *E. coli* 086; 3 *E. coli* 018; 4 *E. coli* 016; 5 *E. coli* 015; 6 *E. coli* 012; 7 *E. coli* 06; 8 *E. coli* 04. (From Di Padova et al. 1995)

To reduce the incidence of possible negative side effects such as the appearance of human anti-mouse antibodies and to increase its half-life, mAb WN1 222-5 was chimerized into a human IgG1 (SDZ 219-800), thus retaining the binding specificity and cross-reactivity of mAb WN1 222-5. Crystallographic analysis of Fab fragments of WN1 222-5 and SDZ 219-800 showed that the two V regions, identical in amino acid sequence, largely overlap (DI PADOVA et al. 1994). As expected, differences are observed in the constant domains. Using purified LPS and free lipid A in enzyme-linked immunosorbent assay, we showed that mAb SDZ 219-800 recognizes all S-form LPS of *E. coli* and *S. enterica*. Reactivity with the known complete core structures of *E. coli* (R1 to R4 and K-12) and *S. enterica* (Ra) was also evident (Fig. 8). The Rc core, expressed in *E. coli* J5, is the minimal structure to which mAb SDZ 219-800 binds. Some residual binding is observed with the truncated core structure of *S. enterica* sv. *minnesota* RcP^-, but no reactivity with Rd and Re structures of *E. coli* and *S. enterica* sv. *minnesota* or free mono- or bisphosphoryl lipid A was detected. In immunoblots of S-form LPS after electrophoresis in polyacrylamide gels mAb SDZ 219-800 reacts not only with the free R component but also with O-side chain substituted core structures, showing that the epitope recognized is accessible on the LPS molecule independently of the presence or absence of the O-specific side chain. This reactivity could be inhibited with an excess of heterologous LPS. These data demonstrate that the epitope is a lateral segment of the core region which has been shown to be structurally rather conserved among Enterobacteriaceae (Fig. 2).

The mAb SDZ 219-800 cross-reacts with all clinical isolates of *E. coli* and *S. enterica* tested and with other Enterobacteriaceae as well. The mAb shows biological activity in vitro as it efficiently inhibits the LPS-induced secretion of IL-6 by mouse peritoneal cells (Table 1). On a molar basis the antibody is significantly more active than polymyxin B (BRUINS et al. 1977). In vivo SDZ 219-800 neutralizes the pyrogenic activity of various enterobacterial S-and R-form LPS and offers protection from the lethal effects of different LPS in D-GalN sensitized mice (Table 2) (DI PADOVA et al. 1993a, b, 1994).

The fact that mAb SDZ 219-800 exhibits high functional neutralizing capacity provides evidence for the accessibility of its epitope on the LPS molecule both in vitro and in vivo. mAb SDZ 219-800 does not bind to lipid A, the toxic component of the LPS molecule. Its potent anti-endotoxic properties in vitro could be based on an indirect steric effect by inhibiting binding of lipid A to its receptor or an antibody-induced change of the toxic conformation of lipid A. In addition to these factors, the neutralizing activity of this mAb in vivo may be explained by a more efficient elimination of antibody-associated LPS through the Fc receptor or the complement receptor. mAb SDZ 219-800 is therefore different from the so-called anti-lipid A IgM antibodies which have so far been tested clinically.

In conclusion, after many years of intense research, murine and chimeric mAbs of IgG class which are widely cross-reactive, and which recognize protective epitopes of the LPS molecule have been identified. The biological properties of these mAbs show that in order to neutralize endotoxicity neither a

Table 1. Monoclonal anti-LPS core antibody SDZ 219-800 inhibits IL-6 secretion of mouse peritoneal macrophages induced by *E. coli* 0111 LPS

Endotoxin *E. coli* 0111 (n*M*)	IL-6 secreted in culture supernatants in the absence or presence of mAb SDZ 219-800[a]		
	Control	MAb SDZ 219-800	
		0.1 n*M*	1n*M*
	ng/ml		
10	23.376	22.612	20.126
1	14.047	10.427	2.225
0.1	1.041	0.684	0.171
0.01	0.267	0.141	0.142
Control	0.125	0.128	0.224

[a]IL-6 in culture supernatants was determined using the B13-29 hybridoma cell line.

Table 2. Inhibition of LPS-induced lethality by monoclonal anti-LPS antibodies in mice[a]

Monoclonal antibody (1 mg/mouse)	Surviving over treated animals after challenge with endotoxin derived from	
	S. enterica sv Abortus equi (1 ng/mouse)	*E. coli* 016 (2 ng/mouse)
WN1 222-5	6/6	5/6
SDZ 219-800	6/6	5/6
HA-1A	0/6	1/6
Control	0/6	1/6

[a]mAb were administered i.v. 2 h before i.v. application of LPS and i.p. administration of D-galactosamine (16 mg/mouse) in C57BL/6 mice.

certain immunoglobulin subclass, such as the IgM class antibody, nor lipid A specificity is required. It is possible, however, that IgM antibodies of the same specificity would express stronger C′-mediated bactericidal properties in addition to similar endotoxin neutralizing capacity. The development and characterization of such mAbs with a broad spectrum of bacterial reactivity therefore offers new opportunities for the immunotherapy of patients suffering from sepsis associated with endotoxemia (Di Padova et al. 1995).

12 Final Remarks

The present chapter discusses the significance of endotoxin as a major sepsis mediator after its release from bacteria and presents new immunological preventive or therapeutic strategies against *circulating* LPS. It should be noted, however, that endotoxin contributes to the pathogenic potential of gram-negative bacteria not only in a released and soluble but also in a *bound* form, i.e., as part of the bacterial outer membrane. Endotoxin constitutes an essential component of this membrane which in turn is essential for bacterial viability. Thus mutants unable to synthesize the lipid A-Kdo domain of LPS cease to grow, and drugs interfering with its biosynthesis are bactericidal. Further, as an exposed membrane structure LPS may not only prevent the activation of complement and thereby contribute to the serum resistance of bacteria but may also act as an inhibitor of phagocytosis and intracellular destruction. In addition, the LPS-containing outer membrane protects gram-negative bacteria against certain antibiotics and bile acids. Therefore, as bacterial membrane components exposed on the cellular surface, endotoxins represent an ideal target to attack pathogenic bacteria by genetical, pharmacological, or immunological strategies. In the future a combination of strategies aimed at LPS both as a soluble toxin and as a cell-bound surface antigen may perhaps become the most promising approach in the fight against gram-negative sepsis.

Acknowledgements. The work described in this paper was supported by the DFG [grants Br 731/9-1, Se 532/2-3, Se 532/4-1, Ho 1259/3-1, Lo 385/4-1, Schu 828/1-1, and the SFB 367 projects B1 (H.B), B2 (G.S., U.M., E.T.R), and C5 (A.J.U)], the BMBF 01KI 9471 [U.S., E.T.R., U.Z., A.J.U.] and the German-Israeli Foundation for Scientific Research and Development (1 169-207.02/94; U.Z., E.T.R.]. The financial aid of the Thyssen Stiftung (R.R.S. "1993/2/84") and the Fonds der Chemischen Industrie (E.T.R., H.D.F.) is greatly acknowledged. We are grateful to Mrs. M. Lohs, F. Richter, and B. Köhler for art and photographic work and to Mrs. R. Mohr and F. Richter for typing this review. We also thank S. Greisman for discussion and valuable suggestions.

References

Anderson J, Melchers F, Galanos C, Lüderitz O (1973) The mitogenic effect of LPS on bone marrow derived mouse lymphocytes. Lipid A as the mitogenic part of the molecule. J Exp Med 137: 943–953

Arditi M, Zhou J, Dorio R, Wie Rong GW, Goyert SM, Kim KS (1993) Endotoxin-mediated endothelial cell injury and activation: role of soluble CD14. Infect Immun 61: 3149–3156

Badger AM, Olivera D, Talmadge JE, Hanna N (1989) Protective effect of SK & F 86002, a novel dual inhibitor of arachidonic acid metabolism, in murine models of endotoxin shock: inhibition of tumor necrosis factor as a possible mechanism of action. Circ Shock 27: 51–61

Baker PJ (1993) Effect of endotoxin on suppressor T cell function. Immunobiology 187: 372–381

Bakouche V, Moreau JL, Lachman LB (1992) Secretion of IL-1: role of protein kinase C. J Immunol 148: 84–91

Baumann H, Gauldie J (1994) The acute phase response. Immunol Today 15: 74–80

Baumgartner J-D, Glauser M-P (1993) Immunotherapy of endotoxemia and septicemia. Immunobiology 187: 464–477

Baumgartner JD, McCutchan JA, van Melle G, Vogt M, Luethy R, Glauser MP, Ziegler EJ, Klamber MR, Muhlen E, Chiolero R, Geroulanos S (1983) Prevention of shock and death in surgical patients by antibody to endotoxin core glycolipid. Lancet 2: 59–63

Baumgartner JD, Heumann D, Gerain J, Weinbreck P, Grau GE, Glauser MP (1990) Association between protective efficacy of anti-lipopolysaccharide (LPS) antibodies and suppression of LPS-induced tumor necrosis factor alfa and interleukin 6: comparison of O-side chain specific antibodies with core LPS antibodies. J Exp Med 171: 889–896

Bažil V, Strominger JL (1991) Shedding as a mechanism of down-modulation of CD14 on stimulated human monocytes. J Immunol 147: 1567–1574

Beckmann F, Moll H, Jäger K-E, Zähringer U (1995) 7-*O*-Carbamoyl-L-glycero-D-manno-heptose: a new core constituent in lipopolysaccharide of Pseudomonas aeruginosa. Carbohydr Res 267: C3–C7

Beekhuizen H, Blokland I, Corsèl-van Tilburg AJ, van Furth R (1991) CD14 contributes to the adherence of human monocytes to cytokine-stimulated endothelial cells. J Immunol 147: 3761–3767

Beekhuizen H, Blokland, van Furth R (1993) Crosslinking of CD14 molecules on monocytes results in a CD11/CD18- and ICAM-1-dependent adherence to cytokine-stimulated human endothelial cells. J Immunol 150: 950–959

Bender A, Sprengler H, Gong J-H, Henke A, Bolte G, Sprengler H-P, Nain M, Gemsa D (1993) The potentiating effect of LPS on tumor necrosis factor-α production by influenza A virus-infected macrophages. Immunobiology 187: 357–371

Bernhagen J, Calandra T, Cerami A, Bucala R (1994) Macrophage migration inhibitory factor is a neuroendocrine mediator of endotoxaemia. Trends Microbiol 2: 198–201

Beutler B, Cerami A (1989) The biology of Cachectin/TNF. A primary mediator of the host response. Annu Rev Immunol 7: 625–655

Bhakdi S, Grimminger F, Suttorp N, Walmrath D, Seeger W (1994) Proteinaceous bacterial toxins and pathogenesis of sepsis syndrome and septic shock: the unknown connection. Med Microbiol Immunol 183: 119–144

Bhat NM, Bieber MM, Chapman CJ, Stevenson FK, Teng NNH (1993) Human antilipid A monoclonal antibodies bind to human B cells and the i antigen on cord red blood cells. J Immunol 151: 5011–5021

Bhat UR, Forsberg LS, Carlson RW (1994) Structure of the lipid A component of Rhizobium leguminosarum bv. phaseoli lipopolysaccharide. J Biol Chem 269: 14402–14410

Bogard WC Jr, Dunn DL, Abernethy K, Kilgariff C, Kung PC (1987) Isolation and characterization of murine monoclonal antibodies specific for gram-negative bacterial lipopolysaccharide: association of cross-genus reactivity with lipid A specificity. Infect Immun 55: 899–908

Bone RC, Fisher CJ Jr, Clemmer TP, Slotman GJ, Metz CA, Balk RA, the Methyl-prednisolone Severe Sepsis Study Group (1987) A controlled clinical trial of high-dose methylprednisolone in the treatment of severe sepsis and septic shock. N Engl J Med 317: 389–393

Bone RC, Fisher CJ Jr, Clemmer TP, Slotman GJ, Metz CA, Balk RA, the Methyl-prednisolone Severe Sepsis Study Group (1989) Sepsis syndrome: a valid clinical entity. Crit Care Med 17: 389–393

Brade H, Brade L, Rietschel ET, (1988) Structure-activity relationships of bacterial lipopolysaccharides (endotoxins). Zentralbl Bakteriol Mikrobiol Hyg [A] 268: 151–179

Brade L, Schramek S, Schade U, Brade H (1986) Chemical, biological, and immunochemical properties of the Chlamydia psittaci lipopolysaccharide. Infect Immun 54: 568–574

Brade L, Kosma P, Appelmelk BJ, Paulsen H, Brade H (1987a) Use of synthetic antigens to determine the epitope specificities of monoclonal antibodies against the 3-deoxy-D-manno-octulosonate region of bacterial lipopolysaccharide. Infect Immun 55: 462–466

Brade L, Brandenburg K, Kuhn H-M, Kusumoto S, Rietschel ET, Brade H (1987b) The immunogenicity and antigenicity of lipid A are influenced by its physicochemical state and environment. Infect Immun 55: 2636–2644

Brade L, Holst O, Brade H (1993) An artificial glycoconjugate containing the bisphosphorylated glucosamine disaccharide backbone of lipid A binds lipid A monoclonal antibodies. Infect Immun 61: 4514–4517

Brandenburg K, Mayer H, Koch MHJ, Weckesser J, Rietschel ET, Seydel U (1993) Influence of the supramolecular structure of free lipid A on its biological activity. Eur J Biochem 218: 555–563

Brandwein SR (1986) Regulation of interleukin-1 production by mouse peritoneal macrophages. J Biol Chem 261: 8624–8632

Braquet P, Touqui L, Shen TY, Vargaftig BB (1987) Perspectives in platelet-activating factor research. Pharmacol Rev 39: 97–112

Braude AI, Douglas H (1972) Passive immunization against the local Schwartzman reaction. J Immunol 108: 505–512

Bruins SC, Stumacher R, Johns MA, McCabe WR (1977) Immunization with R mutants of Salmonella minnesota. II. Comparison of the protective effect of immunization with lipid A and the Re mutant. Infect Immun 17: 16–20

Carr BD, Vicky VM, Schnyder B, Ozmen L, Huang S, Galley P, Heumann D, Aguet M, Ryffel B (1994) Interferon γ receptor deficient mice are resistant to endotoxic shock. J Exp Med 179: 1437–1444
Centers for Disease Control (1990) Increase in national hospital discharge survey cases for septicemia. MMWR 39: 31–34
Chang ZL, Novotney A, Suzuki T, (1990) Phospholipase C and A_2 in tumoricidal activation of murine macrophage-like cell lines. FASEB J 4: A1753
Chen TY, Lei MG, Suzuki T, Morrison DC, (1986) Lipopolysaccharide receptors and signal transduction pathways in mononuclear phagocytes. Curr Top Microbiol Immunol 181: 169–188
Chensue SW, Kunkel SL (1985) Induction of interleukin 1 release by leukotrienes. Fed Proc 44: 1270
Christ WJ, Asano O, Robidoux ALC, Perez M, Wang Y, Dubuc GR, Gavin WE, Hawkins LD, McGuinness PD, Mullarkey MA, Lewis MD, Kishi Y, Kawata T, Bristol JR, Rose JR, Rossignol DP, Kobayashi S, Hishinuma I, Kimura A, Asakawa A, Asakawa N, Katayama K, Yamatsu I (1995) E5531, a pure endotoxin antagonist of high potency. Science 268: 80–83
Chung J, Pelech SL, Blenis J (1991) Mitogen-activated Swiss mouse 3T3 RSK kinase I and II are related to pp44mpk from sea star oocytes and participate in the regulation of pp90rsk activity. Proc Natl Acad Sci USA 88: 4981–4985
Colwell DE, Michalek SM, Brilles DE, Jirillo E, McGhee JR (1984) Monoclonal antibodies to Salmonella lipopolysaccharide: anti-O-polysaccharide antibodies protect C3H mice against challenge with virulent Salmonella typhimurium. J Immunol 133: 950–957
Cordle SR, Donald R, Read MA, Hawiger J (1993) Lipopolysaccharide induces phosphorylation of MAD3 and activation of c-Rel and related NF-κB proteins in human monocytic THP-1 cells. J Biol Chem 268: 11803–11810
Couturier C, Jahns G, Kazatchkine MD, Haeffner-Cavaillon N (1992) Membrane molecules which trigger the production of interleukin-1 and tumor necrosis factor-α by lipopolysaccharide-stimulated human monocytes. Eur J Immunol 22: 1461–1466
Cryz SJ, Wedgwood J, Lang AB, Ruedeberg A Queju, Fürer E, Schaad UB (1994) Immunization of noncolonized cystic fibrosis patients against Pseudomonas aeruginosa. J Infect Dis 169: 1159–1162
Dérijard B, Raingeaud J, Barrett T, Wu I-H, Han J, Ulevitch RJ, Davis R,(1994) Independent human MAP kinase signal transduction pathways defined by MEK and MKK isoforms. Science 267: 682–685
Di Padova FE, Gram H, Barclay R, Kleuser B, Liehl E, Rietschel ET (1993a) New anticore LPS monoclonal antibodies with clinical potential. In: Levin J, Alving CR, Munford RS, Stuetz PL (eds) Bacterial endotoxin: recognition and effector mechanisms. Elsevier, Amsterdam, pp 325–335
Di Padova FE, Brade H, Barclay R, Poxton IR, Liehl E, Schuetze E, Kocker HP, Ramsay G, Schreier MH, McClelland DBL, Rietschel ET (1993b) A broadly cross-protective monoclonal antibody binding to Escherichia coli and Salmonella lipopolysaccharides. Infect Immun 61: 3863–3872
Di Padova FE, Mikol V, Barclay GR, Poxton IR, Brade H, Rietschel ET (1994) Prog Clin Biol Res 388: 85–94
Di Padova FE, Gram H, Barclay R, Poxton IR, Liehl E, Rietschel ET (1996) Monoclonal antibodies to endotoxin core as a new approach in endotoxemia therapy. In: Morrison DC, Ryan JL (eds) Novel therapeutic strategies in the treatment of sepsis. Dekker, New York pp 13–31
Dinarello CA (1994) The interleukin-1 family: 10 years of discovery. FASEB J 8: 1314–1325
Dinarello CA, Bishai J, Rosenwasser LJ, Coceani F (1984) The influence of lipoxygenase inhibitors on the in vitro production of human leucocytic pyrogen and lymphocyte activating factor (interleukin 1). Int J Immunopharmacol 6: 43–50
Doherty DE, Zagarello L, Henson PM, Worthen GS (1989) Lipopolysaccharide stimulates monocyte adherence by effects on both the monocyte and the endothelial cell. J Immunol 143: 3673–3679
Dominioni L, Dionigi R, Zanello M, Chiaranda M, Dionigi R, Acquarolo A, Ballabio A, Sguotti C (1991) Effects of high-dose IgG on survival of surgical patients with sepsis scores of 20 or greater. Arch Surg 126: 236–240
Drysdale PE, Yapundich RA, Shin ML, Shin HS (1987) Lipopolysaccharide-mediated macropahge activation: the role of calcium in the generation of tumoricidal activity. J Immunol 139: 951–956
Dunn DI, Ewald DC, Chandan N, Cerra FB (1986) Immunotherapy of gram-negative bacterial sepsis. A single murine monoclonal antibody provides cross-genera protection. Arch Surg 121: 58–62
Durieux JJ, Vita N, Popescu O, Guette F, Calzadawack J, Munker R, Schmidt RE, Lupker J, Ferrara P, Ziegler-Heitbrock HWL, Labeta MO (1994) The two soluble forms of the lipopolysaccharide receptor, CD14: characterization and release by normal human monocytes. Eur J Immunol 24: 2006–2012

Dziarski R (1989) Corrrelation between ribosylation of pertussis toxin substrates and inhibition of peptidoglycan-muramyl dipeptide- and lipopolysaccharide induced mitogenic stimulation in B lymphocytes. Eur J Immunol 19: 125–130
Dziarski R (1994) Cell-bound albumin is the 70 kDa peptidoglycan-, lipopolysaccharide- and lipoteichoic acid-binding protein on lymphocytes and macrophages. J Biol Chem 269: 20431–20436
Echternacher B, Falk W, Männel DN, Krammer PH (1990) Requirement of endogenous tumor necrosis factor cachectin for recovery from experimental peritonitis. J Immunol 145: 3762–3766
Elekes E, Jakobs D, Schade FU (1993) Suppression of endotoxin mitogenicity of spleen cells by lipoxygenase inhibitors and its reversal by 13-hydroxyoctadecadienoic acid. FEMS Immunol Med Microbiol 6: 13–20
Elsbach P, Weiss J (1993) The bactericidal/permeability-increasing protein (BPI), a potent element in host-defense against gram-negative bacteria and lipopolysaccharide. Immunobiology 187: 417–429
Erwin AL, Munford RS (1992) Processing of LPS by phagocytes. In: Morrison DC, Ryan JL (eds) Molecular biochemistry and cellular biology. CRC Press, Boca Raton, pp 405–430 (Bacterial endotoxic lipopolysaccharide, vol 1)
Flad H-D, Loppnow H, Rietschel ET, Ulmer AJ (1993) Agonists and antagonists for lipopolysaccharide-induced cytokines. Immunobiology 187: 303–316
Flegel WA, Wölpl A, Männel DN, Northoff H (1989) Inhibition of endotoxin-induced activation of human monocytes by human lipoproteins. Infect Immun 57: 2237–2245
Freudenberg MA, Bog-Hansen TC, Back U, Galanos C (1980) Interaction of lipopolysaccharides with plasma high-density lipoprotein in rats. Infect Immun 28: 373–380
Freudenberg MA, Meier-Dieter U, Staehelin T, Galanos C (1991) Analysis of LPS released from Salmonella abortus equi in human serum. Microb Pathogen 10: 93–104
Frey EA, Miller DS, Jahr TG, Sundan A, Bažil V, Espevik T, Finlay BB, Wright SD (1992) Soluble CD14 participates in the response of cell to lipopolysaccharide. J Exp Med 176: 1665–1671
Fujihara Y, Lei M-G, Morrison DC (1993) Characterization of specific binding of a human immunoglobulin M monoclonal antibody to lipopolysaccharide and its lipid A domain. Infect Immun 61: 910–918
Galanos C, Freudenberg MA (1993) Mechanisms of endotoxin shock and endotoxin hypersensitivity. Immunobiology 187: 346–356
Galanos C, Freudenberg MA, Jay F, Nerkar D, Veleva K, Brade H, Strittmatter W (1984) Immunogenic properties of lipid A. Rev Infect Dis 6: 546–552
Galanos C, Lüderitz O, Rietschel ET, Westphal O, Brade H, Brade L, Freudenberg MA, Schade FU, Imoto M, Yoshimura S, Kusumoto S, Shiba T (1985) Synthetic and natural Escherichia coli free lipid A express identical endotoxic activities. Eur J Biochem 148: 1–5
Galanos C, Freudenberg MA, Katschinski T, Salomao R, Mossmann H, Kumazawa Y (1992) Tumor necrosis factor and host response to endotoxin. In: Ryan JL, Morrison DC (eds) Immunopharmacology and pathophysiology. CRC Press, Boca Raton, pp 75–102 (Bacterial endotoxic lipopolysaccharides, vol 2)
Gallay P, Carrel S, Glauser MP, Barras C, Ulevitch RJ, Tobias PS, Baumgartner J-D, Heumann D (1993a) Purification and characterization of murine lipopolysaccharide-binding protein. Infect Immun 61: 378–383
Gallay P, Heumann D, Le Roy D, Barras C, Glauser MP (1993b) Lipolysaccharide-binding protein.as a major plasma protein responsible for endotoxemic shock. Proc Natl Acad Sci USA 90: 9935–9938
Gallay P, Heumann D, Le Roy D, Barras C, Glauser MP (1994a) Mode of action of anti-lipopolysaccharide-binding protein antibodies for prevention of endotoxemic shock in mice. Proc Natl Acad Sci USA 91: 7922–7926
Gallay P, Barras C, Tobias PS, Calaudra T, Glauser MP, Heumann D, (1994b) Lipopolysaccharide (LPS)-binding protein in human serum determines the tumor necrosis factor response of monocytes to LPS. J Infect Dis 170: 1319–1322
Gazzano-Santoro H, Mészáros K, Birr C, Carroll SF, Theofan G, Horwitz AH, Lim E, Aberle S, Kasler H, Parent JB (1994) Competition between $rBPI_{23}$, a recombinant fragment of bactericidal/permeability-increasing protein, and lipopolysaccharide (LPS)-binding protein for binding to LPS and gram-negative bacteria. Infect Immun 62: 1185–1191
Glauser MP, Zanetti G, Baumgartner J-D, Cohen J (1991) Septic shock: pathogenesis. Lancet 338: 732–736
Greenman RL, Schein RMH, Martin MA, Wenzel RP, McIntyre NR, Emmanuel G (1991) A controlled clinical trial of E5 murine monoclonal IgM antibody to endotoxin in the treatment of gram-negative sepsis. JAMA 266: 1097–1102

Greisman SE, Johnston CA (1988) Failure of antisera to J5 and R595 rough mutants to reduce endotoxemic lethality. J Infect Dis 157: 54–64

Greisman SE, Dubuy JB, Woodward CL (1978) Experimental gram-negative bacterial sepsis: reevaluation of the ability of rough mutant antisera to protect mice. Proc Soc Exp Biol Med 158: 482–490

Hailman E, Lichenstein HS, Wurfel MM, Miller DS, Johnson DA, Kelley M, Busse LA, Zukowski MM, Wright SD (1994) Lipopolysaccharide (LPS)-binding protein accelerates the binding of LPS to CD14. J Exp Med 179: 269–277

Han J, Lee JD, Tobias PS, Ulevitch RJ (1993) Endotoxin induces rapid protein tyrosine phosphorylation in 70Z/3 cells expressing CD14. J Biol Chem 268: 25009–25014

Han J, Lee JD, Bibbs L, Ulevitch RJ (1994) A MAP kinase targeted by endotoxin and hyperosmolarity in mammalian cells. Science 265: 808–811

Hauschildt SH, Scheipers P, Bessler WG (1994) Lipopolysaccharide-induced change of ADP-ribosylation of a cytosolic protein in bone-marrow-derived macrophages. Biochem J 297: 17–20

Havell EA (1989) Evidence that tumor necrosis factor has an important role in antibacterial resistance. J Immunol 143: 2894–2899

Haziot A, Chan S, Ferrero E, Low MG, Silber R, Goyer SM (1988) The monocyte differentiation antigen, CD14, is anchored to the cell membrane by a phosphatidylinositol linkage. J Immunol 141: 547–552

Haziot A, Rong GW, Bažil V, Silver J, Goyert SM (1994) Recombinant soluble CD14 inhibits LPS-induced tumor necrosis factor-α production by cells in whole blood. J Immunol 152: 5868–5876

Heine H, Brade H, Kusumoto S, Kusama T, Rietschel ET, Flad H-D, Ulmer AJ (1994) Inhibition of LPS binding on human monocytes by phosphonooxyethyl analogs of lipid A. J Endotox Res 1: 14–20

Heine H, Ulmer AJ, Flad H-D, Hauschildt S (1995) LPS-induced change of phosphorylation of two cytosolic proteins in human monocytes is prevented by inhibitors of ADP-ribosylation. J Immunol 155: 4899–4908

Heumann D, Barras C, Severin A, Glauser MP, Tomasz A (1994) Gram-positive cell walls stimulate synthesis of tumor necrosis factor alpha and interleukin-6 by human monocytes. Infect Immun 62: 2715–2721

Hirata M, Shimomura Y, Yoshida M, Morgan JG, Palings I, Wilson D, Yen MH, Wright SC, Larrick JW (1994) Characterization of a rabbit cationic protein (CAP18) with lipopolysaccharide-inhibitory activity. Infect Immun 62: 1421–1426

Hoess A, Watson S, Siber GR, Liddington R (1993) Crystal structure of an endotoxin-neutralizing protein from horseshoe crab, Limulus anti-LPS factor, at 1.5 Å resolution. EMBO J 12: 3351–3356

Hollingsworth RI, Lill-Elghanian DA (1989) Isolation and characterization of the unusual lipopolysaccharide component, 2-amino-2-deoxy-2-N-(27-hydroxyoctacosanoyl)-3-O-(3-hydroxytetradecanoyl)-gluco-hexuronic acid, and its de-O-acylation product from the free lipid A of Rhizobium trifolii ANU 843. J Biol Chem 264: 14039–14042

Holst O, Brade H (1992) Chemical structure of the core region of lipopolysaccharides. In: Morrison DC, Ryan JL (eds) Molecular biochemistry and cellular biology. CRC Press, Boca Raton, pp 135–170 (Bacterial endotoxic lipopolysaccharides, vol 1)

Hurme M, Serkkola F (1991) Differential activation signals are required for the expression of interleukin-1-α and -β-genes in human moncytes. Scand J Immunol 33: 713–718

Hurme M, Viherluoto J, Nordstrom T (1992) The effect of calcium mobilization on LPS-induced IL-1β production depends on the differentiation stage of the monocytes/macrophages. Scand J Immunol 36: 507–511

Ingalls RR, Golenbock DT (1995) CD11c/CD18, a transmembrane signalling receptor for lipopolysaccharide. J Exp Med 181: 1473–1479

Introna M, Hamilton TA, Kaufmann RE, Adams DO, Bast RCJ (1986) Treatment of murine peritoneal macrophages with bacterial lipopolysaccharide alters expression of c-fos and c-myc oncogenes. J Immunol 137: 2711–2715

Jackway JP, De Franco AL (1986) Pertussis toxin inhibition of B cell and macrophage responses to bacterial lipopolysaccharide. Science 234: 743–746

Jakobs DM, Schade FU (1994) Regulation of endotoxin mitogenicity in murine spleen cells by tumor necrosis factor. J Endotox Res 1: 175–180

Jochum M, Machleidt, Fritz H (1992) Proteolysis-induced pathomechanisms in acute inflammation and related therapeutic approaches In: Sies H, Flohe L, Zimmer G (eds) Molecular aspects of inflammation. Springer Berlin Heidelberg New York, pp 73–92 (42th Mosbach colloquium)

Johnston CA, Greisman SE (1984) Endotoxemia induced by antibiotic therapy: a mechanism for adrenal corticosteroid protection in gram-negative sepsis. Trans Assoc Am Physicians 97: 172–181

Juan TS-C, Hailman E, Kelley MJ, Busse LA, Davy E, Empig CJm Narhi LO, Wright SD, Lichenstein HS (1995) Identification of a lipopolysaccharide binding domain in CD14 between amino acids 57 and 64. J Biol Chem 270: 5219–5224

Kabir S, Rosenstreich DL, Mergenhagen SE (1978) Bacterial endotoxin and cell membranes. In: Jeljaszewic J, Wadstrom T (eds) Bacteria toxins and cell membranes. Academic, London, pp 59–87

Kawahara K, Seydel U, Motohiro M, Hirofumi D, Rietschel ET, Zähringer U (1991) Chemical structure of glycosphingolipids isolated form Sphingomonas paucimobilis. FEBS Lett 292: 107–110

Keppler D, Hagmann W, Rapp S (1987) Role of leukotrienes in endotoxin action in vivo. Rev Infect Dis 5: 580–584

Kirikae T, Kirikae F, Schade FU, Yoshida M, Kondo S, Hisatsune K, Nishikawa S-I, Rietschel ET (1991) Detection of lipopolysaccharide-binding proteins on membranes of murine lymphocyte and macrophage-like cell lines. FEMS Microbiol Immunol 76: 327–336

Kirikae T, Schade FU, Zähringer U, Kirikae F, Brade H, Kusumoto S, Kusama T, Rietschel ET (1993a) The significance of the hydrophilic backbone and the hydrophobic fatty acid regions of lipid A on macrophage binding and cytokine induction. FEMS Immunol Med Microbiol 8: 13–26

Kirikae T, Schade FU, Kirikae F, Rietschel ET, Morrison DC (1993b) Isolation of a macrophage-like cell line defective in binding of lipopolysaccharide (LPS). J Immunol 151: 2742–2752

Kirikae T, Schade FU, Kirikae F, Qureshi N, Takayama K, Rietschel ET (1994) Diphosphoryl lipid A derived from the lipopolysaccharide (LPS) of Rhodobacter sphaeroides ATCC 17023 is a potent competitive LPS inhibitor in murine macrophage-like J774.1 cells. FEMS Immunol Med Microbiol 9: 237–244

Kirkland TN, Ziegler EJ (1984) An immunoprotective monoclonal antibody to lipopolysaccharide. J Immunol 132: 2590–2592

Kirkland TN, Finley F, Leturcq D, Moriarty A, Lee J-D, Ulevitch RJ, Tobias PS (1993) Analysis of lipopolysaccharide binding by CD14. J Biol Chem 268: 24818–24823

Kitchens RL, Munford RS (1995) Enzymatically deacylated lipopolysaccharide (LPS) can antagonize LPS at multiple sites in the LPS recognition pathway. J Biol Chem 270: 9904–9910

Kitchens RL, Ulevitch RJ Munford RS (1992) Lipopolysaccharide (LPS) partial structures inhibit responses to LPS in human macrophage cell line without inhibiting LPS uptake by a CD14-mediated pathway. J Exp Med 176: 485–494

Knirel YA, Rietschel ET, Marre R, Zähringer U (1994) The structure of the O-specific chain of Legionella pneumophila serogroup 1 lipopolysaccharide. Eur J Biochem 221: 239–245

Kovach NL, Yee E, Munford RS, Raetz CRH, Harlan JM (1990) Lipid IVA inhibits synthesis and release of tumor necrosis factor induced by lipopolysaccharide in human whole blood ex vivo. J Exp Med 172: 77–84

Kovacs EJ, Radzioch D, Young HA, Varesio L (1988) Differential inhibition of IL-1 and TNF-α mRNA expression by agents which block second messenger pathways in murine macrophages. J Immunol 141: 3101–3105

Krziwon C, Zähringer U, Kawahara K, Weidemann B, Kusumoto S, Rietschel ET, Flad H-D, Ulmer AJ (1995) Glycosphingolipids from Sphingomonas paucimobilis induce monokine production in human mononuclear cells. Infect Immun 63: 2899–2905

Kuhn H-M, Brade L, Appelmelk BJ, Kusumoto S, Rietschel ET, Brade H (1992) Characterization of the epitope specificity of murine monoclonal antibodies directed against lipid A. Infect Immun 60: 2201–2210

Kusumoto S (1992) Chemical synthesis of lipid A. In: Morrison DC, Ryan JL (eds) Molecular biochemistry and cellular biology. CRC Press, Boca Raton, pp 81–106 (Bacterial endotoxic lipopolysaccharides, vol 1)

Larrick JW, Morgan JG, Palings I, Hirata M, Yen MH (1991) Complementary DNA sequence of rabbit CAP18, a unique lipopolysaccharide binding protein. Biochem Biophys Res Commun 179: 170–175

Lauener RP, Geha RS, Vercelli D (1990) Engagement of the monocyte surface antigen CD14 induces lymphocytic function-associated antigen-1/intercellular adhesion molecule 1-dependent homotypic adhesion. J Immunol 145: 1390–1398

Lee J-D, Kato K, Tobias PS, Kirkland TN, Ulevitch RJ (1992) Transfection of CD14 into 70Z/3 cells dramatically enhances the sensitivity to complexes of lipopolysaccharide (LPS) and LPS binding protein. J Exp Med 175: 1697–1705

Lee J-D, Laydon JT, McDonnell PC, Gallagher TF, Kumar S, Green D, McNulty D, Blumenthal MJ, Heys JR, Landvatter SW, Strickler JE, McLaughlin MM, Siemens IR, Fisher SM, Livi GP, White JR, Adams JL, Young PR (1994) A protein kinase involved in the regulation of inflammatory cytokine biosynthesis. Nature 372: 739–746

Lei M-G, Morrison DC (1988a) Specific endotoxic lipopolysaccharide-binding receptors on murine splenocytes I. Detection of LPS binding sites on splenocytes and splenocyte subpopulation. J Immunol 141: 996–1005

Lei M-G, Morrison DC (1988b) Specific endotoxic lipopolysaccharide binding proteins on murine splenocytes. II. Membrane localization and binding characteristics. J Immunol 141: 1006–1011

Levy O, Weiss J, Zarember K, Ooi CE, Elsbach P (1993) Antibacterial 15-kDa protein isoforms (p15s) are members of a novel family of leukocyte proteins. J Biol Chem 268: 6058–6063

Li P, Allen H, Banerjee S, Franklin S, Herzog L, Johnston C, McDowell J, Paskind M, Rodman L, Salfeld J, Towne E, Tracey D, Wardwell S, Wei F-Y, Wong W, Kamen R, Seshadri T (1995) Mice deficient in IL-1β-converting enzyme are defective in production of mature IL-1β and resistant to endotoxic shock. Cell 80: 401–411

Libby P, Loppnow H, Fleet JC, Palmer H, Li HM, Warner SJC, Salomon RN, Clinton SK (1991) Production of cytokines by vascular cells – an update and implications for atherogenesis. In: Gottlieb AI, Langille BL, Federoff S (eds) Arteriosclerosis-cellular and molecular interactions in the artery wall. Pleunum, New York, pp 161–169 (1st Albsclul symposium)

Little RG, Kelner DN, Lim E, Burke DJ, Conlon PJ (1994) Functional domains of recombinant bactericidal/permeability increasing protein ($rBPI_{23}$). J Biol Chem 269: 1865–1872

Liu MK, Velit PH, Brownsey RW, Reiner NE (1994) CD14-dependent activation of protein kinase C and mitogen-activated protein kinases (p42 and p44) in human monocytes treated with bacterial lipopolysaccharide. J Immunol 153: 2642–2652

Loppnow H, Libby P (1989) Adult human vascular endothelial cells express the IL-6 gene differently in response to LPS and IL-1. Cell Immunol 122: 493–503

Loppnow H, Libby P (1990) Proliferating or IL-1 activated human vascular smooth muscle cells secrete copious IL-6. J Clin Invest 85: 731–738

Loppnow H, Libby P (1992) Functional significance of human vascular smooth muscle cell-derived interleukin 1 in paracrine and autocrine regulation pathways. Exp Cell Res 198: 283–290

Loppnow H, Brade H, Dürrbaum I, Dinarello CA, Kusumoto S, Rietschel ET, Flad H-D (1989) Interleukin 1 induction-capacity of defined lipopolysaccharide partial structures. J Immunol 142: 3229–3238

Loppnow H, Libby P, Freudenberg MA, Kraus JH, Weckesser J, Mayer H (1990) Cytokine induction by lipopolysaccharide (LPS) corresponds to the lethal toxicity and is inhibited by nontoxic Rhodobacter capsulatus LPS. Infect Immun 58: 3743–3750

Loppnow H, Rietschel ET, Brade H, Feist W, Wang M-H, Heine H, Kirikae T, Schönbeck U, Dürrbaum-Landmann I, Grage-Griebenow E, Brandt E, Schade FU, Ulmer AJ, Campos-Portuguez S, Krauss J, Mayer H, Flad H-D (1993) Lipid A precursor Ia and Rhodobacter capsulatus LPS: potent endotoxin antagonists. In: Levin J (ed) Endotoxin research series, vol 2. Excerpta Medica Elsevier, Amsterdam, pp 337–348

Loppnow H, Stelter F, Schönbeck U, Schlüter C, Ernst M, Schütt C, Flad H-D (1995) Endotoxin activates human vascular smooth muscle cells despite lack of expression of CD14 mRNA or endogenous membrane CD14. Infect Immun 63: 1020–1026

Lüderitz O, Freudenberg MA, Galanos C, Lehmann V, Rietschel ET, Shaw D (1982) Lipopolysaccharides of gram-negative bacteria. In: Razin S, Rottem S (eds) Current topics in membranes and transport, vol 17. Academic, New York, pp 79–151

Lüderitz T, Brandenburg K, Seydel U, Roth A, Galanos C, Rietschel ET (1989) Structural and physicochemical requirements of endotoxins for the activation of arachidonic acid metabolism in mouse peritoneal macrophages in vitro. Eur J Biochem 179: 11–16

Lynn WA, Golenbock DT (1992) Lipopolysaccharide antagonists. Immunol Today 13: 271–276

Mamat U, Grimmecke H-D, Schmidt G (1991) Phage-encoded inhibition of O-antigen biosynthesis (Abstr). Immunobiology 183: 251

Mamat U, Rietschel ET, Schmidt G (1995) Repression of lipopolysaccharide biosynthesis in Escherichia coli by an antisense RNA of Acetobacter methanolicus phage Acm 1. Mol Microbiol 15: 1115–1125

Mantovani A, Bussolino F (1991) Endothelium-derived modulators of leukocyte function. In: Gordon JL (eds) Vascular endothelium: interactions with circulating cells. Elsevier, New York, pp 129–140

Marra MN, Wilde CG, Griffith JE, Snable JL, Scott RW (1990) Bactericidal/permeability-increasing protein has endotoxin-neutralizing activity. J Immunol 144: 662–666

Marra MN, Wilde CG, Collins MS, Snable JL, Thornton MB, Scott RW (1992) The role of bactericidal/permeability-increasing protein as a natural inhibitor of bacterial endotoxin. J Immunol 148: 532–537

Mastroeni P, Jannello D, Mastroeni PI (1993) TNFα as a modulator of the interaction between macrophages and intracellular parasites. Eur Bull Drug Res 2: 163–174

Mattern T, Thanhäuser A, Reiling N, Toellner K-M, Duchrow M, Kusumoto S, Rietschel ET, Ernst M, Brade H, Flad H-D, Ulmer AJ (1994) Endotoxin and lipid A stimulate proliferation of human T cells in the presence of autologous monocytes. J Immunol 153: 2996–3004

Mayer H, Campos-Portuguez SA, Busch M, Urbanik-Sypniewska T, Bhat RU (1990) Lipid A variants – or, how constant are the constant regions in lipopolysaccharides? Excerpta Med Int Congr Ser 923: 111–120

McCabe WR (1972) Immunization with R mutants of Salmonella minnesota. I. Protection against challenge with heterologous Gram-negative bacilli. J Immunol 108: 601–610

Mohri M, Spriggs DR, Kufe D (1990) Effects of lipopolysaccharide on phospholipase A2 activity and tumor necrosis factor expression in HL-60 cells. J Immunol 144: 2678–2682

Moll H, Sonesson A, Jantzen E, Marre R, Zähringer U (1992) Identification of 27-oxooctacosanoic acid and heptocosane-1, 27-dioic acid in Legionella pneumophilia. FEMS Microbiol Lett 97: 1–6

Morrison DC (1989) The case for specific lipopolysaccharide receptors expressed on mammalian cells. Microb Pathogen 7: 389–398

Morrison DC (1990) Diversity of mammalian macromolecules which bind to bacterial lipopolysaccharides. Excerpta Med Int Congr Ser 923: 183–189

Morrison DC, Ryan JL (1987) Endotoxins and disease mechanisms. Annu Rev Med 38: 417–432

Morrison DC, Lei M-G, Kirikae T, Chen T-Y (1993) Endotoxin receptors on mammalian cells. Immunobiology 187: 212–226

Müller JM, Ziegler-Heitbrock HWL, Baeuerle PA (1993) Nuclear factor kappa B, a mediator of lipopolysaccharide effects. Immunobiology 187: 233–256

Müller-Loennies S, Holst O, Brade H (1994) Chemical structure of the core region of Escherichia coli J-5 lipopolysaccharide. Eur J Biochem 224: 751–760

Munford RS, Hall CL (1989) Purification of acyloxyacyl hydrolase, a leukocyte enzyme that removes secondary acyl chains from bacterial lipopolysaccharides. J Biol Chem 264: 15613–15619

Munford RS, Hall CL, Lipton JM, Dietschy JM (1982) Biological activity, lipoprotein-binding behavior, and in vivo disposition of extracted and native forms of Salmonella typhimurium lipopolysaccharides. J Clin Invest 80: 877–888

Natanson C, Hoffman WD, Suffredini AF, Eichacker PQ, Danner RL (1994) Selected treatment strategies for septic shock based on proposed mechanisms of pathogenesis. Ann Intern Med 120: 771–783

Nathan CF (1987) Secretory products of macrophages. J Clin Invest 79: 319–323

Nikaido H (1970) Lipopolysaccharide in the taxonomy of gram-negative bacteria. Int J Syst Bacteriol 20: 383–406

Nnalue NA, Lind SM, Lindberg AA (1992) The disacharide L-α-D-heptose-1 → 7-L-α-D-heptose1→ of the inner core domain of Salmonella lipopolysaccharide is accessible to antibody and is the epitope of a broadly reactive monoclonal antibody. J Immunol 149: 2722–2728

Novotney M, Chang ZL, Uchiyama H, Suzuki T (1991) Protein kinase C in tumoricidal activation of murine macrophage-like cell lines. Biochemistry 30: 5597–5604

Nurminen M, Rietschel ET, Brade H (1985) Chemical characterization of Chlamydia trachomatis lipopolysaccharide. Infect Immun 48: 573–575

Parkar BA, McCormick ME, Foster SJ (1990) Leukotrienes do not regulate interleukin 1 production by activated macrophages. Biochem Biophys Res Commun 169: 422–429

Parillo JE (1993) Pathogenic mechanisms of septic shock. N Engl J Med 328: 1471–1477

Pierce L, (1993) HA-1A has a checkered past. Biol World Financial Watch 25: 1–2

Pohlmann TH, Munford RS, Harlan JM (1987) Deacylated lipopolysaccharide inhibits neutrophil adherence to endothelium induced by lipopolysaccharide in vitro. J Exp Med 165: 1393–1402

Pollack M (1990) New therapeutic strategies in gram-negative sepsis and septic shock based on molecular mechanisms of pathogenesis. In: Mandell GL, Douglas RG, Bennet JE (eds) Principles and practice of infectious diseases. Churchill Livingstone, New York, pp 1–18

Price NP, Kelly TM, Raetz CRH, Carlson RW (1994) Biosynthesis of a structurally novel lipid A in Rhizobium leguminosarum: identification and characterization of six metabolic steps leading from UDP-GlcNAc to 3-deoxy-D-manno-2-octulosonic acid$_2$-lipid IV_A. J Bacteriol 176: 4646–4655

Prpic V, Wciel JE, Somers SD, Diguiseppi J, Gomas SL, Pizzo SV, Hamilton TA, Herman B, Adams DO (1987) Effect of bacterial lipopolysaccharide on the hydrolysis of phosphatidylinositol-4,5-bisphosphate in murine peritoneal macrophages. J Immunol 139: 526–533
Pugin J, Schurer-Maly C-C, Leturcq D, Moriarty A, Ulevitch RJ, Tobias PS (1993) Lipopolysaccharide activation of human endothelial and epithelial cells is mediated by lipopolysaccharide binding protein and soluble CD14. Proc Natl Acad Sci USA 90: 2744–2748
Pugin J, Heumann D, Tomasz A, Kravchenko VV, Akamatsu Y, Nishijima M, Glauser MP, Tobias PS, Ulevitch RJ (1994) CD14 is a pattern recognition receptor. Immunity 1: 509–516
Pulverer BJ, Avruch J, Kyriakis JM, Nikolakaki E, Woodgett JR (1991) Phosphorylation of c-jun mediated by MAP kinases. Nature 353: 670–674
Quezado ZMN, Natanson C, Alling DW, Banks SM, Koev CA, Elin RJ, Hosseini JM, Bacher JD, Danner RL, Hoffman WD (1993) A controlled trial of HA-1A in a canine model of gram-negative septic shock. JAMA 269: 2221–2227
Qureshi N, Takayama K, Kurtz R (1991) Diphosphoryl lipid A obtained from the nontoxic lipopolysaccharide of Rhodopseudomonas sphaeroides is an endotoxin antagonist in mice. Infect Immun 59: 441–444
Radziejewska-Lebrecht J, Krajewska-Pietrasik D, Mayer H (1990) Terminal and chain-linked residues of D-galacturonic acid: characteristic constituents of the R-core regions of Proteeae and of Serratia marcescens. Sys Appl Microbiol 13: 214–219
Raetz CRH (1990) Biochemistry of endotoxins. Annu Rev Biochem 59: 129–170
Raetz CRH, Ulevitch RJ, Wright SD, Sibley CH, Ding A, Nathan CF (1991) Gram-negative endotoxin: an extraordinary lipid with profound effects on eukaryotic signal transduction. FASEB J5: 2652–2660
Rietschel ET, Brade H (1992) Bacterial endotoxins. Sci Am 267: 26–33
Rietschel ET, Galanos C (1977) Lipid A antiserum-mediated protection against lipopolysaccharide and lipid A-induced fever and skin necrosis. Infect Immun 15: 34–49
Rietschel ET, Brade H, Brade L, Brandenburg K, Schade U, Seydel U, Zähringer U, Galonos C, Lüderitz O, Westphal O, Labischinski H, Kusumoto S, Shiba T (1987) Lipid A, the endotoxic center of bacterial lipopolysaccharides: relation of chemical structure to biological activity. Prog Clin Biol Res 231: 25–53
Rietschel ET, Kirikae T, Schade FU, Ulmer AJ, Holst O, Brade H, Schmidt G, Mamat U, Grimmecke H-D, Kusumoto S, Zähringer U (1993) The chemical structure of bacterial endotoxin in relation to bioactivity. Immunobiology 187: 169–190
Rietschel ET, Kirikae T, Schade FU, Mamat U, Schmidt G, Loppnow H, Ulmer AJ, Zähringer U, Seydel U, Di Padova F, Schreier M, Brade H (1994) Bacterial endotoxin: molecular relationships of structure to activity and function. FASEB J 218: 217–225
Robinson PJ (1991) Phosphatidylinositol membrane anchors and T-cell activation. Immunol Today 12: 35–41
Rola-Plesczcynski M, Lemaire I (1985) Leukotrienes augment interleukin 1 production by human monocytes. J Immunol 135: 3985–3961
Rola-Plesczcynski M, Gagnon L, Chavaillaz PA (1988) Immune regulation by leukotriene B4. Ann N Y Acad Sci 524: 218–226
Rozalski A, Brade L, Kuhn H-M, Brade H, Kosma P, Appelmelk BJ, Kusumoto S, Paulsen H (1989) Determination of the epitope specificity of monoclonal antibodies against the inner core region of bacterial lipopolysaccharides by use of 3-deoxy-D-manno-octulosonate-containing synthetic antigens. Carbohydr Res 193: 257–270
Satta N, Toti F, Feugeas O, Bohbot A, Dachary-Prigent J, Eschwège V, Hedman H, Freyssinet J-M (1994) Monocyte vesiculation is a possible mechanismen for dissemination of membrane-associated procoagulant activities and adhesion molecules after stimulation by lipopolysaccharide. J Immunol 153: 3245–3255
Saxen H, Nurminen M, Kuusi N, Svenson SB, Mäkelä PH (1986) Evidence for the importance of O antigen specific antibodies in mouse-protective Salmonella outer membrane protein (porin) antisera. Microb Pathog 1: 433–441
Schade FU, Burmeister I, Engel R (1987) Increased 13-hydroxyoctadecadienoic acid content in lipopolysaccharide-stimulated macrophages. Biochem Biophys Res Commun 147: 695–700
Schade FU, Burmeister I, Elekes E, Engel R, Wolter DT (1989a) Mononuclear phagocytes and eicosanoids: aspects of their synthesis and biological activities. Blut 59: 475–485
Schade FU, Ernst M, Reinke M, Wolter DT (1989b) Lipoxygenase inhibitors suppress formation of tumor necrosis factor in vitro and in vivo. Biochem Biophys Res Commun 159: 748–754

Schade FU, Engel R, Jakobs D (1991) Differential protective activities of site specific lipoxygenase inhibitors in endotoxic shock and production of tumor necrosis factor. Int J Immunopharmacol 13: 565–571
Schedel I, Dreikhausen U, Nentwig B, Höckenschnieder M, Rauthmann D, Balikcioglu S, Coldewey R, Diecher H (1991) Treatment of gram-negative septic shock with an immunoglobulin preparation: a prospective, randomized clinical trial. Crit Care Med 19: 1104–1113.
Schletter J, Brade H, Krüger C, Loppnow H, Rietschel ET, Flad H-D, Ulmer AJ (1994) Binding of lipopolysaccharide to an 80 kDa membrane protein of human cells is mediated by serum factors. Immunobiology 191: 273
Schletter J, Brade H, Brade L, Krüger C, Loppnow H, Kusumoto S, Rietschel ET, Flad H-D, Ulmer AJ (1995) Binding of lipopolysaccharide (LPS) to an 80 kD membrane protein of human cells is mediated by soluble CD14 and LPS-binding protein. Infect Immun 63: 2576–2580
Schönbeck U, Brandt E, Flad H-D, Rietschel ET, Loppnow H (1994) S-form lipopolysaccharide induces leukocyte adhesion to human vascular endothelial cells as potent as IL-1: lipid A precursor Ia antagonizes induction of adhesion by LPS. J Endotox Res 1: 4–13
Schönbeck U, Brandt E, Petersen F, Flad H-D, Loppnow H (1995) Interleukin 8 specifically binds to endothelial but not to smooth muscle cells. J Immunol 154: 2375–2883
Schromm AB, Brandenburg K, Rietschel ET, Seydel U (1995) Do endotoxin aggregates intercalate into phospholipid membranes in a nonspecific, hydrophobic manner? J Endotox Res 2: 313–323
Schumann RR (1992) Function of lipopolysaccharide binding protein and CD14, the receptor for LPS/LBP complexes: a short review. Res Immunol 143: 11–15
Schumann RR, Leong SR, Flaggs GW, Gray PW, Wright SD, Mathison JC, Tobias PS, Ulevitch RJ (1990) Structure and function of lipopolysaccharide binding protein. Science 249: 1429–1431
Schütt C, Schumann RR (1993) Der Endotoxinrezeptor CD14. Immun Infekt 21: 36–40
Schütt C, Ringel B, Nausch M, Bažil V, Horejsi V, Neels P, Alzel H, Jonas L, Siegl E, Friemel H, Plantikow A (1988) Human monocyte activation induced by an anti-CD14 monoclonal antibody. Immunol Lett 19: 321–328
Schütt C, Schilling T, Grünwald U, Schoenfeld W, Krüger C (1992) Endotoxin-neutralizing capacity of soluble CD14. Res Immunol 143: 71–78
Seydel U, Labischinski H, Kastowsky M, Brandenburg K (1993) Phase behaviour, supramolecular structure, and molecular conformation of lipopolysaccharide. Immunobiology 187: 191–211
Shear MJ, Turner FC (1943) Chemical treatment of tumors. V. Isolation of the hemorrhage-producing fraction from Serratia marcescens (Bacillus prodigiosus) culture filtrates. J Natl Cancer inst 4: 81–97
Shinomiya H, Hirata H, Nakano M (1991) Purification and characterization of the 65-kDa protein phosphorylated in murine macrophages by stimulation with bacterial lipopolysaccharide. J Immunol 146: 3617–3625
Shnyra A, Hultenby K, Lindberg AA (1993) Role of the physical state of Salmonella lipopolysaccharide in expression of biological and endotoxic properties. Infect Immun 61: 5351–5360
Sonesson A, Moll H, Jantzen E, Zähringer U (1993) Long-chain α-hydroxy-(w-1)-oxo fatty acids and α-hydroxy-1, w-dioic fatty acids are cell wall constituents of Legionella (L. jordanis, L maceachernii and L. micdadei). FEMS Microbiol Lett 106: 315–320
Steel DM, Whitehead AS (1994) The major acute phase reactants: C-reactive protein, serum amyloid P component and serum amyloid A protein. Immunol Today 15: 81–88
Stefanova I, Corcoran ML, Horak EM, Wahl LM, Bolen JB, Horak ID (1994) Lipopolysaccharide induces activation of CD14-associated protein tyrosine kinase p53/56lyn. J Biol Chem 268: 20725–20728
Sturgill TW, Ray LB, Erikson E, Maller JL (1988) Insulin-stimulated MAP-2 kinase phosphorylates and activates ribosomal protein S6 kinase II. Nature 334: 713–718
Süsskind M, Müller-Loennies S, Nimmich W, Brade H, Holst O (1995) Structural investigation on the carbohydrate backbone of the lipopolysaccharide from Klebsiella pneumoniae rough mutant R20/O1$^-$. Carbohydr Res 269: C1–C7
Swierzko A, Brade L, Höffgen E, Paulsen H, Brade H (1993) Specificity of rabbit antisera against the lipopolysaccharide of Salmonella minnesota strain R7 (chemotype Rd_1P^-). FEMS Med Microbiol Immunol 7: 265–270
Swierzko A, Brade I, Brabetz W, Zych K, Paulsen H, Brade H (1994) Monoclonal antibodies against the heptose region of enterobacterial lipopolysaccharides. J Endotox Res 1: 38–44

Takayama K, Qureshi N (1992) Chemical structure of lipid A. In: Molecular biochemistry and cellular biology. Morrison DC, Ryan JL (eds) CRC Press, Boca Raton, pp 43–60 (Bacterial endotoxin lipopolysaccharides, vol 1)

Takayama K, Mitchell DH, Din ZZ, Mukerjee P, Li C, Coleman DL (1994) Monomeric Re lipopolysaccharide from Escherichia coli is more active than the aggregated form in the Limulus amoebocyte lysate assay and in inducing Egr-1 mRNA in murine peritoneal macrophages. J Biol Chem 269: 2241–2244

Teng NNH, Kaplan HS, Herbert JM, Moore C, Douglas H, Wunerlich A, Braude A (1985) Protection against gram-negative bacteremia and endotoxemia with human monoclonal IgM antibodies. Proc Natl Acad Sci USA 82: 1790–1794

Theofan G, Horwitz AH, Williams RE, Liu P-S, Chan I, Birr C, Carroll SF, Mészáros K, Parent JB, Kasler H, Aberle S, Trown PW, Gazzano-Santoro H (1994) An aminoterminal fragment of human lipopolysaccharide-binding protein retains lipid A binding but not CD14-stimulatory activity. J Immunol 152: 3623–3629

Tiegs G, Wendel A (1988) Leukotriene-mediated liver injury. Biochem Pharmacol 37: 2569–2573

Tiegs G, Barsig J, Matiba B, Uhlig S, Wendel A (1994) Potentiation by granulocyte macrophage colony-stimulating factor of lipopolysaccharide toxicity in mice. J Clin Invest 93: 2616–2622

Tobias PS, Ulevitch RJ (1993) Lipopolysaccharide binding protein and CD14 in LPS-dependent macrophage activation. Immunobiol 187: 227–232

Tobias PS, Soldau K, Ulevitch RJ (1986) Isolation of a lipopolysaccharide binding acute phase reactant from rabbit serum. J Exp med 164: 777–793

Tobias PS, Mathison JC, Ulevitch RJ (1988) A family of lipopolysaccharide binding proteins involved in responses to Gram-negative sepsis. J Biol Chem 263: 13479–13781

Tobias PS, Soldau K, Gegner JA, Mintz D, Ulevitch RJ (1995) Lipopolysaccharide binding protein-mediated complexation of lipopolysaccharide with soluble CD14. J Biol Chem 270: 10482–10488

Tracey KJ, Beutler B, Lowry SF, Merryweather J, Wolpe S, Milsark IW, Hariri RJ, Fahey TJ III, Zentella A, Albert JD (1986) Shock and tissue injury induced by recombinant human cachectin. Science 234: 470–474

Trautman M, Hahn H (1985) Antiserum against Escherichia coli J5: a reevaluation of its in vitro and in vivo activity against heterologous gram-negative bacteria. Infection 13: 140–145

Ulevitch RJ (1993) Recognition of bacterial endotoxin by receptor dependent mechanisms. Adv Immunol 53: 267–288

Ulevitch RJ, Johnston AR, Weinstein DB (1981) New function for high density lipoproteins. Isolation and characterization of a bacterial lipopolysaccharide-high density lipoprotein complex formed in rabbit plasma. J Clin Invest 67: 827-837

Ulevitch RJ,Schumann RR, Mathison JC, Leavesley DI, Martin TR, Soldau K, Kline L, Wolfson E, Tobias PS (1990) Endogenous anti-endotoxin mechanisms. In: Baumgartner, Calandra, Carlet (eds) Endotoxin: from pathophysiology to therapeutic approaches. Flammarion, Paris, pp 31–42

Ulmer AJ, Feist W, Heine H, Kirikae T, Kirikae F, Kusumoto S, Kusama T, Brade H, Schade FU, Rietschel ET, Flad H-D (1992) Modulation of endotoxin-induced monokine release in human monocytes by lipid A partial structures inhibiting the binding of ^{125}I-LPS. Infect Immun 60: 5145-5152

Viriyakosol S Kirkland TN (1995) A region of human CD14 required for lipopolysaccharide binding. J Biol Chem 270: 361-368

Vogel SN, Hogan MM (1990) The role of cytokines in endotoxin-mediated host responses. In: Oppenheim JJ, Shevack EM (eds) Immunopharmacology - the Role of Cells and Cytokines in Immunity and Inflammation. Oxford University Press, New York, pp 238–258

Vosbeck K, Tobias PS, Mueller H, Allen RA, Arfors K-E, Ulevitch RJ, Sklar LA (1990) Priming of polymorphonuclear granulocytes by lipopolysaccharides and its complexes with lipopolysaccharide binding protein and high density lipoprotein. J Leukocyte Biol 47: 97–104

Waage A, Halstensen A, Espevik T (1987) Association between tumor necrosis factor in serum and fatal outcome in patients with meningococcal disease. Lancet 1: 355

Wang M-H, Flad H-D, Feist W, Brade H, Kusumoto S, Rietschel ET, Ulmer AJ (1991) Inhibition of endotoxin-induced interleukin-6 production by synthetic lipid A partial structures in human peripheral blood mononuclear cells. Infect Immun 59: 4655–4664

Warren HS, Danner RL, Munford RS (1993) Anti-endotoxin monoclonal antibodies. N Engl J Med 326: 1153–1157

Weidemann B, Brade H, Rietschel ET, Dziarski R, Bažil V, Kusumoto S, Flad H-D, Ulmer AJ (1994) Soluble peptidoglycan-induced monokine production can be blocked by anti-CD14 monoclonal antibodies and by lipid A partial structures. Infect Immun 62: 4709–4715

Weinstein SL, Gold MR, DeFranco AL (1991) Bacterial lipopolysaccharide stimulates protein tyrosine phosphorylation in macrophages. Proc Natl Acad Sci USA 88: 4148–4152

Weinstein SL, Sanghera JS, Lemke K, De Franco AL, Pelech SL (1992) Bacterial lipopolysaccharide induces tyrosine phosphorylation and activation of mitogen-activated protein kinases in macrophages. J Biol Chem 267: 14955–14962

Weinstein SL, June CH, De Franco AL (1993) Lipopolysaccharide-induced protein tyrosine phosphorylation in human macrophages is mediated by CD14. J Immunol 151: 3829–3838

Weintraub A, Zähringer U, Wollenweber H-W, Seydel U, Rietschel ET (1989) Structural characterization of the lipid A component of Bacteroides fragilis strain NCTC 9343 lipopolysaccharide. Eur J Biochem 183: 425–431

Weiss J, Elsbach P, Shu C, Castillo J, Horwitz A, Theofan G (1992) Human bactericidal/permeability-increasing protein and a recombinant NH_2-terminal fragment cause killing of serum-resistant Gram-negative bacteria in whole blood and inhibit tumor necrosis factor release induced by the bacteria. J Clin Invest 90: 1122–1130

Wenzel R, Bone R, Fein A, Quenzero R, Schentag J, Gorelick KJ, Wedel NI, Perl T (1991) Results of a second double-blind, randomized, controlled trial of antiendotoxin antibody E5 in gram-negative sepsis (Abstract). Interscience Conference on Antimicrobial Agents and Chemotherapy, Sept 19 to Oct 2, American Society of Microbiology, Washington p 294

Westphal O, Westphal U, Sommer T (1977) The history of pyrogen research. In: Schlessinger D (ed) Microbioloy – 1977. American Society of Microbiology, Washington, pp 221–238

Wolff M (1904) Beiträge zur Immunitätslehre (Contributions to the teaching of Immunity). Zentralbl Bakteriol Parasitenkd Infektionskr Hyg I Orig 37: 392

Wright SD (1994) Septin, an activity in plasma that enables CD14-dependent recognition of LPS. Prog Clin Biol Res 388: 53–57

Wright SD Levin SM, Jong MTC, Chad Z, Kabbash LG (1989) CR3 (CD11b/CD18) expresses one binding site for Arg-Gly-Asp-containing peptides and a second site for bacterial lipopolysaccharide. J Exp Med 169: 175–183

Wright SD, Ramos RA, Tobias PS, Ulevitch RJ, Mathison JC (1990) CD14, a receptor for complexes of lipopolysaccharide (LPS) and LPS binding protein. Science 249: 1431–1433

Wright SD, Ramos RA, Patel M, Miller DS (1992) A factor in plasma that opsonizes lipopolysaccharide-bearing particles for recognition by CD14 on phagocytes. J Exp Med 176: 719–727

Wurfel MM, Kunitake ST, Lichenstein H, Vane JP, Wright SD (1994) Lipopolysaccharide (LPS)-binding protein is carried on lipoproteins and acts as a cofactor in the ventralization of LPS. J Exp Med 180: 1025–1035

Young LS, Glauser MP (eds) (1991) Gram-negative septicemia and septic shock. Saunders, Philadelphia (Infectious Disease Clinics of North America, vol 5)

Young LS, Gascon R, Alam S, Bermudez LEM (1989) Monoclonal antibodies for treatment of gram-negative infections. Rev Infect Dis 11: 1564–1571

Yu CL, Haskard DO, Cavender D, Ziff M (1986) Effects of bacterial lipopolysaccharides on the binding of lymphocytes to endothelial cell monolayers. J Immunol 136: 569–573

Zähringer U, Lindner B, Seydel U, Rietschel ET, Noaki H, Unger FM, Imoto M, Kusumoto S, Shiba T (1985) Structure of de-O-acylated lipopolysaccharide from the Escherichia coli Re mutant strain F515. Tetrahedr Lett 26: 6321–6324

Zähringer U, Lindner B, Rietschel ET (1994) Molecular structure of lipid A, the endotoxic center of bacterial lipopolysaccharides. Adv Carbohydr Chem Biochem 50: 211–276

Zähringer U, Knirel YA, Helbig JH, Sonesson A, Marre R, Rietschel ET (1995) The lipopolysaccharide of Legionella pneumophila serogroup 1 (strain Philadelphia 1): Chemical structure and biological significance. Prog Clin Biol Res 392: 114–139

Ziegler EJ, Douglas H, Sherman JE, Davis CE, Braude AI (1973) Treatment of E. coli and Klebsiella bacteremia in agranulocytic animals with antiserum to a UDP-Gal epimerase-deficient mutant. J Immunol 111: 433–438

Ziegler EJ, McCutchan JA, Fierer J, Glauser MP, Sadoff JC, Douglas H, Braude A (1982) Treatment of Gram-negative bacteremia and shock with human antiserum to a mutant Escherichia coli. N Engl J Med 307: 1225–1230

Ziegler SF, Wilson CB, Perlmutter RM (1988) Augmented expression of a myeloid specific protein tyrosine kinase gene (hck) after macrophage activation. J Exp Med 168: 1801–1810

Ziegler EJ, Fischer CJ, Sprung CI, Straube RC, Sadoff JC, Foulke GE, Wortel CH, Fink MP, Dellinger RP, Teng NNH, Allen IE, Berger HJ, Knatterud GI, LoBuglio AF, the HA-1A Sepsis Study Group (1991) Treatment of gram-negative bacteremia and septic shock with HA-1A human monoclonal antibody against endotoxin. A randomized, double-blind, placebo-controlled trial. N Engl J Med 324: 429–436

Ziegler-Heitbrock HWL, Ulevitch RJ (1993) CD14: cell surface receptor and differentiation marker. Immunol Today 14: 121–125

Superantigen-Mediated Lethal Shock: The Functional State of Ligand-Reactive T Cells

K. Heeg, T. Miethke, and H. Wagner

1 Superantigens . . . 83

2 The In Vivo Response to SAg: A Complex Pattern of Reactivities . . . 85

3 The First Hours: Hyperactivation of SAg-Reactive T Cells . . . 86

4 Acute Cytokine Release Causes Lethal T Cell Shock . . . 87

5 Induction of Unresponsiveness in SAg-Reactive T Cells: Anergy, Downregulation, and Deletion . . . 89

6 Exhaustive Proliferation of Anergic T Cells . . . 91

7 Effects of SAg on Antigen-Presenting Cells . . . 92

8 SAg-Induced Anergy Defined In Vitro Is Not Correlated with In Vivo Reactivity . . . 93

9 Superantigens: Pathogenic Factor and Tool to Analyze T Cell Function . . . 96

References . . . 97

1 Superantigens

A group of bacterial and viral proteins has been termed superantigens (SAg) because of their unique mechanisms to interact with antigen-presenting cells (APC) and T lymphocytes (Herman et al. 1991a; Marrack and Kappler 1990; Herrmann and MacDonald 1991; Acha-Orbea et al. 1993). SAg are comprised of bacterial exoproducts including the staphylococcal enterotoxins, proteins encoded by viral genomes, and retroviral products such as those of open reading frame region encoded SAg of mouse mammary tumor viruses (Table 1; MacDonald et al. 1988; Marrack et al. 1991,1993; Janeway et al. 1989; Acha-Orbea et al. 1993). All SAg share the distinctive ability to bind to MHC class II molecules as intact molecules, thus bypassing rate-limiting steps during normal antigen processing, including antigen uptake, antigen degradation, and internal binding as peptide fragment to the antigen-presenting groove of class II MHC molecules of APC (Carlsson et al. 1988; Mollick et al. 1989; Fleischer and Schrezenmeier

Institut für Medizinische Mikrobiologie und Hygiene, Technische Universität München, Trogerstr. 9, 81675 München, Germany

Table 1. Superantigens

Source		
Bacterial	*S. aureus*	Staphylococcal enterotoxins A–E
	S. aureus	Toxic shock syndrome toxin 1
	S. pyogneus	Pyrogenic toxins A,C
	M. arthritidis	MAS
Retroviral	MMTV	Mls
Viral	Rabies	nucleocapsid

1988; Torres et al. 1993; Thibodeau et al. 1994; Mottershead et al. 1995). SAg bind directly to regions of the class II molecule that are outside the physiological MHC haplotype-restricted antigen-binding groove (Fig. 1; Dellabona et al. 1990; Herman et al. 1991b; Jardetzky et al. 1994). The SAg binding region seems to be conserved within various MHC haplotypes of mammalian species since SAg bind to murine and rat as well as to human class II molecules (Herrmann and MacDonald 1991; Fraser 1989). Nevertheless a hierarchy of SAg/class II interactions is apparent in that human HLA-DR and mouse H2-IE molecules are superior binders of SAg. On the other side, individual SAg differ in their relative affinity for class II molecules, for example, *Staphylococcus aureus* enterotoxin A (SEA) binds more closely to murine class II that does staphylococcal enterotoxin B (SEB; Cantor et al. 1993). The binding affinity of SAg/class II interactions is in the range of 10^{-5} *M*, that is much lower than the binding constant of antibodies (Seth et al. 1994). As an exception to the rule, the bacterial SAg staphylococcal

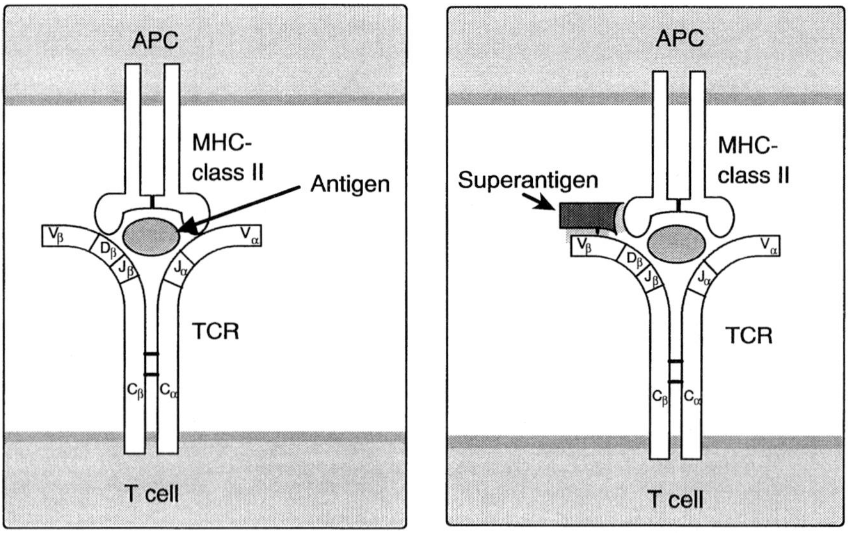

Fig. 1. Recognition of antigen and superantigen

enterotoxin C (SEC) known to bind to class II MHC also appears to bind to VCAM-1 molecules on the cell surface (CANTOR et al. 1993; AVERY et al. 1994). The extent to which other molecules can serve as SAg binding structures is unknown to date. SAg bound to class II molecules and to a lesser extent also free SAg bind to certain regions of the β-chain of the T cell receptor (TCR) that are encoded by Vβ-gene segments (CHOI et al. 1990; IRWIN et al. 1992, 1993; HAYBALL et al. 1994). Consequently a trimolecular complex of TCR, SAg, and MHC is formed which causes cross-linking and immobilization of the TCR and activation of the respective T cell (FLEISCHER and SCHREZENMEIER 1988; SETH et al. 1994; WHITE et al. 1989; KAPPLER et al. 1989). The binding affinities of the SAg–TCR interaction are in the same order of magnitude as those of the SAg–class II interaction (SETH et al. 1994). Thus SAg/TCR binding resembles at least in affinity the binding of isolated TCR to peptide–MHC complexes (MATSUI et al. 1991; WEBER et al. 1992). The binding region of the TCR β-chain for SAg is genetically encoded by Vβ segments of the TCR gene and lies outside the antigen recognition area of the TCR (WHITE et al. 1989; KAPPLER et al. 1989, 1994; HAMAD et al. 1994; HAYBALL et al. 1994). Thus binding of SAg to Vβ TCR is independent of the actual clonal antigen specificity of the TCR in question. Individual SAg interact selectively with certain Vβ encoded regions of the TCR β-chain. Because 25–50 individual Vβ segments make up the human or mouse TCR Vβ gene, an individual SAg reacts with at least 2%–5% of peripheral T cells (HERRMANN and MACDONALD 1991). Thus SAg induce a strong multiclonal yet Vβ-selective T cell activation that includes 1000 times more T cells than a T cell response toward a nominal antigen.

2 The In Vivo Response to SAg: A Complex Pattern of Reactivities

Confrontation of a mature immune system with the bacterial SAg SEB in vivo activates the reactive Vβ8-expressing $CD4^+$ and $CD8^+$ T lymphocytes leading to a complex pattern of concomitant and sequential reactivities that include hyperactivation (MIETHKE et al. 1992b; GAUS et al. 1994), deletion (KAWABE and OCHI 1991; WAHL et al. 1993; HUANG and CRISPE 1993), clonal expansion (MACDONALD et al. 1991; HEEG et al. 1993, 1995; GONZALO et al. 1992), and ligand-specific anergy (KAWABE and OCHI 1990; O'HEHIR and LAMB 1990; RELLAHAN et al 1990; MACDONALD et al. 1991; HEEG et al. 1993, 1995; BASCHIERI et al. 1993; Fig. 2). The net result of SEB-induced T cell activation includes opposite outcomes such as T cell dependent toxic lymphokine syndrome (T cell shock) or ligand-specific, long-lasting T cell anergy. This review follows up the fate of SEB-reactive T cells in vivo after SEB challenge and describes the sequential effects of a SEB-specific yet multiclonal T cell activation in vivo.

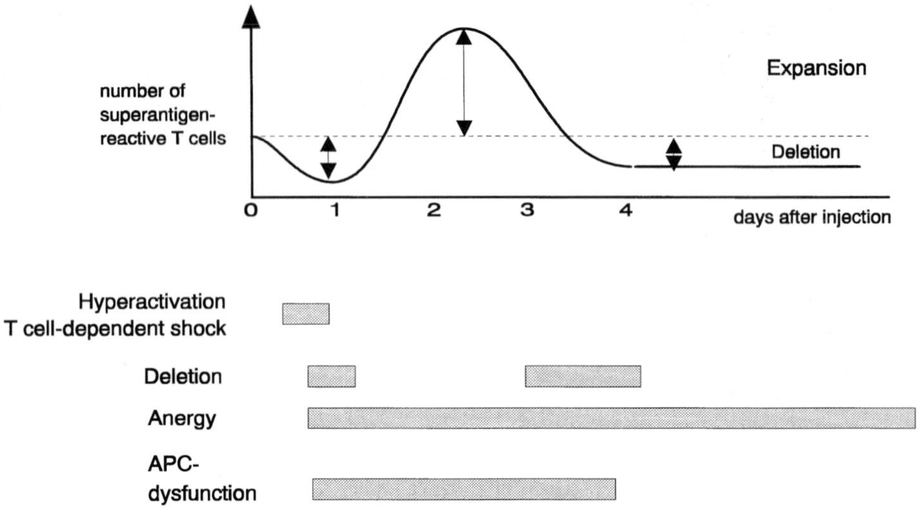

Fig. 2. In vivo response to superantigen

3 The First Hours: Hyperactivation of SAg-Reactive T Cells

SEB injected locally enters within minutes the draining lymph nodes through afferent lymphatic vessels, resulting in high local concentrations that allow binding to class II positive cells [dendritic cells (DC), macrophages (Mϕ), B cells]. Due to the concentration gradient within the lymph node and the low SEB–class II binding affinity SEB transits to the efferent lymphatic vessel, leaving SEB an effective half-time of about 30 min within a lymph node (M. VABULAS and T. MIETHKE, unpublished observations). While SEB disperses thereafter within the body fluids, it accumulates within the kidney. Overall it appears that the effective contact time of SEB with class II positive APC and T cells within a local lymph node is limited.

The short interaction time, however, is sufficient to trigger SAg-induced T cell responses. As a first sign, all SEB-reactive $V\beta8^+$ T lymphocytes in the draining lymph node shed L selectin (Mel-14) molecules from their surface within 10–15 min (MIETHKE et al. 1993c). Loss of Mel-14 is SAg-specific and Vβ-restricted, and lasts for at least 24h (MIETHKE et al. 1993c). Within 30 min Mel-14^- Vβ8 T cells initiate transcription of T cell activation related mRNAs (MIETHKE et al. 1993c; GAUS et al. 1994) that include T cell-specific lymphokines [interleukin (IL)–1α, –2, –4, –6, and –10, interferon (IFN) –γ, lymphotoxin –α, and tumor necrosis factor (TNF) –α]. While $CD4^+$ $V\beta8^+$ T cells transcribe both Th_1- and Th_2-type lymphokines, in $CD8^+$ $V\beta8^+$ T cells mRNA is preferentially found encoding IL-2, IFN-γ and to a lesser extent TNF-α (GAUS et al. 1994). Transcription is followed by translation and systemic release of lymphokines. Peak values for TNF-α are reached within 1–2 h; IL-1, -2, -4, and -6, and IFN-γ peak at 2–4 h; and IL-10 reaches ist maximum after 12–16 h (MIETHKE et al. 1993d; GAUS et al. 1994). The lymphokine content of serum is shut off promptly after reaching peak values, for example, TNF-α and IL-2 serum levels fall to preinjection values within 8 h (MIETHKE etl al. 1992b; GAUS et al. 1994). In this respect it is interesting that blockade of glucocorticoid receptors by RU 486 delays acute downregulation of cytokine production (GONZALO et al. 1993b; LUSSOW et al. 1993). Since glucocorticoids are produced after SAg injection with similar kinetics as cytokines (GONZALO et al. 1993b) (own unpublished data) glucocorticoids may act by suppressing lymphokine-specific mRNA synthesis via interaction with corticoid responsive elements in the promoter region of certain lymphokines (DAYNES and ARANEO 1989). The cytokine profile induced by different SAg is qualitatively not at variance; however, quantitative differences exist. Toxic shock syndrome toxin type 1 (TSST-1), for example, induces per cell higher TNF-α release as SEB (MIETHKE et al. 1993a).

SAg-induced production of lymphokines is strictly dependent on binding to class II molecules and on interaction with T-cells. Cyclosporin A (CsA), which blocks transcription of lymphokine mRNA, completely prevents lymphokine production induced by SEB in vivo (VANIER and PRUD'HOMME 1992; GONZALO et al. 1992; HEEG et al. 1993). In addition, severe combined immunodeficiency (SCID) mice lacking T cells are insensitive toward SEB and fail to respond in vivo (MIETHKE et al. 1992b). Finally, MHC class II deficient mice fail to respond to SAg that exclusively bind to class II molecules (SEB, SEA, TSST-1; AVERY et al. 1994). In these mice only SAg (SEC1) which bind to non-MHC molecule (VCAM-1) are effective. Taken together, the first phase of a SAg-induced response of T cells in vivo is characterized by hyperacute T cell activation of all SAg-reactive T cells, leading to immediate lymphokine mRNA transcription followed by release of T cell dependent cytokines.

4 Acute Cytokine Release Causes Lethal T Cell Shock

SAg-triggered cytokine release leads to wasting of mice, associated with a transient weight loss probably due to systemic action of SAg-induced TNF and lymphotoxin (MARRACK et al. 1990). When high doses (>500 μg per mouse) of SEB are injected, aged mice die within 48 h after injection (AROEIRA et al. 1994). As such the mechanisms of this SAg-induced lethal cytokine syndrome resemble those of the classical endotoxic shock syndrome (Fig. 3; GLAUSER et al. 1991; RIETSCHEL and BRADE 1992). In the latter system bacterial-derived lipopolysaccharides (LPS) activate Mϕ which in turn release high concentrations of TNF-α, IL-1, and IL-6 (GLAUSER et al. 1991; RIETSCHEL and BRADE 1992; FREUDENBERG and GALANOS 1990). These cytokines are the key players in mediating the LPS-induced lethal shock syndrome (GLAUSER et al. 1991; RIETSCHEL and

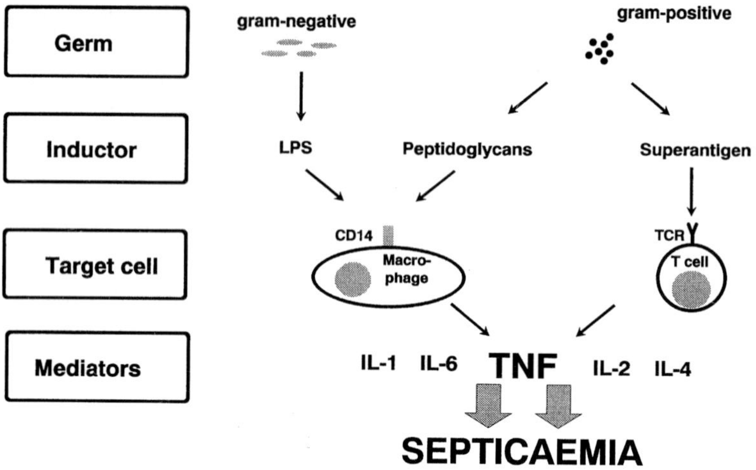

Fig. 3. Pathogenesis of lethal shock induced by endotoxin or superantigen

BRADE 1992). In contrast to primates, mice require relatively high concentrations of LPS for the induction of lethal shock, i.e., they appear endotoxin resistant (LEHMANN et al. 1987). Fortunately, mice can be sensitized for the lethal action of LPS-induced cytokines by simultaneous injection of D-galactosamine (D-GalN) together with LPS (GALANOS et al. 1979; LEHMANN et al. 1987). As a result, LD_{50} doses of LPS are reduced 100-fold, and mice thus acquire an endotoxin-sensible phenotype (GALANOS et al. 1979; LEHMANN et al. 1987).

To determine in vivo the full pathogenic potential of SAg it was deemed mandatory to analyze the consequences of SAg administration in D-GalN sensitized mice. We found that, as for LPS, the LD_{50} values for SAg in D-GalN sensitized mice are 100-fold reduced (1 μg SEB per mouse; MIETHKE et al. 1992b). This in turn allowed detailed analysis of the pathomechanisms of SAg-induced lethality. It was shown that T cells and T cell derived TNF-α are key mediatorys in the SAg-induced lethal shock syndrome since anti-TNF monoclonal antibodies efficiently prevent lethality (MIETHKE et al. 1992b). In addition, the T cell specific immunosuppressive agent CsA is protective (MIETHKE et al. 1992b). Finally, lethal shock cannot be brought about in T cell deficient SCID or nude (nu/nu) mice (MIETHKE et al. 1992b). In addition to D-GalN, at least two further "sensitizers" can synergize with SAg in inducing lethal shock, presumably via different mechanisms. First, RU 486, which blocks the action of glucocorticoids by competitive blockade of the glucocorticoid receptor, enhances SAg-induced lethality by preventing glucocorticoid-mediated downregulation of lymphokine production probably in SAg-activated T cells (GONZALO et al. 1993b). Accordingly, the amount of T cell-derived cytokines increases in RU 486 pretreated mice. Second, N^G-nitro-L-arginine methylester, which blocks action of the inducible nitric oxide synthetase (iNOS), also sensitizes mice for SAg-induced lethal shock and seems to act distally from T cells by blocking nitric oxide dependent protective regulatory mechanisms in the periphery.

Obviously not only endotoxins but also exotoxins (SAg) induce in mice a toxic shock syndrome. Therefore the classical concept of pathogenesis of septic (bacterial) induced lethal shock can now be complemented by SAg-induced T cell activation (Fig. 3). In this unified concept endotoxin (LPS) and exotoxin (SAg) are inducers of toxic responses of host immune cells. While LPS acts on Mϕ that are activated to produce cytokines, SAg interacts with T cells which in turn are triggered to activation and cytokine production. In both situations Mϕ-derived or T cell-derived TNF-α acts as a key mediator for the induction of the lethal septic cascade (Fig. 3). According to this concept, the effector phase but not the induction phase of endotoxin- and exotoxin-mediated shock are similar to each other. On the other hand, recent, reports indicate that even the induction phases are interrelated. For example, LPS and SAg appear to synergize in the induction of lethal shock (Stiles et al. 1993) (unpublished observations). If so, synergistic interactions of endotoxin and exotoxin may condition lethal septicemia.

5 Induction of Unresponsiveness in SAg-Reactive T Cells: Anergy, Downregulation, and Deletion

While the first hours after SAg administration in vivo are characterized by hyperactivation of the responding Vβ8$^+$ T cells, after 4–6 h suppressive effects on T cells become dominant. Ligand-specific unresponsiveness (tolerance) is induced by multiple and consecutive mechanisms. First Vβ8$^+$ T cells from SEB-injected mice transit from a reactive to an anergic T cell phenotype shortly after the hyperreactive period. Vβ8$^+$ CD4$^+$ T cells removed within the first 2 h after SEB injection can be restimulated in vitro with SEB and respond with IL-2 production and subsequently enter cell cycle in vitro (Fig. 4). However, CD4$^+$ Vβ8$^+$ T cells recovered later (>6 h) fail to produce IL-2 upon restimulation with SEB in vitro and thus do not respond to SEB, i.e., these cells display an anergic phenotype (Fig. 4). Anergy is Vβ selective and confined to impaired lymphokine production in vitro. However, virtually all anergic Vβ8$^+$ T lymphocytes express IL-2 receptors (IL-2R) in vivo (Miethke et al. 1993b). In vitro, IL-2R of these anergic T cells are functional since exogenously added IL-2 induces proliferation of these cells (Miethke et al. 1993b; Heeg et al. 1993, 1995). Hence induction of anergy is SAg-reactive T cells is an immediate process that follows an initial state of hyperreactivity. It appears to be triggered directly by the TCR and clearly is not linked to cell cycle progression. T cell unresponsiveness is due not only to anergy. In vivo, anergic Vβ8$^+$ IL-2R$^+$ T cells proceed within hours through distinct levels of T cell unresponsiveness (tolerance) that may ultimately lead to T cell deletion via apoptosis. This stepwise process includes Vβ-selective TCR downregulation followed by loss of coreceptors (CD4/CD8, CD2). Apoptosis

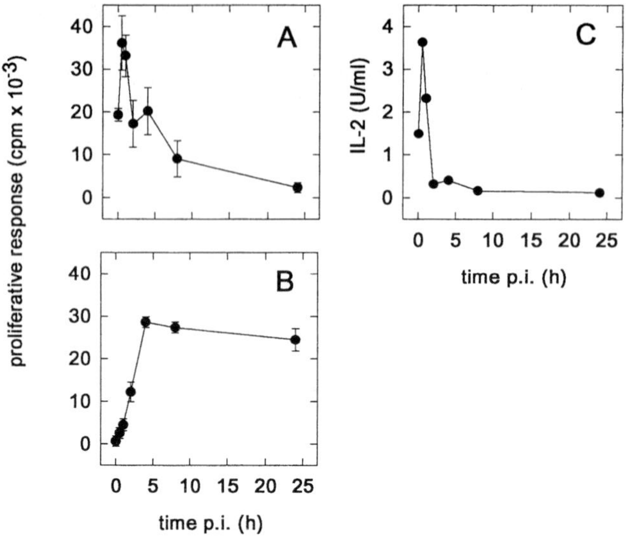

Fig. 4. SEB induces anergy. Lymph node cells from SEB-injected mice were restimulated in vitro after the indicated time periods with SEB (**A, C**) or IL-2 (**B**). Proliferative responses (**A, B**) or IL-2 production (**C**) were recorded

occurs only in a subset of anergic Vβ8^{+} T cells. TCR downregulation may be transient, since TCR^{-} T cells regain TCR expression, at least when transferred to cultures (Miethke et al. 1994). Since not all anergic Vβ T cells proceed to the level of tolerance defined by deletion (apoptosis), multiple levels of tolerance are observed at the same time (Miethke et al. 1994). It is not known how progression to later stages of tolerance induction is controlled. Recent experiments suggest that the dose of antigen that interacts with the responder T cells determines the level of tolerance induced. Accordingly, low concentrations of SEB induce only anergy while higher concentrations of SEB also induce TCR downregulation or even deletion (apoptosis). Experiments in TCR-transgenic mice support this hypothesis. When T cells from TCR-transgenic mice were confronted in vivo with their antigen whose expression was under control of an inducible promoter, a similar antigen dose/response relationship was observed (Schönrich et al. 1992; Hämmerling et al. 1991; Arnold et al. 1993).

Induction of apoptosis of SAg-reactive T cells is not inhibited by CsA (Vanier and Prud'homme 1992; Heeg et al. 1993), suggesting that neither T cell-derived cytokines nor Ca^{2+}-sensitive signal steps within T cells are involved in this process. In contrast, linomide can suppress apoptosis, but the mechanisms of its action are not defined (Gonzalo et al. 1993a). In fas-defective lpr/lpr mice SAg-induced apoptosis can be observed (unpublished observations), suggesting that besides fas/fas ligand-induced apoptosis other signal pathway can be utilised to induce apoptosis in SAg-reactive T cells.

According to the two-signal concept of T cell activation, the second signal provided by APC critically controls whether T cells are activated or anergized. (BRETSCHER and COHN, 1970; LAFFERTY and WOOLNOUGH 1977). Since SEB binds to class II positive cells irrespective of whether these cells are costimulus-competent APC (DC, activated B cells) or costimulus-incompetent APC (resting B cells), it has been suggested that SEB bound to incompetent APC induces anergy and possibly even deletion of the SAg-reactive T cells. Experiments addressing this concept have been controversial. B cell knock-out mice, which lack B cells, were once shown to be indistinguishable from wildtype type strains in deleting SAg-reactive T cells after SEB administration in vivo (MIETHKE et al. 1995). On the other hand, concomitant injection of LPS (which polyclonally activates B cells and thus converts B cells to competent APC) prevents SAg-induced deletion (VELLA et al. 1995). However, in the latter system an effect of LPS-triggered macrophage activation with subsequent cytokine release needs to be evaluated. Interestingly, apoptosis of SEB-reactive T cells can be induced in vitro under certain conditions even by costimulus-competent DC (unpublished observations). In this system transient presentation of high concentration of SEB by DC is required to induce apoptosis. We believe that a TCR signal induced by high concentrations of antigen rather than the availability of costimulatory signals controls induction of apoptosis.

6 Exhaustive Proliferation of Anergic T Cells

When TCR downregulation or T cell deletion has occurred in a subset of SAg-reactive T cells, the residual anergic Vβ8$^+$ T cells enter cell cycle and expand clonally (Fig. 2) in vivo (MACDONALD et al. 1991; GONZALO et al. 1992; HEEG et al. 1993). The frequency of Vβ8$^+$ T cells, which initially fell to about half of control values 14–20 h after SAg challenge, then increases dramatically. Within 2 days after SEB injection T cell areas in spleen and lymph node enlarge. SEB-reactive CD8$^+$ and CD4$^+$ Vβ8$^+$ T cells then comprise 50% of all T lymphocytes in these secondary lymphatic organs. At the time of maximal expansion CD8$^+$ Vβ8$^+$ exhibit SEB-specific cytolytic activity when tested directly ex vivo (KAWABE and OCHI 1990; HERRMANN et al. 1992). A pathophysiological role of these cytotoxic T cells (CTL) in vivo is unlikely since the antigen, SEB, appears no longer to be available. Expansion of Vβ8$^+$ T cells ceases after 72–96 h, and a mayor proportion of the expanded T cells again undergo apoptosis (KAWABE and OCHI 1991; MACDONALD et al. 1991; GONZALO et al. 1992; VANIER and PRUD'HOMME 1992; HEEG et al. 1993). As a result the number of CD8$^+$ Vβ8$^+$ T cells falls to preinjection values, while the frequency of CD4$^+$ Vβ8$^+$ T cells is reduced to half of control values.

At present it is not known whether clonal expansion is driven by T cell derived lymphokines or by cell associated ligands. At the times of expansion SEB-reactive T cells are already anergic, and serum lymphokine levels are undetectable (MIETHKE et al. 1992b; HEEG et al. 1993). Moreover, IL-2R expression of the expanding cells has ceased (MIETHKE et al. 1992a, 1993b). On the other hand, CsA completely blocks Vβ8⁺ T cell expansion (HEEG et al. 1993). $CD8^+$ T cell growth but not that of $CD4^+$ T cells can be brought about in CsA-treated mice by supplementation with exogenous added IL-2 (HEEG et al. 1993). Since cytokine production by SEB-reactive T cells and their transient IL-2R expression overlap in time, we favor the hypothesis that $V\beta8^+$ T cells becomes triggered to expansion during the initial period of a SEB-response, i.e., the time point preceding anergy induction. According, expanding T cells would starve from cytokines (exhaustion), and cytokine deprivation would initiate the second wave of clonal deletion. Recently, however, another explanation for clonal deletion has emerged. It was shown that $CD8^+$ CTL can be stimulated by $CD4^+$ T cells in a Vβ-restricted manner in vivo and in vitro, i.e., CTL appear to recognize Vβ-derived peptides on class Ib molecules of $CD4^+$ T cells (JIANG et al. 1995). These experiments suggest that regulatory $CD8^+$ CTL may be involved in the second deletion phase of SEB-reactive T cells in vivo.

7 Effects of SAg on Antigen-Presenting Cells

Although it has been shown that SAg can activate Mϕ via binding to class II molecules (GROSSMAN et al. 1990; CHATILA and GEHA 1993), cytokines produced by T cells during the initial hyperreactive phase seem considerably to govern APC function. It is known that bactericidal Mϕ function is correlated with the expression of iNOS (NATHAN and XIE 1994). TNF-α is a powerful inductor of iNOS (CUNHA et al. 1994). We therefore analyzed iNOS expression after SEB administration. iNOS expression is observed during the first day after SEB injection, peaking at 16–24 h postinjection (unpublished observations). Moreover, resistance to lethal intraperitoneal infection with *Listeria monocytoges* is enhanced after SEB injection (unpublished observations). These data suggest that SEB-induced iNOS is functional and protective in vivo. In addition to TNF-α, other SEB-induced cytokines may alter Mϕ and APC function. For example, we observed high serum levels of IL-10 peaking 16–20 h upon challenge of mice with SEB. Since IL-10 exerts suppressive effects on APC (DING et al. 1993; D'ANDREA et al. 1993; ENK et al. 1993), we tested the ability of APC from SEB-injected mice to induce in vitro an allograft response. We found that APC function is transiently reduced 24–48 h after SEB injection (unpublished observations). Reduced APC function was not correlated with the expression of costimulatory molecules (B7) on APC. On the other hand, APC function was retained when mice were injected with SEB and CsA. We therefore conclude that T cell derived CsA-sensible cytokines (probably IL-10) are key mediators in suppressing APC function transiently. As a consequence during this phase the in vivo response toward third-party SAg (e.g., TSST-1) is completely suppressed (MIETHKE et al. 1993b). In addition, IL-10 protects mice from SAg-induced lethal shock (BEAN et al. 1993). At first glance this unresponsiveness resembles LPS-induced tolerance. However, the third-party SAg-specific T cells are fully reactive in vitro when stimulated with normal APC. Thus in the course of a SAg-response a transient desensitization to SAgs is induced due to a loss of APC function of SAg-presenting cells. Desensitization is therefore not Vβ-restricted and embodies SAg independently of their intrinsic T cell specificity.

8 SAg-Induced Anergy Defined In Vitro Is Not Correlated with In Vivo Reactivity

When the second wave of apoptosis has passed, the number of $CD8^+$ $V\beta8^+$ T cells normalizes while the frequency of $CD4^+$ $V\beta8^+$ T cells remains reduced to about half of control values. $V\beta8^+$ T lymphocytes are now unresponsive to rechallenge in vitro with SEB and retain this anergic phenotype for weeks (KAWABE and OCHI 1990, 1991; O'HEHIR and LAMB 1990; RELLAHAN et al. 1990; MACDONALD et al. 1991; VANIER and PRUD'HOMME 1992; MIGITA and OCHI 1993). Detailed analyses have shown that upon restimulation anergic $CD4^+$ and $CD8^+$ T cells reexpress IL-2R in vivo and in vitro but fail to produce lymphokines such as IL-2 (HEEG et al. 1993, 1995; GAUS et al. 1994). Limiting-dilution analyses have shown that virtually all $V\beta8^+$ T cells can be triggered to IL-2R-expression and proliferation in the presence of exogenous IL-2 (HEEG et al. 1995). Thus SEB-induced anergy is characterized by a selective loss of lymphokine production while other TCR-mediated signal pathways such as IL-2R expression are fully retained. On the other hand, distinct fucntional activities of anergic $CD4^+$ T cells (e.g., some forms of B cell help) seem to be maintained in vivo (OTTEN and GERMAIN 1991; LUSSOW and MACDONALD 1994; BANDEIRA et al. 1993). Further, SEB-induced anergy appears to be reversible since infection with the nematode *Nippostrongylus brasiliensis* changes the phenotype of anergic $CD4^+$ T cells which regain the ability to produce IL-4 (RÖCKEN et al. 1992).

When we evaluated SEB-induced unresponsiveness of $V\beta8^+$ T cells in vivo, the results were at variance to those obtained in vitro. While in vitro T cells from SEB-injected mice failed to produce IL-2 upon restimulation, the very same cells produced large amounts of IL-2 in vivo (HEEG et al. 1995; GAUS et al. 1994). Compared to primary stimulation with SEB in vivo, qualitatively and quantitatively the same combination of cytokines was produced with the exception that IFN-γ peak levels were five fold increased, and IL-10 peak levels were reached after only 4h (GAUS et al. 1994). Polymerase chain reaction analyses indicated that the $V\beta8^+$ T cell subset produces these cytokines upon in vivo rechallenge (HEEG et al. 1995). As a corollary, when SEB-tolerized mice were sensitized with D-GalN, the lethal shock syndrome was provoked by SEB rechallenge (HEEG et al. 1995; GAUS

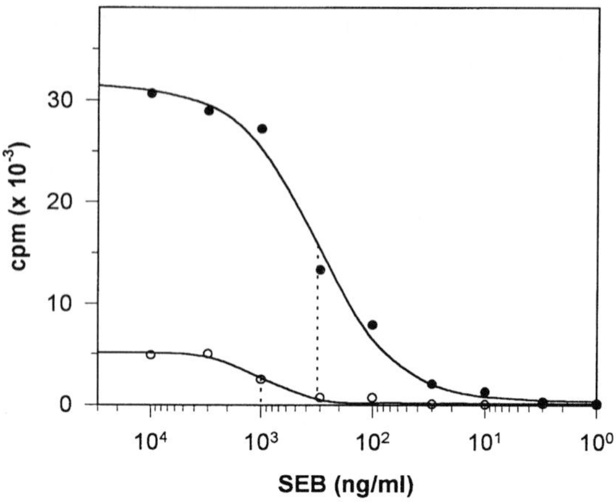

Fig. 5. Quantitative analysis of T cell triggering by SEB. $CD4^+$ T cells from normal (*closed circles*) or anergized (*open circles*) were stimulated with SEB concentrations as indicated on splenic feeder cells. Proliferative responses were recorded 5 days later. *Dashed line*, SEB concentration for half-maximal growth

et al. 1994). How can the paradox of in vitro anergy as opposed to in vivo reactivity be explained?

To answer this question we resorted to in vitro analyses and took advantage of the fact that SEB binds to class II positive cells without being processed. This in turn allows us to titrate in vitro the amount of SEB required to trigger a give T cell response, i.e., to determine the TCR threshold (BHARDWAJ et al. 1993). Further, this approach allows determination of the relative avidity of a given T cell population towards SEB (BHARDWAJ et al. 1993; HEEG and WAGNER 1995). Figure 5 details a representative analysis of this type. SEB (or SEA) titrations result in sigmoid-shaped dose-response curves that can be characterized by the amount of ligand (antigen) that induces half-maximal responses (SD_{50}). In the course of these studies it became evident that SD_{50} values depend critically on intrinsic properties of APC. For example, DC (as APC) require up to 20-fold less SEB for induction of proliferative responses of normal T cells than do splenocytes (Table 2). Resting B cells fail to induce a proliferative response at all, while activated B cells are almost as effective as DC (HEEG and WAGNER 1995). MHC

Table 2. SD_{50} values (ng/ml) for stimulation with SEB on various APC

		Spleen		Resting B cells		Activated B cells		Dendritic cells	
Responder cells	Anti-CD28	Proliferation	SD_{50}	Proliferation	SD_{50}	Proliferation	SD_{50}	Proliferation	SD_{50}
Normal	–	++	≈250	(+)	≈700	+++	≈30	+++	≈10
	+	+++	≈200	++	≈500	n.d.		n.d.	
Anergic	–	(+)	>1000	ϕ	>1000	++	≈250	+++	≈100
	+	++	≈1000	(+)	>1000	n.d.		n.d.	

class II expression and expression of a costimulatory molecules (B7) on the APC correlated with their efficacy to stimulate a response (HEEG and WAGNER 1995). In addition, cross-linking of CD28 molecules on the responding T cells by mAb substitutes, at least in part, for lacking B7 molecules on splenic APC or resting B cells without influencing SD_{50} values (Table 2, HEEG and WAGNER 1995).

In applying this technique we observed that anergic T cells (from SEB-pre-injected mice) could be reactivated provided they were challenged with higher concentrations of SEB. SD_{50} values were at least ten fold higher than those of normal T cells (Table 2). Increased SD_{50} values were evident with all types of APC tested (Table 2) and thus represent an intrinsic characteristic of anergic cells. Note that with splenic APC and resting B cells the high ligand density necessary for activation cannot be reached. Consequently $V\beta8^+$ T cells appear anergic. The critical ligand density, however, can be reached with DC or activated B cells. Thus stimulation with SEB with these APC allow reactivation of anergic cells and therefore discloses a reactive phenotype (HEEG and WAGNER 1995). Taken together, these experiments indicate that manifestation of an anergic phenotype of $V\beta8^+$ T cells depends on intrinsic and extrinsic factors. Elevated intrinsic TCR threshold (SD_{50}) values of SEB-anergized T cells can be overcome by stimulation with professional APC (extrinsic). As a result anergy can be turned into responsiveness by the use of professional APC and high ligand densities.

The observed elevation of SD_{50} values in anergic T cells can be explained by at least two mechanisms, which appear to be mutually exclusive: either the activation threshold for anergic versus normal T cells differs, or the relative avidity of the anergic T cell population toward SEB is lower than that in normal T cells. Since SEB causes apoptosis of a large proportion of SEB-reactive T cells, one could argue that the surviving T cells exhibit a lower TCR affinity, i.e, the T cells expressing high-affinity TCR for SAg would be preferentially deleted. Consequently only low-affinity TCR would escape deletion. In attempts to discriminate between these alternatives we tested the ligand dose required to induce a sustained TCR-initiated function of anergic cells, i.e., the induction of functional IL-2R (Table 3). It was shown that the SD_{50} values for induction of IL-2R expression on anergic T cells a identical to those of normal T cells (Table 3). Since the level of TCR expression is not reduced in anergic T cells, these results suggest that the relative TCR avidity of the anergic T cell population is not appreciably reduced compared to normal T cells. We therefore conclude that not TCR avidity but TCR threshold for triggering IL-2 gene activation is altered in

Table 3. TCR threshold (ng/ml) is elevated in anergic T cells

	SD_{50} for expression of functional IL-2R (APC: activated B cells + exogenous IL-2)	SD_{50} for induction of IL-2 production (APC: activated B cells)
Normal T cells	≈ 10	≈ 30
Anergic T cells	≈ 10	≈ 900

anergic cells. In fact, anergic T cells require 20- to 30-fold more ligand to trigger IL-2 production (Table 3). Hence, the refractoriness of anergic T cells is relative and can be characterised by a selectively elevated TCR threshold for induction of IL-2 production. Elevated TCR threshold can be overcome by raising the ligand concentration or by efficient stimulation with competent APC. Induction of responsiveness in SEB-anergic mice is clearly due to SEB stimulation with professional APC, thus converting anergy into responsiveness.

9 Superantigens: Pathogenic Factor and Tool to Analyze T Cell Function

Collectively, the diverse immunological responses towards SAg demonstrate their analytical and pathophysiological power and justify the tremendous interest which they have spawned in recent years. First, SAg have allowed delineation of the pathophysiology of T cell dependent shock syndrome that is triggered by hyperacute Vβ-selective T cell activation and release of T cell derived cytokines. Accordingly, a unified concept for the pathogenesis of septicemia can be formulated in which T cell derived (SAg) or Mϕ-derived (endotoxin) TNF-α act as key mediators for the induction of subsequent lethal shock. Future analyses of the interconnected action of endotoxin and SAg will allow delineation of synergy mechanisms between SAg and LPS which may be of relevance to understand the pathophysiology of septicemia in man. Second, SAg represent a tool for analyzing the mechanisms that lead to activation versus tolerization of the responding T cells. The Vβ-selective activation of T cells made it possible to follow up the fate of SAg-reactive T-cells in vivo and in vitro. It is puzzling that after an initial phase of hyperactivation distinct levels of tolerance become induced in vivo. The use of SAg will help to define the role of APC and costimulatory signals in inducing activation versus tolerization. Because of their unique presentation pathway SAg will be informative in analyzing the dose-dependent tolerization/activation signals in T cells in vivo and in vitro. Futhermore, since SAg activate a large fraction of peripheral T cells, it will now be feasible to analyze at the molecular level the signaling pathways leading either to activation or inactivation. Finally, there is an ongoing debate as to whether SAg are involved in the pathogenesis of certain human diseases in addition (Kotzin et al. 1994), including Kawasaki syndrome (Pietra et al. 1994), rheumatoid arthritis (Paliard et al. 1991), and even AIDS (Pantaleo et al. 1994; Chen et al. 1994; Lewis et al. 1994; Dadaglio et al. 1994).

References

Acha-Orbea H, Held W, Waanders GA, Shakhov AN, Scarpellino L, Lees RK, MacDonald HR (1993) Exogenous and endogenous mouse mammary tumor virus superantigens. Immunol Rev 131: 5–25

Arnold B, Schönrich G, Hämmerling GJ (1993) Multiple levels of peripheral tolerance. Immunol Today 14: 12–14

Aroeira LS, Williams O, Lozano EG, Martinez AC (1994) Age-dependent changes in the response to staphylococcal enterotoxin B. Int Immunol 6: 1555–1560

Avery, AC, Markowitz, JS, Grusby MJ, Glimcher LH, Cantor H (1994) Activation of T cells by superantigen in class II-negative mice. J Immunol 153: 4853–4861

Bandeira A, Mengel J, Burlen-Defranoux O, Coutinho A (1993) Proliferative T cell anergy to Mls-1a does not correlate with in vivo tolerance. Int Immunol 3: 923–931

Baschieri S, Lees RK, Lussow AR, MacDonald HR (1993) Clonal anergy to staphylococcal enterotoxin B in vivo: selective effects on T cell subsets and lymphokines. Eur J Immunol 23: 2661–2666

Bean AG, Freiberg RA, Andrade S, Menon S, Zlotnik A (1993) Interleukin 10 protects mice against staphylococcal enterotoxin B-induced lethal shock. Infect Immun 61: 4937–4939

Bhardwaj N, Young JW, Nisanian AJ, Baggers J, Steinman RM (1993) Small amounts of superantigen, when presented on dendritic cells, are sufficient to initiate T cell responses. J Exp Med 178: 633–642

Bretscher P, Cohn M (1970) A theory of self-nonself discrimination. Science 169: 1042–1049

Cantor H, Crump AL, Raman VK, Liu H, Markowitz JS, Grusby MJ, Glimcher LH (1993) Immunoregulatory effects of superantigens: interactions of staphylococcal enterotoxins with host MHC and non-MHC products. Immunol Rev 131: 27–42

Carlsson R, Fischer H, Sjögren HO (1988) Binding of staphylococcal enterotoxin A to accessory cells is a requirement for its ability to activate human T cells. J Immunol 140: 2484–2488

Chatila T, Geha RS (1993) Signal transduction by microbial superantigens via MHC class II molecules. Immunol Rev 131: 43–59

Chen ZW, Kou ZC, Shen L, Regan JD, Lord CI, Halloran M, Lee-Parritz D, Fultz PN, Letvin NL (1994) An acutely lethal simian immunodeficiency virus stimulates expansion of Vβ7- and Vβ14-expressing T lymphocytes. Proc Natl. Acad Sci USA 91: 7501–7505

Choi YW, Herman A, DiGiusto D, Wade T, Marrack P, Kappler JW (1990) Residues of the variable region of the T-cell-receptor beta-chain that interact with S. aureus toxin superantigens. Nature 346: 471–473

Cunha FQ, Assreuy J, Moss DW, Rees D, Leal LMC, Moncada S, Carrier M, O'Donnel CA, Liew FY (1994) Differential induction of nitric oxide synthase in various organs of the mouse during endotoxaemia: role of TNF-α and IL-1-β. Immunology 81: 211–215

D'Andrea A, Aste Amezaga M, Valiante NM, Ma X, Kubin M, Trinchieri G (1993) Interleukin 10 (IL-10) inhibits human lymphocyte interferon gamma-production by suppressing natural killer cell stimulatory factor/IL-12 synthesis in accessory cells. J Exp Med 178: 1041–1048

Dadaglio G, Garcia S, Montagnier L, Gougeon ML (1994) Selective anergy of V beta 8(+) T cells in human immunodeficiency virus-infected individuals. J Exp Med 179: 413–424

Daynes RA, Araneo BA, (1989) Contrasting effects of glucocorticoids on the capacity of T cells to produce the growth factors interleukin 2 and interleukin 4. Eur J Immunol 19: 2319–2325

Dellabona P, Peccoud J, Kappler JW, Marrack P, Benoist C, Mathis D (1990) Superantigens interact with MHC class II molecules outside of the antigen groove. Cell 62: 1115–1121

Ding L, Linsley PS, Huang LY, Germain RN, Shevach EM (1993) IL-10 inhibits macrophage costimulatory activity by selectively inhibiting the up-regulation of B7 expression. J Immunol 151: 1224–1234

Enk AH, Angeloni VL, Udey MC, Katz SI (1993) Inhibition of Langerhans cell antigen-presenting function by IL-10. J Immunol 151: 2390–2398

Fleischer B, Schrezenmeier H (1988) T cell stimulation by staphylococcal enterotoxins. Clonally variable response and requirement for major histocompatibility complex class II molecules on accessory or target cells. J Exp Med 167: 1697–1707

Fraser JD (1989) High-affinity binding of staphylococcal enterotoxins A and B to HLA-DR. Nature 339: 221–223

Freudenberg MA, Galanos C (1990) Bacterial lipopolysaccharides: structure, metabolism and mechanisms of action. Int Rev Immunol 6: 207–221

Galanos C, Freudenberg MA, Reutter W (1979) Galactosamine-induced sensitization of the lethal effects of endotoxin. Proc Natl Acad Sci USA 76: 5939–5943

Gaus H, Miethke T, Wagner H, Heeg K (1994) Superantigen-induced anergy of Vβ8$^+$ CD4$^+$ T cells induces functional but nonproliferative T cells in vivo. Immunology 83: 333–340

Glauser MP, Zanetti G, Baumgartner JD, Cohen J (1991) Septic shock: pathogenesis. Lancet 338: 732–736

Gonzalo JA, Moreno de Alboran I, Ales Martinez JE, Martinez C, Kroemer G (1992) Expansion and clonal deletion of peripheral T cells induced by bacterial superantigen is independent of the interleukin-2 pathway. Eur J Immunol 22: 1007–1011

Gonzalo JA, Gonzalez-Garcia A, Kalland T, Hedlund G, Martinez-AC, Kroemer G (1993a) Linomide, a novel immunomodulator that prevents death in four models of septic shock. Eur J Immunol 23: 2372–2374

Gonzalo JA, Gonzalez-Garcia A, Martinez C, Kroemer G (1993b) Glucocorticoid-mediated control of the activation and clonal deletion of peripheral T cells in vivo. J Exp Med 177: 1239–1246

Grossman D, Cook RG, Sparrow JT, Mollick JA, Rich RR (1990) Dissociation of the stimulatory activities of staphylococcal enterotoxins for T cells and monocytes. J Exp Med 172: 1831–1841

Hamad ARA, Herman A, Marrack P, Kappler JW (1994) Monoclonal antibodies defining functional sites on the toxin superantigen staphylococcal enterotoxin BJ Exp Med 180: 615–621

Hayball JD, Robinson JH, O'Hehir RE, Verhoef A, Lamb JR, Lake RA (1994) Identification of two binding sites in staphylococcal enterotoxin B that confer specificity for TCR Vβ gene products. Int Immunol 6: 199–211

Hämmerling GJ, Schönrich G, Momburg F, Auphan N, Malissen M, Malissen B, Schmitt-Verhulst AM, Arnold B (1991) Non-deletional mechanisms of peripheral and central tolerance: studies with transgenic mice with tissue-specific expression of a foreign MHC class I antigen. Immunol Rev 122: 47–67

Heeg K, Wagner H (1995) Induction of responsiveness in superantigen-induced anergic T cells: role of ligand density and costimulatory signals. J Immunol 155: 83–92

Heeg K, Bendings S, Miethke T, Wagner H (1993) Induction of unresponsiveness to the superantigen staphylococcal enterotoxin B: cyclosporin A resisitant split unresponsiveness unfolds in vivo without preceding clonal expansion. Int Immunol 5: 929–937

Heeg K, Gaus H, Griese D, Bendigs S, Miethke T, Wagner H (1995) Superantigen-reactive T cells that display an anergic phenotype in vitro appear functional in vivo. Int Immunol 7: 105–114

Herman A, Kappler JW, Marrack P, and Pullen AM (1991a) Superantigens: mechanism of T-cell stimulation and role in immune responses. Annu Rev Immunol 9: 745–772

Herman A, Labrecque N, Thibodeau J, Marrack P, Kappler JW, Sekaly RP (1991b) Identification of the staphylococcal enterotoxin A superantigen binding site in the beta 1 domain of the human histocompatibility antigen HLA-DR. Proc Natl Acad Sci USA 88: 9954–9958

Herrmann T and MacDonald HR (1991) T cell recognition of superantigens. Curr Top Microbiol Immunol 174: 21–38

Herrmann T, Baschieri S, Lees RK, MacDonald HR (1992) In vivo responses of CD4$^+$ and CD8$^+$ cells to bacterial superantigens. Eur J Immunol 22: 1935–1938

Huang L, Crispe IN (1993) Superantigen-driven peripheral deletion of T cells. Apoptosis occurs in cells that have lost the alpha/beta T cell receptor. J Immunol 151: 1844–1851

Irwin MJ, Hudson KR, Fraser JD, Gascoigne NRJ (1992) Enterotoxin residues determining T-cell receptor V beta binding specificity. Nature 359: 841–843

Irwin MJ, Hudson KR, Ames KT, Fraser JD, Gascoigne NRJ (1993) T-cell receptor β-chain binding to enterotoxin superantigens. Immunol Rev 131: 61–78

Janeway CA, Yagi J, Conrad PJ, Katz ME, Jones B, Vroegop S, Buxser S, (1989) T-cell responses to Mls and to bacterial proteins that mimic its behavior. Immunol Rev 107: 61–88

Jardetzky TS, Brown JH, Gorga JC, Stern LJ, Urban RG, Chi Y, Stauffacher C, Strominger JL, Wiley DC (1994) Three-dimensional structure of a human class II histocompatibility molecule complexed with superantigen. Nature 368: 711–718

Jiang H, Ware R, Stall A, Flaherty L, Chess L, Pernis B (1995) Murine CD8$^+$ T cells that specifically delete autologous CD4$^+$ T cells expressing Vβ8 TCR: a role of the Qa-1 molecule. Immunity 2: 185–194

Kappler JW, Kotzin BL, Herron L, Gelfand EW, Bigler RD, Boylston A, Carrel S, Posnett DN, Choi YW, Marrack P (1989) V beta-specific stimulation of human T cells by staphylococcal toxins. Science 244: 811–813

Kappler JW, White J, Kozono H, Clements J, Marrack P (1994) Binding of a soluble aβ T-cell receptor to superantigen/major histocompatibility complex lignands. Proc Natl Acad Sci USA 91: 8462–8466
Kawabe Y, Ochi A (1990) Selective anergy of V beta 8^+, $CD4^+$ T cells in Staphylococcus enterotoxin B-primed mice. J Exp Med 172: 1065–1070
Kawabe Y, Ochi A (1991) Programmed cell death and extrathymic reduction of Vbeta8^+ $CD4^+$ T cells in mice tolerant to Staphylococcus aureux enterotoxin B. Nature 349: 245–248
Kotzin BL, Leung DYM, Kappler JW, Marrack P (1994) Superantigens and their potential role in human disease. Adv Immunol 54: 99–166
Lafferty KJ, Woolnough J (1977) The origin and mechanism of the allograft reaction. Immunol Rev 35: 231–262
Lehmann V, Freudenberg MA, Galanos C (1987) Lethal toxicity of lipopolysaccharide and tumor necrosis factor in normal and D-galactosamine-treated mice. J Exp Med 165: 657–663
Lewis DE, Ng Tang DS, Adu-Oppong A, Schober W, Rodgers JR (1994) Anergy and apoptosis in $CD8^+$ T cells from HIV-infected persons. J Immunol 153: 412–420
Lussow AR, MacDonald HR (1994) Differential effect of superantigen-induced "anergy" on priming and effector stages of a T cell-dependent antibody response. Eur J Immunol 24: 445–449
Lussow AR, Crompton T, Karapetian O, MacDonald HR (1993) Peripheral clonal deletion of superantigen-reactive T cells is enhanced by cortisone. Eur J Immunol 23: 578–581
MacDonald HR, Schneider R, Lees RK, Howe RC, Acha-Orbea H, Festenstein H, Zinkernagel RM, Hengartner H (1988) T-cell receptor V beta use predicts reactivity and tolerance to Mlsa-encoded antigens. Nature 332: 40–45
MacDonald HR, Baschieri S, Lees RK (1991) Clonal expansion precedes anergy and death of V beta 8^+ peripheral T cells responding to staphylococcal enterotoxin B in vivo. Eur J Immunol 21: 1963–1966
Marrack P, Kappler JW (1990) The staphylococcal enterotoxins and their relatives. Science 248: 705–711
Marrack P, Blackman MA, Kushnir E, Kappler JW (1990) The toxicity of staphylococcal enterotoxin B in mice is mediated by T cells. J Exp Med 171: 455–464
Marrack P, Kushnir E, Kappler JW (1991) A maternally inherited superantigen encoded by a mammary tumor virus. Nature 349: 524–526
Marrack P, Winslow GM, Choi YW, Scherer MT, Pullen AM, White J, Kappler JW (1993) The bacterial and mouse mammary tumor virus superantigens; two different families of proteins with the same functions. Immunol Rev 131: 79–92
Matsui K, Boniface JJ, Reay PA, Schild H, Fazekas de St Groth B, Davis MM (1991) Low affinity interaction of peptide-MHC complexes with T cell receptors. Science 254: 1788–1791
Miethke T, Gaus H, Wahl C, Heeg K, Wagner H (1992a) T-cell-dependent shock induced by a bacterial superantigen. Chem Immunol 55: 172–184
Miethke T, Wahl C, Heeg K, Echtenacher B, Krammer PH, Wagner H (1992b) T cell-mediated lethal shock triggered in mice by the superantigen staphylococcal enterotoxin B: critical role of tumor necrosis factor. J Exp Med 175: 91–98
Miethke T, Duschek K, Wahl C, Heeg K, Wagner H (1993a) Pathogenesis of the toxic shock syndrome: T cell mediated lethal shock caused by the superantigen TSST-1. Eur J Immunol 23: 1494–1500
Miethke T, Wahl C, Heeg K, Wagner H (1993b) Acquired resistance to superantigen-induced T cell shock. J Immunol 150: 3776–3784
Miethke T, Wahl C, Holzmann B, Heeg K, Wagner H (1993c) Bacterial superantigens induce rapid and TCR Vβ-selective downregulation of L-selectin (gp90-Mel14) in vivo. J Immunol 151: 6777–6782
Miethke T, Wahl C, Regele D, Gaus H, Heeg K, Wagner H (1993d) Superantigen mediated shock: a cytokine release syndrome. Immunobiology 189: 270–284
Miethke T, Wahl C, Gaus H, Heeg K, Wagner H (1994) Exogenous superantigens acutely trigger distinct levels of peripheral T cell tolerance/immunosuppression: dose-response relationship. Eur J Immunol 24: 1893–1902
Miethke T, Wahl C, Heeg K, Wagner H (1995) Bacterial superantigens induce T cell unresponsiveness in B cell deficient mice. Eur J Immunol 25: 3187–3190
Migita K, Ochi A (1993) The fate of anergic T cells in vivo J Immunol 150: 763–770
Mollick JA, Cook RG, Rich RR (1989) Class II MHC molecules are specific receptors for staphylococcus enterotoxin A. Science 244: 817–820
Mottershead DG, Hsu PN, Urban RG, Strominger JL, Huber BT (1995) Direct binding of the Mtv7 superantigen (Mls-1) to soluble MHC class II molecules. Immunity 2: 149–154
Nathan C, Xie Q (1994) Nitric oxide synthases: Roles, tolls, and controls. Cell 78: 915–918

O'Hehir RE, Lamb JR (1990) Induction of specific clonal anergy in human T lymphocytes by Staphylococcus aureus enterotoxins. Proc Natl Sci USA 87: 8884–8888

Otten GR, Germain RN (1991) Split anergy in a $CD8^+$ T cell: receptor-dependent cytolysis in the absence of interleukin-2 production. Science 251: 1228–1231

Paliard X, West SG, Lafferty JA, Clements JR, Kappler JW, Marrack P, Kotzin BL (1991) Evidence for the effects of a superantigen in rheumatoid arthritis. Science 253: 325–329

Pantaleo G, Demarest JF, Soudeyns H, Graziosi C, Denis F, Adelsberger JW, Borrow P, Saag MS, Shaw GM, Sekaly RP, Fauci AS (1994) Major expansion of $CD8^+$ T cells with a predominant Vβ usage during the primary immune response to HIV. Nature 370: 463–467

Pietra BA, de Inocenico J, Giannini EH, Hirsch R (1994) TCR Vβ family repertoire and T cell activation markers in Kawasaki disease. J Immunol 153: 1881–1888

Rellahan BL, Jones LA, Kruisbeek AM, Fry AM, Matis LA (1990) In vivo induction of anergy in peripheral V beta 8^+ T cells by staphylococcal enterotoxin B. J Exp Med 172: 1091–1100

Rietschel ET, Brade H (1992) Bacterial endotoxins. Sci Am 267: 54–61

Röcken M, Urban JF, Shevach EM (1992) Infection breaks T-cell tolerance. Nature 359: 79–82

Schönrich G, Momburg F, Malissen M, Schmitt-Verhulst AM, Malissen B, Hämmerling GJ, Arnold B (1992) Distinct mechanisms of extrathymic T cell tolerance due to differential expression of self antigen. Int Immunol 4: 581–590

Seth A, Stern LJ, Ottenhoff THM, Engel I, Owen MJ, Lamb JR, Klausner RD, Wiley DC (1994) Binary and ternary complexes between T cell receptor class II MHC and superantigen in vitro. Nature 369: 324–327

Stiles BG, Bavari S, Krakauer T, Ulrich RG (1993) Toxicity of staphylococcal enterotoxins potentiated by lipopolysaccharide: major histocompatibility complex class II molecule dependency and cytokine release. Infect Immun 61: 5333–5338

Thibodeau J, Labrecque N, Denis F, Huber BT, Sekaly RP (1994) Binding sites for bacterial and endogenous retroviral superantigens can be dissociated on major histocompatibility complex class II molecules. J Exp Med 179: 1029–1034

Torres BA, Griggs ND, Johnson HM (1993) Bacterial and retroviral superantigens share a common binding region on class II molecules. Nature 364: 152–154

Vanier LE, Prud'homme GJ (1992) Cyclosporin A markedly enhances superantigen-induced peripheral T cell deletion and inhibits anergy induction. J Exp Med 176: 37–46

Vella AT, McCormack JE, Linsley PS, Kappler JW, Marrack P (1995) Lipoplysaccharide interferes with the induction of peripheral T cell death. Immunity 2: 261–270

Wahl C, Miethke T, Heeg K, Wagner H (1993) Clonal deletion as direct consequence of an in vivo T cell response to bacterial superantigen. Eur J Immunol 23: 1197–1200

Weber S, Traunecker A, Oliveri F, Gerhard W, Karjalainen K (1992) Specific low-affinity recognition of major histocompatibility complex plus peptide by soluble T-cell receptor. Nature 356: 793–796

White J, Herman A, Pullen AM, Kubo RT, Kappler JW, Marrack P (1989) The V beta-specific superantigen staphylococcal enterotoxin B: stimulation of mature T cells and clonal deletion in neonatal mice. Cell 56: 27–35

Pathogenesis of Sepsis Syndrome: Possible Relevance of Pore-Forming Bacterial Toxins

S. Bhakdi[1], I. Walev[1], D. Jonas[1], M. Palmer[1], U. Weller[1], N. Suttorp[1], F. Grimminger[2], and W. Seeger[2]

1	Introduction	101
2	*S. aureus* α-Toxin, *E. coli* Hemolysin, and Streptolysin-O: Three Prototypes of Pore-Forming Toxins	103
2.1	*S. aureus* α-Toxin	105
2.2	*E. coli* Hemolysin	106
2.3	Streptolysin-O	106
3	Consequences of Pore Formation in Target Cells	107
3.1	Cell Death	107
3.2	Secondary Cellular Reactions	107
3.2.1	Reactions Provoked by Egress of Macromolecules and Lysosomal Contents from the Cells	108
3.2.2	Ca^{2+} -Dependent Reactions	108
3.2.3	Short-Circuiting of G Protein Dependent Signaling Pathways	110
3.2.4	Reactions Provoked by Selective Flux of Monovalent Ions	111
4	Organ Dysfunction and Lethal Effects	113
5	Synergism Between Pore Formers and Other Toxins	114
6	Pore-Forming Toxins: Candidates for Immune Intervention?	114
7	Conclusion	115
References		116

1 Introduction

This review focuses on a group of bacterial products whose very existence is known to only a minority of clinicians, and whose potential significance as inducers of the sepsis syndrome has eluded the attention of most microbiologists. This is unfortunate because pore-forming bacterial toxins possess all the properties for contributing to the pathogenesis of local and systemic inflammatory reactions. Because pore formers generally are highly immunogenic proteins, the prospects for immune intervention are described that may eventually be of

[1]Institute of Medical Microbiology and Hygiene, University of Mainz, Hochhaus am Augustusplatz, 55101 Mainz, Germany
[2]Zentrum für Innere Medizin, Universität Gießen, Klinikstr. 36, 35392 Gießen, Germany

benefit to patients. The subject is therefore of interest not only from a theoretical but also from a practical point of view.

Pore-forming toxins are produced and secreted by bacteria as water-soluble proteins. They bind via interaction with cell surface receptors or via nonspecific absorption to lipid bilayers. After binding, the polypeptide domains insert into the target membrane to create water-filled, transmembrane pores. This process is often triggered by the oligomerization of toxin molecules in the membrane, with the formation of high molecular weight polymers that can be visualized in the electron-microscope as arc- or ring-shaped structures. It is thought that the oligomerization step provides the energy required to induce conformational changes that drive the membrane-inserting polypeptide domains into the lipid bilayer (Bhakdi and Tranum-Jensen 1987, 1988, 1991). Inhibition of oligomerization has indeed found to be an effective means to abolish the cytotoxic activity of several pore formers.

The size of transmembrane pores generated by these toxins varies between extremes. On the one hand, certain toxins such as the α-toxin of *Staphylococcus aureus* generate very small pores in the membranes of nucleated cells that are permissive only for monovalent ions (Walev et al. 1993, 1995). On the other hand, streptolysin-O and related toxins can within seconds produce pores of enormous dimensions that lead to rapid egress of proteins from the cytoplasm (Buckingham and Duncan 1983; Bhakdi et al. 1985; Alouf and Geoffrey 1991). Other toxins such as those belonging to the RTX family (with *Escherichia coli* hemolysin as the prototype) produce lesions of intermediate size (1–2 nm) that allow the flux of Ca^{2+} and nucleotides but not that of proteins through the membranes (Bhakdi et al. 1986). As is shown below, the functional size of toxin pores is a very significant factor determining the spectrum of cellular reactions that can be elicited in attacked target cells. These short- and long-range secondary effects are, in turn, probably of prime pathophysiological significance. In considering the biological effects of a pore former it is therefore always important to be aware of the respective pore size.

The list of pore formers is long and is continuously growing. Since the identification of *S. aureus* α-toxin as the first bacterial pore former only 14 years ago (Füssle et al. 1981), the majority of medically relevant pathogens have been shown to produce at least one pore-forming toxin. It is also becoming clear that these exotoxins tend to fall into categories or families of structurally and functionally related molecules. The two main families identified to date are (a) the cholesterol-binding toxins elaborated by gram-positive organisms, with streptolysin-O, pneumolysin, and perfringolysin being well-studied prototypes (Alouf and Geoffrey 1991) and (b) *E. coli* hemolysin with related toxins that have been designated the RTX (for repeat in toxin) family (Welch 1991). Another emerging family is that of the vibrio cytolysins. At a molecular level the cytolysin of *Aeromonas hydrophila* has been studied in some detail (Parker et al. 1994). Table 1 summarizes some of the major, medically relevant pore-forming toxins known today.

Table 1. Prototypes of pore-forming bacterial exotoxins

Producing Organism	Designation	Salient Features
Gram-positive bacteria[a]		
Staphylococcus aureus	α-toxin	Binds to unidentified specific sites on target cells; forms voltage-independent pores of ~1.0 nm diameter; *Identified susceptible human cells*: platelets, monocytes, lymphocytes, endothelial cells
Streptococcus pyogenes A	Streptolysin-O	Prototype of polymerizing toxin that forms large pores (up to 30 nm diameter); binds to cholesterol; all mammalian cells examined are susceptible
Gram-negative bacteria[b]		
Escherichia coli	*E. coli* hemolysin	Receptor problem unclarified, may bind nonspecifically to lipid bilayers; forms voltage-dependent, cation-selective pores of 1–2 nm diameter; potent cytocidal effects on all white blood cells and endothelial cells; triggers G protein dependent processes at low concentrations
Serratia marcescens	*Serratia* hemolysin	Forms voltage-independent pores of 1 nm diameter
Aeromonas hydrophilia	Aerolysin	Produces oligomeric pores of 1 nm diameters

[a]At least 14 other related and probably functionally similar toxins are produced by gram-positive organisms, including *Listeria monocytogenes, Streptococcus pneumoniae,* and *Bacillus cereus.*
[b]At least 10 other related toxins are produced by gram-negative organisms, including *Proteus* spp., *Morganella morganii, Pasteurella haemolytica, Actinobacillus pleuropneumoniae,* Bordetella pertussis Other pore-forming toxins are produced by *Vibrio cholerae* and *Vibrio parahaemolyticus* and *Klebsiella.*

2 *S. aureus* α-Toxin, *E. coli* Hemolysin, and Streptolysin-O: Three Prototypes of Pore-Forming Toxins

These three toxins have been extensively studied at the molecular level in this and other laboratories. They represent important prototypes, each with quite characteristic functional properties, and it is useful to review briefly some salient aspects of their structure and function.

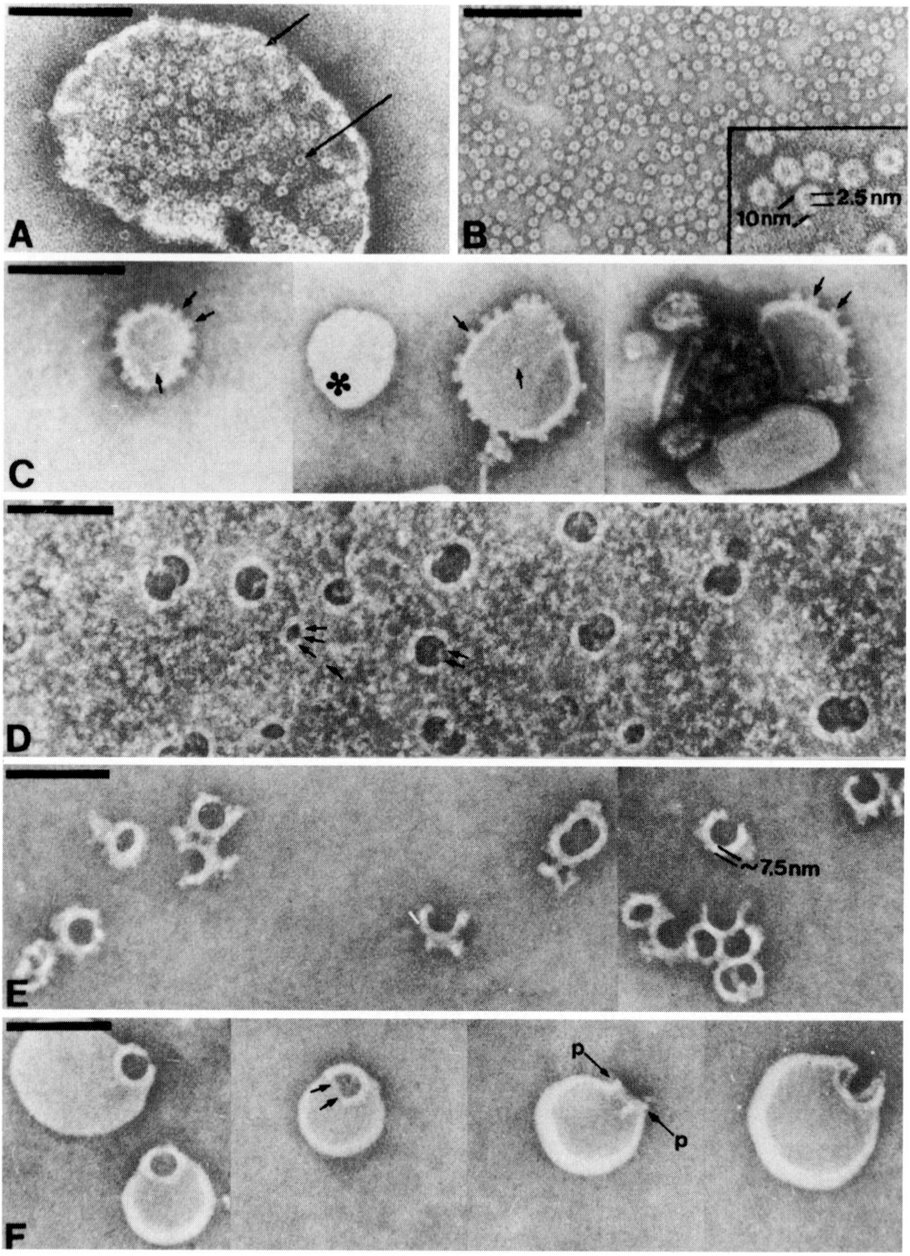

Fig. 1. A Negatively stained fragment of rabbit erythrocyte lysed with staphylococcal α-toxin. Numerous 19-nm ring-shaped structures are seen over the membrane (*arrows*). **B** Isolated toxin hexamers in detergent solution. **C** Lecithin liposomes carrying reincorporated α-toxin hexamers. The hexamers are seen as stubs along the edge of the liposomal membrane and as rings over the membrane (*arrows*). Characteristically, liposomes that escape incorporation of the toxin are

2.1 *S. aureus* α-Toxin

The toxin is secreted as a water-soluble monomer of M_r 34 000 devoid of cysteine residues. β-Sheet structures are the predominant secondary structural feature. The protein binds to as yet unidentified receptor molecules on susceptible cells in a temperature-independent manner (CASSIDY and HARSHMAN 1976; HILDEBRAND et al. 1991). Binding occurs via a conformational binding site that may involve both N- and C-terminus of the molecule. Membrane-bound toxin molecules collide in the membrane plane via lateral diffusion and form oligomers (BHAKDI and TRANUM-JENSEN 1991; Fig. 1) that appear to be heptamers (GOUAX et al. 1994). Oligomerization is accompanied by the ATP-independent insertion of the central molecular domain (encompassing approximately residues 124–138) into the bilayer. This has been delineated by the demonstration of the following: (a) Polarity-sensitive fluorescent molecules bound to these residues shift from a hydrophilic to a hydrophobic environment as toxin oligomerization occurs (VALEVA et al., 1996). (b) Biotin attached to residue 130 becomes inaccessible to streptavidin added to the external membrane face (PALMER et al. 1993a). (c) Alterations in this central region also alter the pore-forming properties of the toxin (PALMER et al. 1993b; WALKER et al. 1993; WALKER and BAYLEY 1994). (d) Derivatization of a cysteine residue placed at position 121 blocks the functional pore. (e) Removal of the blocking molecule opens the pore (VALEVA et al., 1996). Unexpectedly, it was found that the membrane-inserting region of α-toxin is not primarily hidden in the native molecule, but that it is freely accessible to modifying reagents and proteases (PALMER et al. 1993a).

When assessed in planar lipid membranes and erythrocytes, the size of α-toxin pores is approximately 1–1.5 nm, i.e., large enough to accommodate divalent ions and nucleotides (BHAKDI et al. 1984; MENESTRINA 1986). Unexpectedly, the size measured in susceptible nucleated cells such as keratinocytes, monocytes, and fibroblasts is slightly but significantly smaller. In these cases we have found that Ca^{2+} does not freely permeate the lesions (WALEV et al. 1993, 1995). Hence, when applied at physiological concentrations, α-toxin generates pores that are selective for monovalent ions in many cell targets.

impermeable to the stain. **D** Negatively stained erythrocyte membrane lysed by streptolysin-O showing numerous curved rods 25–100 nm long and approximately 7.5 nm wide with inner radius of curvature of 13–16 nm. Most rods are approximately semicircular, often joined in pairs at their ends. Dense accumulations of stain are seen at the concave side of the rods. When these do not form closed profiles, the stain deposit is partly bordered by a "free" edge of the erythrocyte membrane (*arrows*). **E** Negative staining of isolated streptolysin-O oligomers, showing, numerous curved rod structures identical to those found in toxin-treated membranes. **F** Purified streptolysin-O complexes reincorporated into cholesterol-free lecithin liposomes. The toxin oligmers form holes in the liposomes (*unlabeled arrows*); *p*, lesion seen in profile. *Scale bars*, 100 nm in all frames. **B–F** Sodium silicotungstate was used as negative stain; **A** uranylacetate (From BHAKDI and TRANUM-JENSEN 1988)

2.2 *E. coli* Hemolysin

The family of RTX toxins encompasses at least 14 members of homologous proteins elaborated by gram-negative organisms (WELCH 1991). *E. coli* hemolysin is the best studied member thereof, and it has several singular properties that probably extend to other toxins in this family. *E. coli* hemolysin is a single-chain polypeptide devoid of cysteine with a M_r of 107 000. The toxin contains a hydrophobic stretch of amino acids in the N-terminal half of the molecule. It requires acylation with fatty acids at two lysine residues to obtain pore-forming activity (ISSARTEL et al. 1991; STANLEY et al. 1994). The following properties are remarkable. First, all mammalian cells tested thus far are susceptible to toxin action, but certain cell types including granulocytes, monocytes, endothelial cells, and renal epithelial cells are particularly vulnerable. The molecular basis for this phenomenon is unknown. Binding studies conducted in our laboratory in collaboration with V. Koronakis and C. Hughes speak against the presence of specific binding sites on cells that would account for their susceptibility. It appears that binding occurs quite unspecifically, possibly via interaction of the hydrophobic toxin domain with lipid. This interaction does not depend on the presence of covalently bound fatty acids, and toxin mutants lacking the acylation sites can therefore still bind to target cells. Remarkably, the binding step is accompanied by triggering of G protein dependent reactions that have been well studied in human granulocytes. These include the stimulation of the respiratory burst, secretion of vesicular constituents, and production platelet-activating factor, processes that are related to inositol 1,4,5-triphosphate (IP_3) generation, and that can be partially inhibited by pertussis toxin (GRIMMINGER et al. 1991b; BHAKDI and MARTIN 1991). The events underlying the potent triggering of IP_3 related events is not known. They are not related to pore formation, and they can be evoked by nonacylated toxin mutants that create no pores. We speculate that the hydrophobic toxin domain interacts directly with elements of the G protein cascade to evoke the known spectrum of reactions that are normally provoked by the binding of chemokines to their receptors.

Pores created by *E coli* hemolysin are permissive for calcium and nucleotides such as ATP and GTP, but are too small to permit egress from the proteins from the cells (BHAKDI et al. 1986, 1989a; MENESTRINA et al. 1987). The pore-forming domains of *E. coli* hemolysin have not yet been identified.

2.3 Streptolysin-O

Toxins of this family are single-chain polypeptides of M_r 50 000–70 000, containing one cysteine residue whose chemical modification generally leads to inactivation. A stretch of hydrophobic amino acids is located near the cysteine in the highly conserved C-terminus of the protein. This hydrophobic sequence is involved in the primary binding event to cholesterol molecules in target membranes. For these toxins to bind efficiently to membranes the lipid bilayers must have a high content of cholestrol (over 20 mol% of lipid; Bernheimer 1974; Alouf 1980; Ohno-Iwashita et al. 1988; Alouf and Geoffrey 1991). Virtually all mammalian cells are susceptible to their action, whereas bacteria are not. The membrane-bound toxin monomers are not cytotoxic; pore formation occurs when they aggregate laterally with each other to form noncovalent polymers (Fig. 1). The polymers are heterogeneous in size and probably accommodate 25–80 toxin molecules. In analogy to α-toxin, polymerization is accompanied by the insertion of as yet unidentified domains into the bilayer with formation of very large pores (Buckingham and Duncan 1983; Bhakdi et al. 1985). Mutant toxin preparations that have lost the ability to polymerize are also devoid of pore-forming properties.

3 Consequences of Pore Formation in Target Cells

3.1 Cell Death

Cell death is the most obvious and inevitable consequence of transmembrane pore formation if a lesion is not removed or repaired. Death ensues because the cells are rapidly depleted of ATP, and because they are unable to counteract the deleterious effects of ionic disequilibrium. Cell death can have immediate detrimental consequences. Necrosis of tissues generates niches for bacterial survival and multiplication. Death of phagocytes fosters microbial persistence and invasion (Bhakdi et al. 1989a). Endothelial cells are generally highly susceptible towards the action of pore formers; their death can lead to major perturbations in the microcirculation, for example, in the pulmonary vasculature (Seeger et al. 1990). Damage to renal epithelial cells evoked, for instance, by *E. coli* hemolysin is probably directly responsible for renal dysfunction that is typical of acute pyelonephritis (Keane et al. 1987). The cytotoxic action of pneumolysin and streptolysin-O on epithelial cells lining the round window membrane may contribute to sensineural hearing loss during acute otitis media caused by pneumococci because of breakdown of the round window membrane permeability barrier (Engel et al. 1995). There is also increasing evidence that the cytotoxic action of pore formers on epithelial cells of the gastrointestinal tract contributes to pathogenesis of diarrhea. These examples serve to underline the fact that cytotoxicity itself can cause organ dysfunction. It is noteworthy that, although most pore-forming toxins have been designated hemolysins because of their easily detectable hemolytic action in vitro, their in vivo significance is certainly due in the main to their action on susceptible nucleated cells and platelets.

3.2 Secondary Cellular Reactions

It cannot be sufficiently emphasized that secondary reactions mounted by cells under attack by a pore former are of prime importance. It is useful to differentiate four main categories or reactions that are provoked by the different actions of pore-forming toxins.

3.2.1 Reactions Provoked by Egress of Macromolecules and Lysosomal Contents from the Cells

Such reactions are evoked, for example, by streptolysin-O and other toxins of this group. Biologically active molecules may be released from the cytoplasm that then act on cells in the immediate or less immediate surroundings. A simple example is the release of basic fibroblast growth factor from fibroblasts that are killed by streptolysin-O (Walev et al., unpublished). Such events surely contribute to cellular reactions in the infected sites. Another example is the massive liberation of lysosomal enzymes by the cells permeabilized with streptolysin-O. This may be the reason for the observed cleavage and release of interleukin-6 receptor and CD14 from cells bearing these molecules. Release of interleukin-6 receptor and CD14 may be highly significant events in the pathogenesis of septic shock because of transignaling events that render cells primarily devoid of the receptor to become susceptible to this interleukin (Walev et al., unpublished).

3.2.2 Ca^{2+}-Dependent Reactions

These occur when Ca^{2+}-permissive pores are generated that are too small to permit simultaneous leakage of cytoplasmic proteins. Many cellular mechanisms remain functionally intact over a short time period that respond to the rapid influx

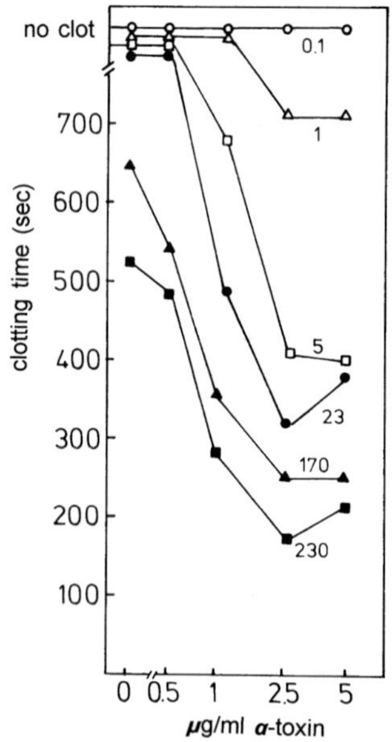

Fig. 2. Reduction in clot times of recalcified human plasma by *S. aureus* α-toxin. Citrated plasma samples containing the depicted numbers of platelets (per nanoliter) were given 12 m*M* Ca^{2+} and α-toxin at the given final concentrations. Marked reductions clot times were noted at toxin concentrations of 1–2.5 µg/ml, dependent on the presence of platelets in the plasma sample. *No clot,* lack of clot formation within 900 s. (From Bhakdi et al. 1988)

of Ca^{2+} ions. Some well-studied phenomena include secretion, activation of phospholipases with generation of eicosanoids, and contraction of cytoskeletal elements. Thus by stimulating secretion of procoagulatory substances from platelets α-toxin exerts potent procoagulatory effects (BHAKDI et al. 1988) when added to human blood (Fig. 2). *E. coli* hemolysin provokes the secretion of granular constituents from human granulocytes; the uncontrolled liberation of elastase may be pathophysiologically relevant (BHAKDI et al. 1989a). *E. coli* hemolysin potently induces generation of biologically active lipid mediators in target endothelial cells and granulocytes (KÖNIG et al. 1986; GRIMMINGER et al. 1990a,b, 1991a,b; SEEGER et al. 1989; SUTTORP et al. 1985, 1993). Both α-toxin and *E. coli* hemolysin provoke cytoskeletal dysfunction in endothelial cells, and cellular contraction leads to the formation of intercellular gaps that cause leakage of macromolecules across a confluent monolayer (Fig. 3; SUTTORP et al. 1988, 1990). Such processes certainly contribute to organ dysfunction that can be

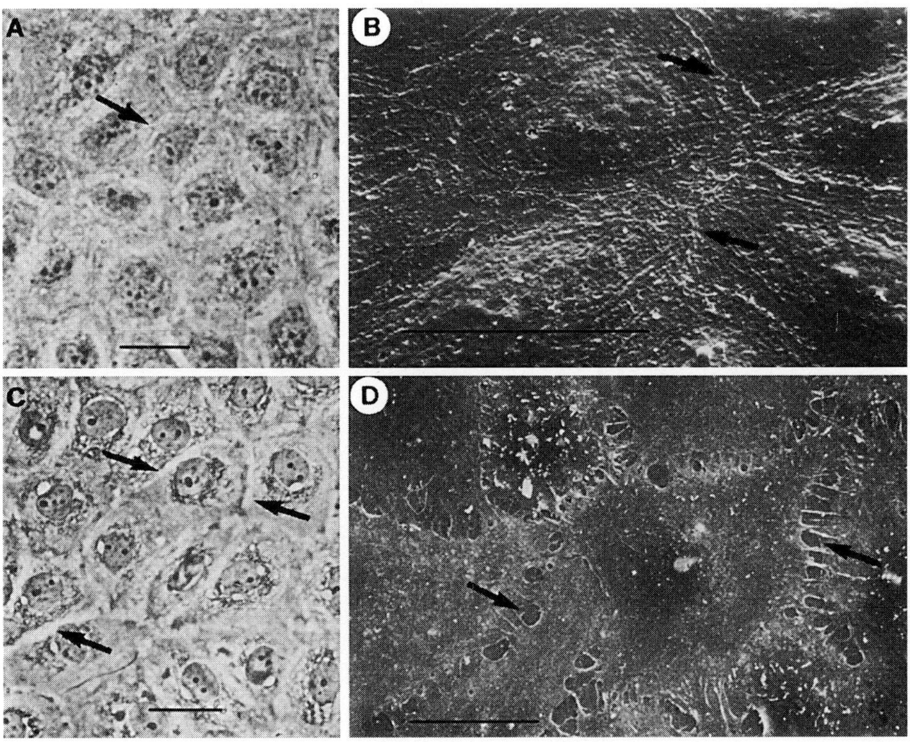

Fig. 3A–D. Intercellular gap formation in cultured endothelial cells induced by a pore-forming toxin. **A,** and **B** Sealed endothelial monolayer of pig pulmonary artery grown on gelatine-coated cover slips and exposed for 90 min to a hydrostatic pressure of 10 cm H_2O. Examination by phase contrast (**A**) and scanning electron microscopy **(B)** shows that virtually all interendothelial junctions (intercellular space) are closed. **C,** and **D** Endothelial monlayer exposed for 90 min to 1 μg/ml staphylococcal α-toxin. Note large interendothelial gaps (→) that are seen by both phase-contrast **(C)** and scanning electron microscopy **(D)**. *Bar*, 20 μm. (From SUTTORP et al. 1988)

provoked in isolated organ systems, and that are observed when animals are challenged with these toxins.

3.2.3 Short-Circuiting of G Protein Dependent Signaling Pathways

This property is characteristic of *E. coli* hemolysin and probably of other RTX toxins. For unknown reasons the initial interaction of *E. coli* hemolysin with the

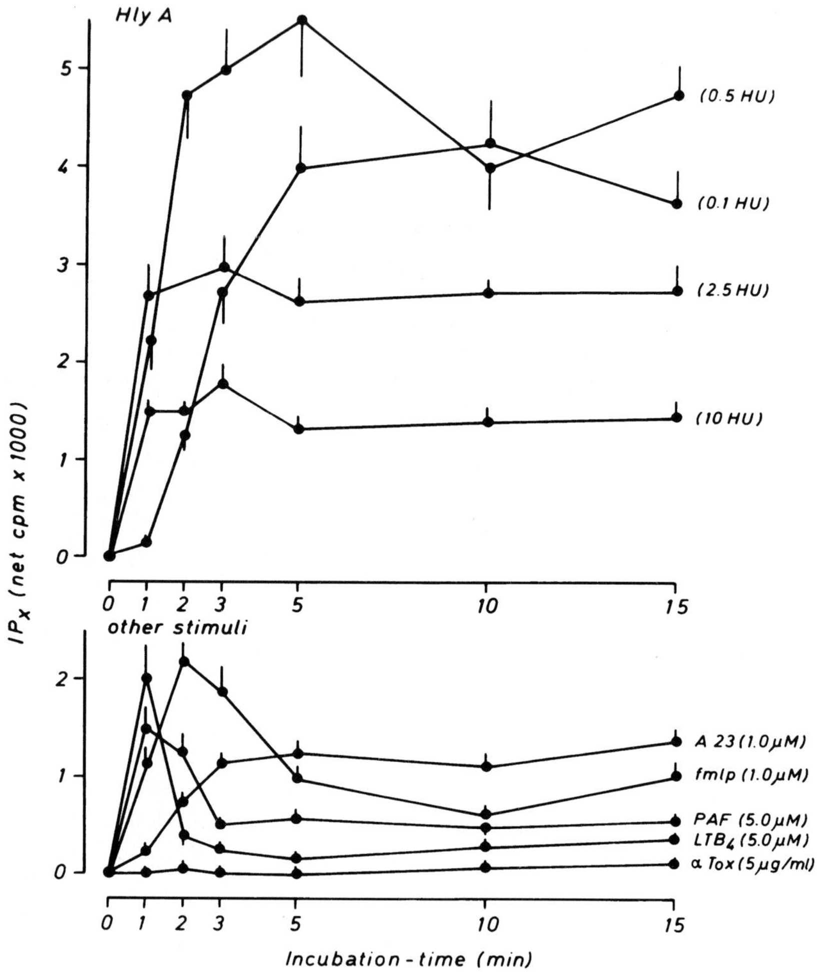

Fig. 4. Time course of neutrophil inositol phosphate generation evoked by *E. coli hemolysin* (H*ly A*) in comparison with various other stimuli. Human neutrophils (10^7) prelabeled with [^{3}H]inositol were incubated with various concentrations of Hly A (*top*) or other stimuli at optimum concentrations (*bottom*) for various time periods. Extracted inositol phosphates were separated by anion-exchange chromatography. Ip_x, IP_3, IP_2, and IP_1, corrected for baseline levels in nonchallenged cells (net cpm; baseline range between 540 and 1100 cpm). Means ± SEM of five (*top*) and four (*bottom*) independent experiments are given. (From Grimminger et al. 1991b)

target membrane is linked to the triggering of G protein dependent pathways and IP_3 generation (Fig. 4). Granulocytes challenged with toxin molecules that can bind but not form pores respond with reactions typical of those normally provoked by chemokines, i.e., secretion and production of reactive oxygen metabolites. Indeed, *E. coli* hemolysin has been identified as the most potent proteinaceous inducer of phosphoinositide hydrolysis known to date. As outlined above, we currently speculate that the triggering of these G protein dependent processes is related to the spontaneous insertion of the hydrophobic domain of *E. coli* hemolysin into the bilayer.

3.2.4 Reactions Provoked by Selective Flux of Monovalent Ions

These have recently been discovered in the course of our studies with α-toxin, which produces small pores in target membranes of many nucleated cells that are not Ca^{2+} permissive. Two striking phenomena have been observed. First, programmed cell death (apoptosis) is triggered in T-lymphocytes (Jonas et al. 1994). An important requisite is that the membranes are not permeable for Ca^{2+}: when α-toxin is administered at high concentrations, Ca^{2+} -permissive lesions are generated, and the extent of programmed cell death (PCD) diminishes (Fig. 5). Remarkably, PCD can be totally inhibited if cells are suspended in KCl rather than in Nacl. Hence internucleosomal DNA degradation ensuing through

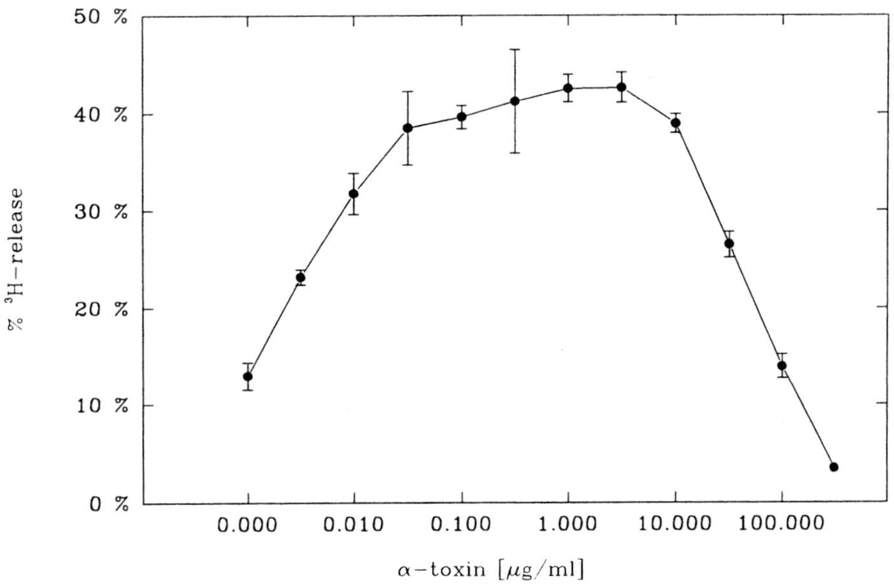

Fig. 5. Dose-independent induction of DNA fragmentation in T-lymphocytes by staphylococcal α-toxin. T-cells were activated with phythemagglutinin and labeled with [^{3}H]thymidine. They were treated with α-toxin at the depicted concentrations and release of [^{3}H]thymidine was measured after 4 h. Note the bell-shaped dose-dependent curve of DNA degradation evoked by the toxin. (From Jonas et al. 1994)

DNAse activation appears to somehow be coupled to perturbance of Na^+/K^+ homeostasis.

A second process of potential significance is the activation of interleukin-converting enzyme (ICE) in human monocytes (Walev et al. 1995). This leads to rapid processing and secretion of interleukin-1β (Fig. 6). Recent studies have shown that ICE is controlled directly or indirectly by K^+. Thus any agents that cause cellular K^+ depletion also activate ICE. Conversely, the activation of ICE can be abrogated or suppressed in permeabilized cells by their transfer to K^+-rich media. Since an ICE-related protease may be involved in triggering apoptosis, the parallel findings made with α-toxin with regard to the triggering of PCD and the activation of interleukin-1β secretion may be linked. The study of secondary reactions evoked by pore-forming toxins has thus uncovered a novel phenom-

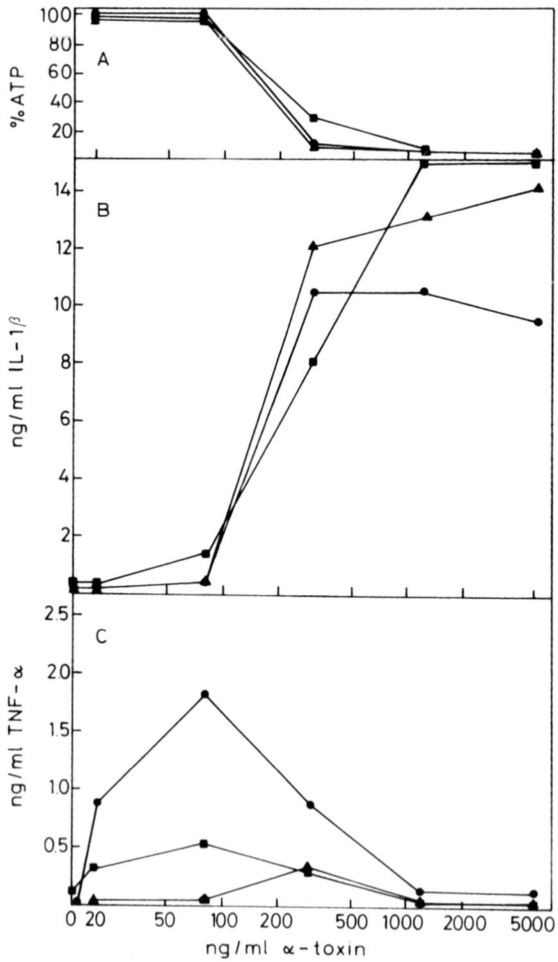

Fig. 6A–C. Action of α-toxin on monocytes in the presence of 10% autologous serum. **A** Cellular ATP content, expressed as a percentage of untreated control levels. **B**, **C** Concentrations of interleukin-1β IL=1β; **B** and tumor necrosis factor-α (TNF-α **C**) in the cell supernatants. The incubation period with α-toxin was 6 h at 37 °C. Each plot depicts the results obtained from one individual. (From Bhakdi et al. 1989c)

enon in cell biology: the role of K^+ in the regulation of vitally important cellular proteases (WALEV et al. 1995).

As a postscript, it is noteworthy that a limited number of small pores created by α-toxin can be repaired in certain nucleated cells such as fibroblasts. The repair process fails at lower pH (< 7) and seems to involve closure of the pores in the membranes (WALEV et al. 1994). Another interesting aspect pertains to the action of pore-forming toxins within cells that have ingested toxin-producing micro-organisms. As has been elegantly shown in the case of listeriolysin, this pore former enables the bacteria to escape from endolysosomes by the disruption of the lysosomal membrane (PORTNOY et al. 1992).

Taken together, secondary reactions by pore-forming toxins are thus very diverse, and it is important to realize that they are probably responsible for an array of short- and long-range effects in the macro-organism. The spectrum of these cellular reactions are indeed at least as wide as those elicited by endotoxins.

4 Organ Dysfunction and Lethal Effects

Since pore formers can evoke functional disturbances in target cells, it is easy to understand that they will also elicit acute organ dysfunction and are lethal in experimental animals. In these respects they are no less potent than endotoxins. Indeed, in some experimental models such as in the isolated lung they are much more toxic than their lipopolysaccharide counterparts. Detailed investigations have been performed in the latter model with both α-toxin (SEEGER et al. 1984, 1990) and *E. coli* hemolysin (GRIMMINGER et al. 1990a, b; SEEGER et al. 1991). Each of these toxins rapidly evoked profound pathophysiological alterations in the pulmonary microvasculature. This subject has been reviewed in sufficient detail recently (BHAKDI et al. 1994a), and it suffices here to say that subnanomolar concentrations of either α-toxin or *E. coli* hemolysin administered to the recirculating perfusion fluid leads to development of irreversible pulmonary edema. The underlying mechanisms are complex but encompass the direct toxic action on endothelial cells and the production of eicosanoids that in turn foster pulmonary arterial hypertension. Overall, it is evident that pore-forming exotoxins have the propensity to evoke acute pulmonary failure such as is typically found in patients presenting with adult respiratory distress syndrome during severe bacterial infections.

That pore-forming toxins are lethal when administered low amounts has long been known for α-toxin and streptolysin-O (ALOUF 1980; BERNHEIMER 1974). Experiments with *E. coli* hemolysin have extended this to the group of RTX toxins. In the case of α-toxin and *E. coli* hemolysin it is noteworthy that animals succumb to the action of both pore formers with symptoms that may be explained from the in vitro observations. For example, α-toxin provokes acute thrombocytopenia and pulmonary edema in rabbits and monkeys (BHAKDI et al. 1989b). Rabbits challenged with *E. coli* hemolysin present with acute leukopenia and hemorrhagic pulmonary edema (VAGTS et al. 1993). These facts support the view that attack by pore formers on susceptible cells, coupled with detrimental cellular reactions, can lead to multiorgan failure and death.

5 Synergism Between Pore Formers and Other Toxins

Importantly, there may be synergism between endotoxin and pore-forming toxins. It has recently been demonstrated that the priming of pulmonary cells in isolated lungs with endotoxin does potentiate vascular abnormalities in response to α-toxin and *E. coli* hemolysin (WALMRATH et al. 1993, 1994). Such synergism and the resulting vascular abnormalities may be relevant to the pathogenesis of organ failure in systemic infections. We anticipate that similar synergisms are operative between superantigens (FLEISCHER 1991) and pore formers.

6 Pore-Forming Toxins: Candidates for Immune Intervention?

Pore-forming toxins usually fall within a narrow range of related protein families exhibiting antigenic cross-reactivity. Therefore development of vaccines against the major toxins is basically feasible. Many toxins have been cloned, and it is easy to generate inactive, site-directed mutants. In the case of α-toxin, for example, the replacement of a single amino acid residue can result in immunogenic but perfectly nontoxic proteins that elicit high antibody titers in animals (JURSCH et al. 1994; BHAKDI et al. 1994b). High-titered human antibodies against α-toxin have also been produced with use of a conventional toxoid. These antibodies are capable of neutralizing detrimental effects of α-toxin in vitro and in vivo (BHAKDI et al. 1989b). Apart from the possibility of passive immunization with such antibodies in patients with severe infections, a theoretical option is to vaccinate patients who are to undergo high-risk operations, for example, with a combination of potentially relevant vaccines. Such a prophylactic regimen would probably be cheap and safe and may also prove to be effective.

7 Conclusion

In contrast to endotoxins, pore-forming bacterial exotoxins continue to receive little attention as possible contributors to the pathogenesis of sepsis syndrome and septic shock. This is inappropriate since they are produced by the majority of important bacterial pathogens, and their toxic action has been extensively documented in defined cell systems, isolated organs, and animals models (e.g., WELCH et al. 1981; O'REILLY et al. 1986; PATON et al. 1993). Mutant bacterial strains that have selectively lost their capacity to produce important pore formers such as α-toxin, *E. coli* hemolysin, and pneumolysin also exhibit marked reduction in virulence in various animal models. Antibodies against α-toxin have been shown to significantly protect animals against various infections with *S. aureus*. Considering the wide spectrum of cellular disturbances that pore-forming toxins can evoke in target cells, these findings are in no way unexpected.

Unfortunately, pore-forming toxins are very difficult to quantify in biological fluids and tissues, Quantitation of the proteins is feasible only if methods can be devised to release them from the target cells. At present no satisfactory solution to the problem is available. In this regard the state of the art corresponds to the situation in clinical endotoxin research approximately 10 years ago, when the problem of endotoxin measurements had not yet been solved. Furthermore, pore-forming toxins probably do not have to be released in measurable quantities into body fluids to exert their action. Very few molecules are generally required to generate a membrane lesion, and attack on a cell may occur within a small, circumscribed areas as bacteria gain contact with their target. For example, seeding toxin inducing *E. coli* onto monocytes at the ration of only 1:1 causes death of the monocytes in the absence of detectable toxin in the cell supernatants (BHAKDI et al. 1990). Similarly, perfusion of isolated lungs with hemolytic *E. coli* also caused massive derangements in pulmonary vasculature in the absence of detectable circulating toxin (SEEGER et al. 1991). For these reasons it will prove difficult to obtain conclusive evidence that pore-forming toxins contribute to the development of sepsis syndrome. Nevertheless, the suggestive evidence at the present is already quite overwhelming, and it is highly appropriate to consider these very widespread bacterial products as probable contributors to sepsis and septic shock.

Acknowledgements. Our studies have received the continued support of the Deutsche Forschungsgemeinschaft, SFB 311, Immunpathogenese.

References

Alouf JE (1980) Streptococcal toxins (streptolysin O, streptolysin S, erythrogenic toxin). Pharmacol Ther 11: 661–717

Alouf JE, Geoffrey C (1991) The family of the antigenically-related, cholesterol-binding ("sulphydryl-activated") cytolytic toxins. In: Alouf JE, Freer JH (eds) Sourcebook of bacterial protein toxins. Academic New York, pp 147–186

Bernheimer AW (1974) Interactions between membranes and cytolytic bacterial toxins. Biochim Biophys Acta 344: 27–50

Bhakdi S, Martin E (1991) Superoxide generation by human neutrophils induced by low doses of Escherichia coli hemolysin. Infect Immun 59: 2955–2962

Bhakdi S, Tranum-Jensen J (1987) Damage to mammalian cells by proteins that form transmembrane pores. Rev Physiol Biochem Pharmacol 107: 147–223

Bhakdi S, Tranum-Jensen J (1988) Damage to cell membranes by pore-forming bacterial cytolysins. Prog Allergy 40: 1–43

Bhakdi S, Tranum-Jensen J (1991) Alpha-toxin of Staphylococcus aureus. Microbiol Rev 55: 733–751

Bhakdi S, Muhly M, Füssle R (1984) Correlation between toxin binding and hemolytic activity in membrane damage by staphylococcal alpha-toxin. Infect Immun 46: 318–323

Bhakdi S, Tranum-Jensen J, Sziegoleit A (1985) Mechanism of membrane damage by streptolysin-O. Infect Immun 47: 52–60

Bhakdi S, Mackman N, Nicaud JM, Holland IB (1986) Escherichia coli hemolysin may damage target cell membrane by generating transmembrane pores. Infect Immun 52: 63–69

Bhakdi S, Muhly M, Mannhardt U, Hugo F, Klapettek K, Mueller Eckhardt C, Roka L (1988) Staphylococcal alpha toxin promotes blood coagulation via attack on human platelets. J Exp Med 168: 527–542

Bhakdi S, Greulich S, Muhly M, Eberspächer F, Becker H, Thiele A, Hugo F (1989a) Potent leukocidal action of Escherichia coli hemolysin mediated by permeabilization of target cell membranes. J Exp Med 169: 737–754

Bhakdi S, Mannhardt H, Ronneberger R, Hungerer KD (1989b) Human hyperimmune globulin protects against the cytotoxic action staphylococcal alpha-toxin in vitro and in vivo. Infect Immun 57: 3214–3220

Bhakdi S, Muhly M, Korom S, Hugo F (1989c) Release of interleukin-1ß associated with potent cytocidal action of staphylococcal alpha-toxin on human monocytes. Infect Immum 57: 3512–3519

Bhakdi S, Muhly M, Korom S, Schmidt G (1990) Effects of Escherichia coli hemolysin on human monocytes. J Clin Invest 85: 1746–1753

Bhakdi S, Weller U, Walev I, Martin E, Jonas D, Palmer M (1993) A guide to the use of pore-forming toxins for controlled permeabilization of cell membranes. Med Microbiol Immunol 182: 167–175

Bhakdi S, Grimminger F, Suttorp N, Walmrath D, Seeger W (1994a) Proteinaceous bacterial toxins and pathogenesis of sepsis syndrome and septic shock: the unknown connection. Med Microbiol Immunol 183: 119–144

Bhakdi S. Jursch R, Bröker M, Ronneberger H, Hungerer K-D (1994b) Functionally inactive S. aureus alpha-toxin containing a single amino acid substitution: potential usefulness as a vaccine. Behring Inst Mitt 95: 80–84

Buckingham L, Duncan JL (1983) Approximate dimensions of membrane lesions produced by streptolysin-S and streptolysin-O. Biochim Biophys Acta 729: 115–122

Cassidy P, Harshman S (1976) Biochemical properties of staphylococcal alpha-toxin. Biochim Biophys Acta 332: 413–423

Engel F, Blatz R, Kellner J, Palmer M, Weller U, Bhakdi S (1995) Breakdown of the round window membrane permeability barrier evoked by streptolysin O: possible etiologic role in development of sensorineural hearing loss in acute otitis media. Infect Immun 63: 1305–1310

Fleischer B (1991) T lymphocyte-stimulating microbial toxins as "superantigens." Med Microbiol Immunol 180: 53–58

Füssle R, Bhakdi S, Sziegoleit A, Tranum-Jensen J, Kranz T, Wellensiek HJ (1981) On the mechanism of membrane damage by S. aureus alpha-toxin. J Cell Biol 91: 83–94

Gouax JE, Braha O, Hobaugh MR, Song L, Cheley S, Shustak C, Bayley H (1994) Subunit stoichiometry of staphylococcal alpha-hemolysin in crystals and on membranes: a heptameric transmembrane pore. Proc Natl Acad Sci USA 91: 12828–12831

Grimminger F, Walmrath D, Birkemeyer D, Bhakdi S, Seeger W (1990a) Burst of leukotriene- and HETE-generation elicited by low dose Escherichia coli hemolysin in rabbit lungs. Infect Immun 58: 2659–2663

Grimminger F, Thomas M, Obernitz R, Walmarth D, Bhakdi S, Seeger W (1990b) Inflammatory lipid mediator generation elicited by viable hemolysin-forming Escherichia coli in lung vasculature. J Exp Med 172: 1115–1125

Grimminger F, Scholz C, Bhakdi S, Seeger W (1991a) Subhemolytic doses of Escherichia coli hemolysin evoke large quantities of lipoxygenase products in human neutrophils. J Biol Chem 266: 14262–14269

Grimminger F, Sibelius U, Bhakdi S, Suttorp N, Seeger W (1991b) Escherichia coli hemolysin is a potent inductor of phosphoinositide hydrolysis and related metabolic responses in human neutrophils. J Clin Invest 88: 1531–1539

Hildebrand A, Pohl M, Bhakdi S (1991) Staphylococcus aurus α-toxin dual mechanisms of binding to target cells. J Biol Chem 266: 17195–17200

Issartel J-P, Koronakis V, Hughes C (1991) Activation of Escherichia coli prohaemolysin to the mature toxin by acyl carrier protein-dependent fatty acylation. Nature 351: 759–761

Jonas D, Walev I, Berger T, Liebetrau M, Palmer M, Bhakdi S (1994) Small transmembrane pores created by staphylococcal alpha-toxin in T lymphocytes evoke internucleosomal DNA-degradation. Infect Immun 62: 1304–1312

Jursch R, Hildebrand A, Hobon G, Kehoe M, Ward R, Bhakdi S (1994) Histidine residues near the N terminus of staphylococcal alpha-toxin as reporters of regions that are critical for oligomerization and pore formation. Infect Immun 62: 2249–2256

Keane WF, Welch R, Gekker G, Peterson PK (1987) Mechanism of Escherichia coli alpha-hemolysin induced injury to isolated renal tubular cells. Am J Pathol 126: 350–357

König B, König W, Scheffer J, Hacker J, Goebel W (1986) Role of Escherichia coli alpha-hemolysin and bacterial adherence in infection: requirement for release of inflammatory mediators from granulocytes and mast cells. Infect Immun 54: 886–892

Menestrina G (1986) Ionic channels formed by Staphylococcus aureus alpha-toxin: voltage-dependent inhibition by divalent and trivalent cations. J Membr Biol 90: 177–190

Ohno-Iwashita Y, Ywamoto M, Mitsui K, Ando S, Nagai K (1988) Protease nicked θ-toxin of *Clostridium perfringens*, a new membrane probe with no catalytic effect reveals two classes of cholesterol as toxin-binding sites on sheep erythrocytes. Eur J Biochem 176: 95–101

O'Reilly M, Azavedo JCS, Kennedy S, Foster TJ (1986) Inactivation of the alpha-hemolysin gene of Staphylococcus aureus 8325-5 by site-directed mutagenesis and studies on the expression of its haemolysins. Microb Pathog 1: 125–138

Palmer M, Jursch R, Weller U, Valeva A, Hilgert K, Kehoe M, Bhakdi S (1993a) Staphylococcus aureus alpha-toxin. J Biol Chem 268: 11959–11962

Palmer M, Weller U, Meßner M, Bhakdi S (1993b) Altered pore-forming properties of proteolytically nicked staphylococcal alpha-toxin. J Biol Chem 268: 11963–11967

Parker MW, Buckley JT, Postma JPM, Tucker AD, Leonard K, Pattus F, Tsernoglou D (1994) Structure of the Aeromonas toxin proaerolysin in its water-soluble and membrane-channel states. Nature 367: 292–295

Paton JC, Andrew PW, Boulnois GJ, Mitchell TJ (1993) Molecular analysis of the pathogenicity of Streptococcus pneumoniae: the role of pneumococcal proteins. Annu Rev Microbiol 47: 89–115

Portnoy DA, Chakraborty T, Goebel W, Cossart P (1992) Molecular determinants of Listeria monocytogenes pathogenesis. Infect Immun 60: 1263–1267

Seeger W, Bauer M, Bhakdi S (1994) Staphylococcal alpha-toxin elicits hypertension in isolated rabbit lungs. Evidence for thromboxane formation and the role of extracellular calcium. J Clin Invest 74: 849–858

Seeger W, Walter H, Suttorp N, Bhakdi S (1989) Thromboxane-mediated hypertension and vascular leakage evoked by low doses of Escherichia coli hemolysin in rabbit lungs. J Clin Invest 84: 220–227

Seeger W, Birkemeyer RG, Ermet L, Suttorp N, Bhakdi S, Duncker HR (1990) Staphylococcal alpha-toxin induced vascular leakage in rabbit lungs. Lab Invest 63: 341–349

Seeger W, Obernitz R, Thomas M, Walmrath D, Holland IB, Grimminger F, Eberspächer B, Hugo F, Suttorp N, Bhakdi S (1991) Lung vascular injury after administration of viable hemolysin-forming Escherichia coli in isolated rabbit lungs. Am Rev Respir Dis 143: 797–805

Stanley P, Packman LC, Koronakis V, Hughes C (1994) Fatty acylation of two internal lysine residues required for the toxic activity of Escherichia coli hemolysin. Science 266: 1992–1996

Suttorp N, Habben E (1988) Effect of staphylococcal alpha-toxin on intracellular Ca^{2+} in polymorphonuclear leukocytes. Infect Immun 56: 2228–2234

Suttorp N, Seeger W, Dewein E, Bhakdi S, Roka L (1985) Staphylococcal alpha-toxin-induced PG12 production in endothelial cells: role of calcium. Am J Physiol 248: C127–C135

Suttorp N, Hessz T, Seeger W, Wilke A, Koob R, Lutz F, Drenckhahn D (1988) Bacterial exotoxins and endothelial permeability for water and albumin. Am J Physiol 255: C369–C376

Suttorp N, Flöer B, Seeger W, Schnittler H, Bhakdi S (1990) Effects of E coli hemolysin on endothelial cell function. Infect Immun 58: 3796–3801

Suttorp N, Fuhrmann M, Tannert-Otto S, Grimminger F, Bhakdi S (1993) Pore-forming bacterial toxins potently induce release of nitric oxide in porcine endothelial cells. J Exp Med 178: 337–341

Vagts D, Dienes H-P, Barth PJ, Ronneberger H, Hungerer K-D, Bhakdi S (1993) In vivo effects of intravascularly applied Escherichia coli hemolysin: dissociation between induction of granulocytopenia and lethality in monkeys. Med Microbiol Immunol 182: 1–12

Valeva A, Weisser A, Walker B, Kehoe M, Bayley H, Bhakdi S, Palmer M (1996) Molecular architecture of a toxin pore: a 15-residue sequence lines the transmembrane channel of staphylococcal alpha-toxin. EMBO J 15 (in press)

Walev I, Martin E, Jonas D, Mohamadzadeh M, Müller-Klieser W, Kunz L, Bhakdi S (1993) Staphylococcal alpha-toxin kills human keratinocytes by permeabilizing the plasma membrane for monovalent ions. Infect Immun 61: 4972–4979

Walev I, Palmer M, Martin E, Jonas D, Weller U, Höhn-Bentz H, Husmann M, Bhakdi S (1994) Recovery of human fibroblasts from attack by the pore-forming alpha-toxin of Staphylococcus aureus. Microb Pathog 17: 187 201

Walev I, Reske K, Palmer M, Valeva A, Bhakdi S (1995) Potassium-inhibited processing of IL-1β in human monocytes. EMBO J 14: 1607–1614

Walker B, Bayley H (1994) A pro-forming protein with a protease-activated trigger. Prot Eng 7: 91–97

Walker B, Krishnasastry M, Bayley H (1993) Functional complementation of staphylococcal alpha-hemolysin fragments. Overlaps, nicks, and gaps in the glycine-rich loop. J Biol Chem 268: 5285–5292

Walmrath D, Griebner M, Kolb B, Grimminger F, Bhakdi S, Galanos C, Schade U, Seeger W (1993) Endotoxin primes rabbit lungs for enhanced pulmonary hypertension, but not vascular leakage, in response to staphylococcal α-toxin. Am Rev Respir Dis 148: 1179–1184

Walmrath D, Pilch J, Scharmann M, Grimminger F, Seeger W (1994) Severe ventilation-perfusion mismatch in perfused rabbit lungs evoked by sequential intravascular challenge with endotoxin and E. coli hemolysin. J Appl Physiol 74: 1972–1980

Welch RA (1991) Pore-forming cytolysins of gram-negative bacetria. Mol Microbiol 5: 521–529

Welch RA, Dellinger EP, Minshew B, Falkow S (1981) Haemolysin contributes to virulence of extra-intestinal E. coli infections. Nature 294: 665–667

The Cellular Response in Sepsis

D. PAJKRT and S.J.H. VAN DEVENTER

1 Introduction 119
2 Leukocytes 120
2.1 Neutrophilic Granulocytes 120
2.2 Mononuclear Cells 122
3 Blood Platelets 123
4 The Vascular Endothelium 124
4.1 Coagulation and Fibrinolysis 124
4.2 Neutrophil–Endothelial Cell Interactions 126
4.3 NO Production 129
5 Conclusion 130
References 131

1 Introduction

The introduction of new medical technologies has resulted in the invasive and more aggressive treatment of diseases that in the past would have remained untreated. As a consequence the incidence of invasive bacterial infections has increased. Sepsis, which can be characterized as an ill-regulated host inflammatory reaction during infection, is a major cause of mortality in critically ill patients. Sepsis is not a disease but encompasses a set of clinical and laboratory findings that can be observed in critically ill patients and are known to be associated with high mortality. Despite various attempts to define sepsis (also known as sepsis syndrome, severe sepsis and systemic inflammatory response syndrome) to date the diagnostic and therapeutic implications of such definitions remain unclear. Recently completed large intervention studies using monoclonal antibodies or recombinant proteins aimed at blocking the inflammatory response in sepsis have not resulted in increased survival, and in some cases such interventions may even be harmful. These data have added to the uncertainty in respect to the clinical relevance of definitions of sepsis and have raised doubt

Center for Hemostasis, Thrombosis, Atherosclerosis, and Inflammation Research, Academic Medical Centre, Meibergdreef 9, 1105 AZ Amsterdam, The Netherlands about the future of immunotherapy in sepsis. It is of considerable interest that, in contrast to the failure of clinical applications of immunotherapy, animal models have demonstrated impressive results of immunotherapy which clearly indicate "proof of principle" for various intervention strategies, such as interleukin-1 receptor antagonist (IL-1Ra), anti-tumor necrosis factor (TNF) antibodies, and platelet-activating factor (PAF) antagonists.

In contrast to the evolving knowledge of soluble mediators, surprisingly little is known about the cellular response in clinical sepsis. Such knowledge is essential because the outcome of sepsis eventually depends on the result of interactions of cells of the immune system and nonimmune cells. A clear example of such interaction is the binding of leukocytes to the inflamed endothelium, which is necessary for successful clearance of local bacterial infections but also may result in damage to the lungs. This chapter reviews data on abnormal cellular responses in sepsis. Although sepsis is regularly complicated by damage to the parenchyma of various organs, this review is restricted to the cells that are thought to be most intimately involved in systemic inflammatory responses, i.e., leukocytes, endothelial cells, and blood platelets.

2 Leukocytes

Exposure to bacterial components such as endotoxin results in the production of various endogenous inflammatory proteins, cytokines, and lipid mediators that initiate an inflammatory response. Several animal studies have identified proinflammatory cytokines as important components of the host defense response, in part by their ability to activate leukocytes, and by guiding recruitment of primed neutrophils to local sites of inflammation. In this respect monocyte- and macrophage- derived TNF and IL-1β, the chemokines, and endothelial cell bound PAF have been characterized as crucial players by providing either chemotactic stimuli and/or regulating the expression of adhesion molecules and their ligands.

2.1 Neutrophilic Granulocytes

Neutrophils are distinguished from other white blood cells by their ubiquitous availability and by their rapid recruitment. Neutrophils are the first phagocytic cells to arrive at sites of local infection. In sepsis and in experimental endotoxemia early neutropenia is usually followed by late neutrophilia. The neutropenia in sepsis results from sequestration of the neutrophils into several organs such as lungs, liver, and spleen (Grisham et al. 1988). It is generally assumed that following sequestration the activation of neutrophils by endotoxin or cytokines contributes to the local injury by causing the release of reactive oxygen species and proteinases. Neutrophil-induced endothelial cell damage is a major pathogenic mechanism in the pathogenesis of pulmonary damage (adult respiratory response syndrome, ARDS) in sepsis, and in ischemia-reperfusion injury. In addition to the classical proinflammatory cytokines, granulocyte-colony stimulating factor (G-CSF) and granulocyte-macrophage colony-stimulating factor (Gm-CSF) can prime neutrophils and may thus influence the resulting recruitment of the neutrophils (BALAZOVICH et al. 1991; DE HAAS et al. 1994).

Neutrophils are major effectors of microbial killing. Opsonized bacteria are recognized by neutrophil Fc receptors and by complement receptors and subsequently engulfed by enclosure within the neutrophil cell membrane. The thus formed phagosome fuses with the intracellular primary and secondary granules, and through a NADPH-oxidase dependent mechanism a reduction of oxygen to superoxide takes place. This superoxide is transformed into hydrogen peroxidase which eventually kills the micro-organism. Several other proteins such as myeloperoxidase and lysozyme which are readily available in the granules of neutrophil granulocyte also contribute to microbial and fungal killing. Neutrophils also have potent nonoxidative ways of killing bacteria, which include the endotoxin-binding proteins CAP-18 and bactericidal permeability protein (BPI). The latter protein is a 55-kDa molecule contained within the neutrophil azurophilic granules. In sepsis and experimental endotoxemia the circulating BPI concentration may increase threefold, nonetheless remaining too low to have significant endotoxin neutralizing effects (VON DER MÖHLEN et al. 1995b). Recombinant aminoterminal fragments of BPI ($rBPI_{23}$ and $rBPI_{21}$) have been shown potently to neutralize the biological effects of endotoxin in animal models of sepsis and in experimental endotoxemia in human volunteers (VON DER MÖHLEN et al. 1995a). In contrast to the relatively low circulating BPI levels, BPI can be present at considerable concentrations in local abscesses and within the inflamed bowel and would therefore be expected to exert its major effects at the local tissue level (MARRA et al. 1994; OPAL et al. 1994; Monajemi et al., in press).

Endotoxin is known to prime neutrophils and to stimulate the oxidative burst (FOREHAND et al. 1989). However, continuous stimulation of neutrophils causes functional alterations that include the loss of lysosomal granules and shedding of $Fc\gamma$ receptors from the cell membrane. Possibly as a consequence of continuous stimulation, under pathological circumstances such as burns and injury both the neutrophil oxidative burst and chemotaxis become impaired (ALEXANDER and WIXSON 1970), and this results in an increased susceptibility for secondary infections. Indeed, in patients with septic ARDS circulating neutrophils have been reported to have a diminished capacity for respiratory bursts and adherence to pulmonary microvascular endothelial cells (FEIN et al. 1991). Although inhibition of neutrophil/endothelial interactions would seem an interesting target for (immuno) therapy in sepsis, and several interventions indeed have shown efficacy in preventing ARDS in ischemia-reperfusion injury, such therapy may harm the infected patient. The crucial importance of neutrophils for effective clearance of bacteria has been demonstrated in several animal models of localized infection in which blockade of neutrophil recruitment by administration of anti-CD18 antibodies caused decreased endotoxin clearance and increased mortality (Nathanson, personal communication)

Few studies have investigated the potential harmful effects of anti-selectin antibodies in models of infection, but because of the apparent redundancy of the various selectins less interference with neutrophil recruitment and little effect on bacterial phagocytosis would be anticipated.

In conclusion, neutrophils are thought to be involved in the organ damage that may occur in critically ill patients. Interference with neutrophil recruitment by targeting either CD11b or CD18 may prevent pulmonary damage in ischemia-reperfusion injury. Such interventions may potently interfere with the clearance of loci of bacterial infection and thus seem contraindicated in most patients with sepsis.

2.2 Mononuclear Cells

The main targets of endotoxin are monocytes and macrophages and many clinical symptoms observed during sepsis are mediated by monocytes and macrophage-derived cytokines. Several proteins present in serum under normal circumstances mediate the interaction between endotoxin and monocytes/macrophages. Lipopolysaccharide (LPS)-binding protein (LBP) is an acute-phase protein synthesized by hepatocytes that binds the endotoxin lipid A moiety. LBP is important in deaggregating circulating endotoxin particles and presenting endotoxin to the GPI-anchored CD14 on neutrophils and monocytes. It should be noted that, although LPS binding to CD14 is undisputed, other endotoxin signal transduction pathways exist because monocytes from patients suffering from paroxysmal nocturnal hemoglobulinuria that lack CD14 on their outer membrane can be activated by LPS. On the other hand, the absence of LBP or the presence of excess soluble CD14 (sCD14) significantly reduces the ability of LPS to induce cytokine synthesis.

It should be noted that various other membrane-expressed molecules, including the scavenger receptor (or scavenger receptor related protein) and the CD11b/CD18 complex can also bind LPS (Wright et al. 1989). Binding of endotoxin to these molecules may be important for endotoxin clearance but has no important role in inducing cytokine synthesis. The control of LPS binding to membrane-bound CD14 is complex because even in healthy individuals relatively high levels of sCD14 are present in the serum (up to 3 μg/ml). An increase in sCD14 levels can be found in a variety of conditions such as severe burns, after trauma, rheumatoid arthritis, and psoriasis. Although the mechanism of elaboration seems to differ, in sepsis sCD14 levels have been reported to correlate with C-reactive protein. It should be noted that sCD14 is able to recognize LPS/LBP complexes and to present these to cells that lack the membrane-bound CD14 (Frey et al. 1992). On the other hand, an excess concentration of sCD14 is able to bind and neutralize endotoxin-LBP complexes and prevent activation of monocytes and macrophages (Schütt et al. 1992). Further complexity is added by the recent observation that LBP may also function to transport endotoxin molecules into high-density lipoprotein (HDL) particles (WURFEL et al. 1994). Hence the circulating concentration of HDL particles would be important for endotoxin removal, low levels driving the reaction toward CD14 binding. Indeed, addition of reconstituted human HDL to human blood decreases the amount of cytokines release after endotoxin stimulation (CASAS et al. 1995). Furthermore, treatment with rHDL has been shown to increase survival in animal models of endotoxemia (HUBSCH et al. 1993; CUÉ et al. 1994).

It has recently been recognized that in patients suffering from the sepsis syndrome a new subset of $CD14^{+}/CD16^{+}$ monocytes appears within the circulation. In healthy humans the concentrations of these cells are extremely low (9%). This amount can increase up to 50% in some septic patients, and a significant correlation has been shown between increased $CD14^{+}/CD16^{+}$ monocytes and the circulating IL-6 concentrations. In addition, these double-positive monocytes express decreased CD11b and CD33 antigen, suggesting a state of advanced differentiation. It is presently unclear whether these double-positive monocytes are responsible for the increased IL-6 production, or whether increased IL-6 levels are responsible for the changed phenotype of the monocytes in sepsis (FINGERLE et al. 1993). Future studies need to elucidate the question of whether monocytes present in the circulation of septic patients change in their adhesion molecule expression, and which monocyte subpopulation is responsible for the increased cytokine production in sepsis.

3 Blood Platelets

Disseminated intravascular coagulation (DIC) is a frequent complication in septic patients. Coagulation activation, platelet activation, and platelet aggregation may all contribute to the pathogenesis of DIC. In contrast to conventional teaching, it is now known that platelets can specifically adhere to intact endothelium when activated by either TNFα, IL-1, or endotoxin (GRAU and LOU 1993). Although thrombocytopenia in sepsis may be multifactorial, including decreased synthesis and immune complex mediated depletion, adhesion of platelets to intact endothelium may be one of the causes. Transient thrombocytopenia also may be observed in endotoxin-challenged volunteers, in which situation other causes are unlikely. The adherence of platelets to intact endothelium induces contact-dependent functional changes that are thought to be mediated by platelet membrane-expressed IL-1 (HAWRYLOWICZ et al. 1991), which is importantly upregulated by thrombin (BEVILACQUA et al. 1984). Of considerable interest is the finding that platelets may also express CD11a/CD18 and ICAM-1, which have been shown to be important for the binding to endothelial cells (PIGUET et al. 1993; MCCAFFERY and BERRIDGE 1986). Blocking the CD11a/CD18 complex with anti-LFA-1 monoclonal or anti-ICAM-1 antibodies

tested, It should further be noted that circulating monocytes from septic patients also express tissue factor and hence importantly contribute to coagulation activation.

One of the main inhibitors of thrombin formation is protein C. Protein C is activated by thrombomodulin after binding thrombin at the endothelial surface. The expression of thrombomodulin by endothelial cells is decreased by proinflammatory cytokines, both by a suppression of thrombomodulin mRNA transcription and by shedding of membrane-expressed thrombomodulin (Conway and Rosenberg 1988). In sepsis thrombomodulin shedding indeed occurs, as reflected by an increase of the circulating thrombomodulin concentration. The most dramatic example of DIC in sepsis is meningococcal septic shock, which is characterized by a severe reduction of protein C levels (Fijnvandraat et al. 1994), and it has recently been reported that the circulating thrombomodulin concentration increases in this disease. However, in a recent study the protein C concentration as expressed by antigen and activity was correlated very well, which suggests decreased protein C synthesis, or increased clearance, rather than a reduction in protein C activation by thrombomodulin, as the cause of the coagulopathy in meningococcal septic shock (Fijnvandraat et al. 1995).

Another important function of the vascular endothelium is modulation of fibrinolysis. In vitro, endothelial cells can be induced to release both tissue plasminogen activator (t-PA) and plasminogen activator inhibitor (PAI-1). In vivo, injection of endotoxin into healthy volunteers causes an initial increase in t-PA release, which is followed by a rise of PAI-1 plasma levels. These changes have been demonstrated to result in an initial increase and subsequent reduction in fibrin formation, as reflected by D-dimer plasma concentrations (Suffredini et al. 1989). It should be noted that the observed changes in fibrinolysis resulted from a single injection of a rather low dose of endotoxin, whereas septic patients are usually exposed to endotoxin, TNF, or IL-1 for a prolonged period. Moreover, in severe DIC in sepsis the depletion of various coagulation proteins may become rate limiting. Nevertheless, plasma PAI-1 concentrations in sepsis have been demonstrated to be correlated well with survival, suggesting a causal role for impaired fibrinolysis in sepsis.

In conclusion, sepsis importantly alters the normal anticoagulant function of the vascular endothelium. In fact, some endothelial surfaces may become strongly procoagulant, causing local fibrin deposition and therefore interfering with tissue perfusion.

4.2 Neutrophil–Endothelial Cell Interactions

The recruitment of inflammatory cells to local tissue sites is necessary for a normal host defense and results from interactions of surface molecules expressed by leukocytes and the vascular endothelium. This process is importantly modulated by several small cytokines (chemokines) and PAF. In sepsis, neutrophils adhering to the endothelium or transmigrating into tissues

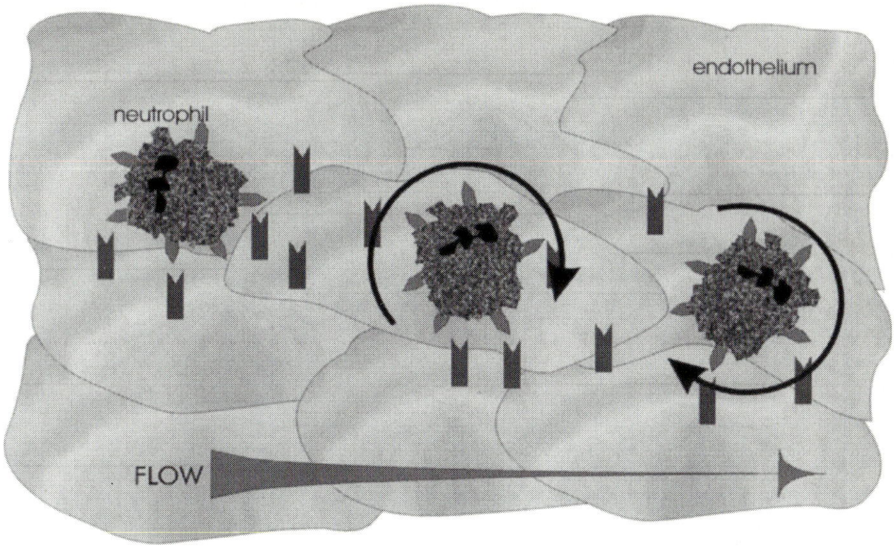

Fig. 2. Neutrophil rolling. In normal conditions few neutrophils are in contact with the vascular endothelium. Inflamed endothelial cells express adhesion molecules of the selectin family (E-selectin, P-selectin) which induces rolling of a substantial fraction of the circulating neutrophil pool

are considered to be involved in organ failure through the production of reactive oxygen species and proteinases. The endothelial adhesion molecules consist of the selectins (E-selectin, formerly known as ELAM-1; P-selectin, formerly known as GMP-140) and molecules that belong to the immunoglobulin superfamily (ICAM-1, ICAM-2, VCAM-1). Elegant in vivo studies using novel imaging techniques have demonstrated that within the microvasculature a minority of the circulating neutrophils is in transient contact with the endothelium. This process, known as "rolling" is a result of the opposed forces of shear stress and endothelial "stickiness," which in large part depends on the expression of the selectin-group of adhesion molecules. The physiological role of unstimulated neutrophil "rolling" remains to be defined, but it may serve to facilitate the normal circulation of phagocytic cells through noninflamed tissues. In inflammatory states the rolling of neutrophils is known to increase dramatically, which depends in part on increased expression of P-selectin, and it has recently been demonstrated that neutrophils may roll not only over endothelial cells but also over already adhering neutrophil monolayers (Fig. 2). At local inflammatory sites rolling neutrophils come to a halt, forming tight adhesions with endothelial cells. This tight endothelial-neutrophil interaction is a consequence of an interaction of the neutrophil β_2-integrins, in particular CD11b/CD18, with ICAM-1 (Fig. 3).

This CD11b/CD18 complex is expressed constitutively on the surface of leukocytes but is normally in an "inactive" state. Transformation to the active state occurs following stimulation by chemotactic factors such as endothelial-

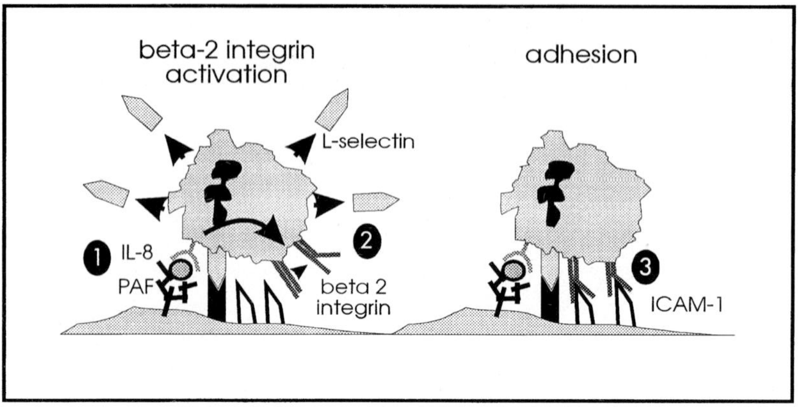

Fig. 3. Neutrophil adhesion. Rolling of neutrophils is not sufficient for thight adhesion and hence does not lead to transmigration. Close contact of neutrophils with the vascular endothelium causes activation of β_2-integrins (*2*), which is in part mediated by IL-8 and PAF (*1*), that bind to ICAM-1 on endothelial cells (*3*). In this process neutrophils shed L-selectin

bound IL-8, G-CSF, and PAF. The endothelial-neutrophil interaction is further facilitated by increased expression of ICAM-1 that results from stimulation by endotoxin or cytokines. The important role of β_2 integrins for the adherence of neutrophils to the endothelium cells is underscored by the sometimes dramatic clinical consequences of the leukocyte adhesion deficiency syndrome, a disease caused by absent expression of all three members of the leukocyte integrins. Several investigators have shown that blocking the integrins with monoclonal antibodies can reduce aggregation and adherence to endothelial cells (Arnaout 1990). Neutrophil CD11b/CD18 expression becomes highly elevated in low-dose endotoxemia in volunteers (unpublished), and it is now understood that this increased expression is induced at least partly by TNFα, for the CD11/CD18 upregulation can be reduced with anti-TNFα monoclonal antibodies (Windsor et al. 1993). In contrast to the considerable knowledge of neutrophil-endothelial cell interactions in vivo, the data on neutrophil-endothelial cell interactions in sepsis are scant. Increased sequestration of neutrophils into the lungs of patients with septic ARDS has been demonstrated using lung scans with ^{111}In-labeled neutrophils (Warshawski et al. 1986), but more detailed descriptions of adhesion receptor activation at the alveolar endothelium are still missing.

Clearly, neutrophils do not transmigrate as long as they remain bound to the luminal site of endothelial cells, and this implies that binding must be reversible. The exact molecular mechanisms involved in transmigration of neutrophils through endothelial cells remain to be characterized, but in vivo studies using HUVEC have demonstrated a role for PAF and IL-8. In this respect the observations that endothelial cells express a membrane-bound PAF, or PAF-like activity, and that IL-8 is associated with endothelial cells by heparan sulfate, seem to be of importance. Moreover, endothelial cell production of IL-8 is polarized, resulting in a preferential basal secretion and thus a concentration gradient that promotes neutrophil transmigration.

In conclusion, neutrophil-endothelial cell interactions have been fairly well characterized at the molecular level. Obviously it is difficult to study these interactions in septic patients, and indeed very few studies have addressed this problem.

4.3 NO Production

One of the main determinants of vascular tone in both health and disease is nitric oxide (NO). Endothelial cells maintain vascular tone by constitutive production of NO as a result of synthesis from arginine resulting from the activity of a constitutive NO synthetase (NOS; Fig. 4.). Endothelial cells are capable of synthesis of arginine from citrulline by means of an incomplete urea cycle and citrulline transport into endothelial cells is negatively influenced by glutamine. Hence, although infusion of large amounts of arginine may cause a drop in blood pressure, the changes in circulating amino acids may affect NO synthesis in a more complex manner. Moreover, various cytokines may induce other types of NOS (collectively known as iNOS), thus enhancing NO production (MONCADA et al. 1991). No doubt increased NO production in sepsis occurs, and inhibition of NO

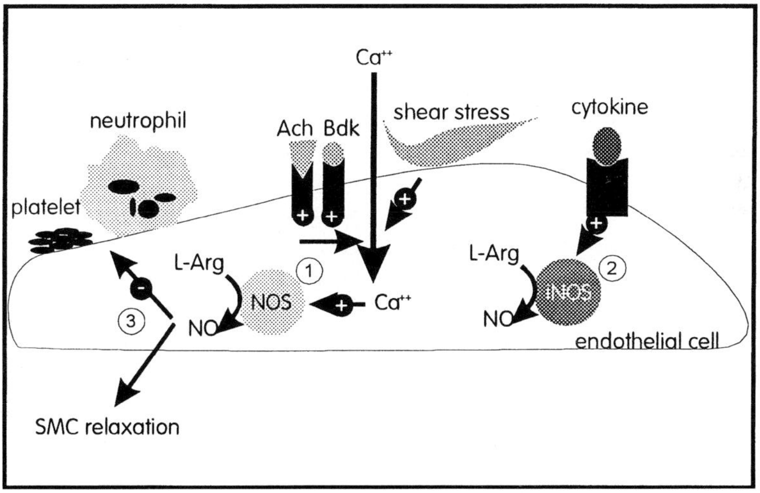

Fig. 4. NO production by endothelial cells in sepsis. Constitutive NO production from L-arginine by endothelial cells is dependent on the function of NOS (*1*). The activity of NOS is Ca-dependent and regulated by shear stress, acetylcholine, and bradykinin. Cytokines can induce another NOS, commonly known as iNOS (*2*). NO is an important regulator of vascular tone, and increased NO production is thought to be a major pathogenic factor for the development of irreversible hypotension in sepsis. However, complete inhibition of NO may not be beneficial in sepsis because NO inhibits both CD18-dependent neutrophil adhesion and adhesion of blood platelets to endothelial cells (*3*)

synthesis by the administration of NOS blockers may reverse hypotension, albeit at the expense of decreased tissue perfusion. It is beyond the scope of this chapter to extensively discuss the role of NO in sepsis, but we would like to stress that increased endothelial NO production may have a protective role. For example, inhibition of NO production has been shown to facilitate adhesion of platelets to endothelial cells in vitro (Katusic 1994), and the abnormally high rate of adhesion of neutrophils to canine coronary arteries following ischemia-reperfusion injury was reversed by the administration of an NO-donating compound. In fact, severe myocardial ischemia may occur in endoxemic animals after blockade of NO production, and this may exaggerate the late-phase hypotension that has been observed in such models.

5 Conclusion

Mononuclear cells, neutrophils, and the vascular endothelium all become activated in sepsis. Although it is popular to think of neutrophils and monocytes as the major effector cells in sepsis, the vascular endothelium being a prime target, organ damage in sepsis should be considered the result of complex cellular interactions. In this respect it is clear that the endothelium may become an important proinflammatory surface by enabling neutrophil adhesion and by facilitating intravascular coagulation activation. Although the induction of various cytokines in sepsis is now relatively well characterized, there is an important lack of knowledge of the cell activation processes in critically ill patients. For example, it remains unclear what cells contribute to cytokine production in sepsis, and no clinically useful methods exist that would enable the characterization of neutrophil-endothelial interactions in septic patients. Hence most knowledge of cell activation processes in sepsis is derived from data on soluble adhesion molecules and circulating cells, which may differ importantly from the resident or adhering populations. Thus far, coagulation activation and impaired fibrinolysis are among the best characterized consequences of activation of monocytes and the vascular endothelium, and inhibition of tissue factor and administration of protein C seem to be promising therapeutic strategies. Inhibition of neutrophil–endothelial cell adhesion events is now feasible, but the results of such strategies in experimental sepsis cast considerable doubt about their efficacy in reducing mortality. Finally, NO has been characterized as an important endothelial cell derived gas that is involved in the pathogenesis of irreversible septic shock. Although inhibition of NO may be an attractive strategy in septic shock, such an intervention may have adverse effects by increasing endothelial-neutrophil interactions and by promoting neutrophil adhesion.

References

Alexander JW, Wixson D (1970) Neutrophil dysfunction and sepsis in burn injury. Surg Gynecol Obstet 130: 431–438

Arnaout MA (1990) Structure and function of the leukocyte adhesion molecules CD11/CD18. Blood 75: 1037–1050

Balazovich KJ, Almeida HI, Boxer LA (1991) Recombinant G-CSF and GM-CSF prime human neutrophils for superoxide production through different signal transduction mechanisms. J Lab Clin Med 18: 576–584

Bauer KA, ten Cate H, Barzegar S, Spriggs DR, Sherman ML, Rosenberg RD (1989) Tumor necrosis factor infusions have a procoagulant effect on the hemostatic mechanisms of humans. Blood 74: 165–172

Bevilacqua MP, Pober JS, Majeau GR, Cotran RS, Gimbrone MA Jr (1984) Interleukin-1 (IL-1) induces biosynthesis and cell expression of procoagulant activity in human vascular endothelial cells. J Exp Med 160: 618–623

Brox JH, Osterud B, Bjorklid E, Fenton JW II (1984) Produktion and availability of thromboplastin in endothelial cells: the effects of thrombin, endotoxin and platelets. Br J Haematol 57: 239–246

Casas A, Hubsch AP, Rogers BC, Doran JE (1995) Reconstituted high-density lipoprotein reduces LPS stimulated TNFα. J Surg Res (in press)

Conway EM, Rosenberg RD (1988) Tumor necrosis factor suppresses transcription of the thrombomodulin gene in endothelial cells. Mol Cell Biol 8: 5588–5592

Cué JI, DiPiro JT, Brunner LJ, Doran JE, Blankenship ME, Mansberger AR, Hawkins ML (1994) Reconstituted high density lipoprotein inhibits physiologic and tumor necrosis factor-α response to lipopolysaccharide in rabbits. Arch Surg 129: 193–197

De Haas M, Kerst JM, van der Schoot CE, Hack CE, Calafat J, Nuijen JH, Roos D, van Oers RHJ, von dem Borne (1994) rG-CSF administration to healthy volunteers: immediate effects on circulating neutrophils. Blood 84: 3885–3894

Fein AM, Grant MM, Niederman MS, Kantrowitz N (1991) Neutrophil-endothelial cell interaction in critical illness. Chest 99: 1456–1462

Fijnvandraat K, Peters M, Derkx HHF, van Deventer SJH, ten Cate JW (1994) Endotoxin-induced coagulation activation and protein C reduction in meningococcal septic shock. Prog Clin Biol Res 388: 247–254

Fijnvandraat K, Derkx HHF, Peters M, Bijlmer R, Sturk A, van Deventer SJH, ten Cate JW (1995) Coagulation activation and tissue necrosis in meningococcal septic shock: severely reduced protein C predicts a high mortality. Haemost Thromb 73: 15–20

Fingerle G, Pforte A, Passlick B, Blumenstein M, Ströbel M, Ziegler-Heitbrock HWL (1993) The novel subset of CD14+/CD16+ blood monocytes is expanded in sepsis patients. Blood 82: 3170–3176

Forehand JR, Pabst MJ, Phillips WA, Johnston RB Jr (1989) Lipopolysaccharide priming of human neutrophils for an enhanced respiratory burst. Role of intracellular calcium. J Clin Invest 83: 74–83

Frey EA, Miller DS, Gullstein Jahr T, Sundan A, Bazil V, Espevik T, Finlay BB, Wright SD (1992) Soluble CD14 participates in the response of cells to lipopolysaccharide. J Exp Med 176: 1665–1671

Grau GE, Lou L (1993) TNF in vascular pathology: the importance of platelet-endothelial interactions. Res Immunol 144: 355–363

Grisham MB, Everse J, Janssen HF (1988) Endotoxemia and neutrophil activation in vivo. Am J Physiol 254: H1017–H1022

Hawrylowicz CM, Howells GL, Feldmann M (1991) Platelet-derived interleukin-1 induces human endothelial adhesion molecule expression and cytokine produktion. J Exp Med 174: 785–790

Hubsch AP, Powell F, Lerch PG, Doran JE (1993) A reconstituted, apolipoprotein A-1 containing lipoprotein reduces tumor necrosis factor release and attenuates shock in endotoxemic rabbits. Circ Shock 40: 14–23

Kaplanski G, Porat R, Aiura K, Erban JK, Gelfand JA, Dinarello CA (1993) Activated platelets induce endothelial secretion of interleukin-8 in vitro via an interleukin-1-mediated event. Blood 81: 2492–2495

Katusic ZS (1994) Nitric oxide: regulator of P-selectin expression. Gastroenterology 107: 1199–1201

Marra MN, Scott RW, McKelligon BM, Opal SM (1994) Bactericidal/permeability-increasing protein (BPI) levels exceed lipopolysaccharide binding protein (LBP) in fluids in infected sites (Abstr). Clin Res 42: 152A

McCaffery PJ, Berridge MV (1986) Expression of leukocyte functional molecule (LFA-1) on mouse platelets. Blood 67: 1757–1764

Moncada S, Palmer RM, Higgs EA (1991) Nitric oxide: physiology, pathophysiology and pharmacology. Pharmacol Rev 43: 109–142

Opal SM, Palardy JE, Marra MN, Fisher CJ Jr, McMelligon BM, Scott RW (1994) Relative concentration of endotoxin-binding proteins in body fluids during infection. Lancet 344: 429–431

Piguet PF, Vesin C, Ryser JE, Senaldi G, Grau GE, Tacchini-Cottier F (1993) An effector role of platelets in the systemic and local lipopolysaccharide-induced toxicity in mice, mediated by a CD11a- and CD54-dependent interaction with the endothelium. Infect Immun 61: 4182–4187

Prydz H, Pettersen KS (1988) Synthesis of thromboplastin (tissue factor) by endothelial cells. Haemostasis 18: 215–223

Schütt C, Schilling T, Grunwald U, Schönfeld W, Krüger C (1992) Endotoxin-neutralizing capacity of soluble CD14. Res Immunol 143: 71–78

Suffredini AF, Harpel PC, Parillo JE (1989) Promotion and subsequent inhibition of plasminogen activator after administration of intravenous endotoxin to normal subjects. N Engl J Med 320: 1165–1172

Van der Poll T, Büller HR, ten Cate H et al. (1990) Activation of coagulation after administration of tumor necrosis factor to normal subjects. N Engl J Med 322: 1622–1627

Van der Poll T, Levi M, Hack CE, ten Cate H, van Deventer SJH, Eerenberg AJM, de Groot ER, Jansen J, Gallati H, Büller HR, ten Cate JW, Aarden LA (1994a) Elimination of interleukin 6 attenuates coagulation activation in experimental endotoxemia in chimpanzees. J Exp Med 179: 1253–1259

Van der Poll T, Levi M, van Deventer SJH, ten Cate H, Haagmans BL, Biemond BJ, Büller HR, Hack CE, ten Cate JW (1994b) Differential effects of anti-tumor necrosis factor monoclonal antibodies on systemic inflammatory responses in experimental endotoxemia in chimpanzees. Blood 83: 446–451

Von der Möhlen MAM, Kimmings AN, Wedel NI, Mevissen MLCM, Jansen J, Friedmann N, Lorenz TJ, Nelson BJ, White ML, Bauer R, Hack CE, Eerenberg A, van Deventer SJH (1995a) Inhibition of endotoxin-induced cytokine release and neutrophil activation in humans using recombinant bactericidal/permeability-increasing protein ($rBPI_{23}$). J Infect Dis 172: 144–151

Von der Möhlen MAM, Wortel CH, Obradov D, Tobias PS, Hack CE, ten Cate JW, van Deventer SJH (1995b) Endotoxin-binding proteins in patients with severe gram-negative sepsis. In: Faist (ed) Inflammation and tissue injury: basic mechanisms and pathophysiology. Springer, Berlin Heidelberg New York (Immune consequences of trauma, shock and sepsis, mechanisms and therapeutic approaches, vol 1) (in press)

Warshawski FJ, Sibbald WJ, Driedger AA, Cheung H (1986) Abnormal neutrophil-pulmonary interaction in the adult respiratory distress syndrome: qualitative and quantitative assessment of ulmonary kinetics with "in vivo" indium neutrophil scintigraphy. Am Rev Respir Dis 133: 797–804

Windsor ACJ, Walsh CJ, Mullen PG, Cook DJ, Fisher BJ, Blocher CR, Leeper-Woodford SK, Sugerman HJ, Fowler AA III (1993) Tumor necrosis factor-α blockade prevents neutrophil CD18 receptor upregulation and attenuates acute lung injury in porcine sepsis without inhibition of neutrophil oxygen radical generation. J Clin Invest 91: 1459–1468

Wright SD, Levin SM, Jong MTC, Chad Z, Kabbash LG (1989) CR3 (CD 11b/CD18) expresses one binding site for Arg-Gly-Asp-containing peptides and a second site for bacterial lipopolysaccharide. J Exp Med 169: 175–183

Wurfel MM, Kunitake ST, Lichtenstein H, Kane JP, Wright SD (1994) Lippopolysaccharide (LPS)-binding protein as a cofactor in the neutralization of LPS. J Exp Med 180: 1025–1035

Yamazaki H, Fujimoto T, Suzuki H, Akamatsu N, Katagiri Y, Yamaguchi A, Tanoue K (1992) Interaction of platelets and bloodvessel-vascular injury induced by platelet activation in vivo. Jpn Circ J 56: 178–186

Cytokines as Mediators in the Pathogenesis of Septic Shock

C.A. Dinarello

1 Introduction 134

2 Biological Effects of IL-1 and TNF Relevant to Septic Shock 136
2.1 Local and Systemic Effects 136
2.2 Synergistic Actions of IL-1 and TNF 136
2.3 Expression of Various Genes in Cells Exposed to IL-1 and TNF 138
2.4 Effects Mediated by Prostanoids 138
2.5 Effects Mediated by Nitric Oxide 140
2.6 Effect on Receptor Expression 141
2.7 IL-1 and TNF Infusion Mimics Septic Shock 141
2.8 Effects on the Vascular Wall 142
2.9 Comparison with IL-6 142

3 Experimental Sepsis Affected by Reducing Production of IL-1 and TNF 144
3.1 Corticosteroids 144
3.2 Inhibition of Lipoxygenase 144
3.3 Inhibition by Dietary ω-3 Fatty Acids 144
3.4 Effect of Cyclo-oxygenase Inhibitors on the Production of IL-1 and TNF 145
3.5 Inhibition of IL-1 and TNF Processing Enzymes 146
3.6 Cytokine-Suppressing Anti-inflammatory Drugs 148

4 The Anti-inflammatory Cytokine Network in Sepsis 148
4.1 IL-4, IL-10, IL-13, and TGFβ 149
4.2 IL-6 and Other Ligands for the gp 130 Signal Transducer 150
4.3 Interferons 151

5 Naturally Occurring Inhibitors of IL-1 152
5.1 IL-1 Receptor Antagonist 152
5.2 IL-1Ra Production in Sepsis 152
5.3 IL-1Ra in Experimental Endotoxemia in Humans 153
5.4 Clinical Trials of IL-1Ra in Septic Shock 154
5.5 Soluble IL-1 Receptors 155
5.6 Soluble IL-1 Receptor Type I in Human Studies 155

6 Reducing the Activity of TNF 156
6.1 Soluble TNF Receptors 156
6.2 Neutralizing Antibodies to TNF 157

References 158

Department of Medicine, Division of Infectious Diseases, University of Colorado, Health Science Center, 4200 East Ninth Avenue, Denver, CO 80262, USA

1 Introduction

During septic shock the host produces several proinflammatory cytokines which have been implicated as playing a critical role in the pathogenesis of the disease. The production of these cytokines is initiated by the organisms themselves (phagocytosis) or by soluble products of the organisms. For example, the lipopolysaccharide endotoxins (LPS) of gram-negative bacteria, the protein exotoxins of gram-positive bacteria, and the cell-wall glycopeptides such as teichoic acids and muramyl peptides. Of course, LPS is by far the most potent soluble product of bacteria which induces cytokine production, and therefore most information about cytokine induction is derived from studies using LPS in vitro and in vivo. However, it is important to recognize that the cytokine production in septic shock is neither specific nor unique. The cytokines which contribute to pathological changes in septic shock are not unique to infection. Multiple trauma, ischemia-reperfusion injury, acute transplant rejection, antigen-specific immune responses, and various acute inflammatory states (acute hepatitis and pancreatitis) initiate the same cytokine cascade and result in both systemic and local inflammatory processes.

However, special consideration exists for septic shock since no other disease is associated with such high mortality, despite our ability to provide patients with septic shock with appropriate antibiotics and supportive therapy. Gene deletion, neutralizing antibody studies, and specific receptor blockade of cytokines have revealed their pivotal role in the pathogenesis of spetic shock, at least in animal studies. This chapter reviews the cytokines most likely to participate in the pathological events of septic shock and discusses the mechanisms by which these cytokines contribute to the shock syndrome. In addition, we consider the evidence that nature did not leave us without counterregulatory mechanisms to oppose the effects of proinflammatory cytokines. Anti-inflammatory cytokines and specific naturally occurring inhibitors of cytokines are produced in health as well as in disease. These inhibitors likely reduce the severity of inflammation induced by proinflammatory cytokines. Therefore an emerging concept is that in a given individual, the net biological response of pro-and anti-inflammatory cytokines (sometimes called "balance" in this chapter) affects the outcome of a particular disease process. The systemic inflammation of septic shock is no exception to this concept. The preexisiting status of the patient may affect the nature of the pro-and anti-inflammatory cytokine response. Genetic factors also play an important role in this balance.

In infectious diseases the micro-organism induces a cytokine profile which is distinct from that induced during a response to foreign-tissue antigens. For example, during bacterial infections there is little or no production of interferon (IFN)-γ or interleukin (IL)-2 whereas these cytokines are prominent components of the cytokine profile during transplant rejection and immunologically mediated diseases. In both situations and tumor necrosis factor (TNF) are produced and function primarily as proinflammatory cytokines.

Biologically IL-1 and TNF are closely related, although the structure and receptors for IL-1 and TNF are clearly distinct. IL-1 and TNF are active in the low pico- and femtomolar range. Based on short-term blockade of IL-1 and TNF receptors in humans and animals and recent data on IL-1β and TNFα deficient mice, there is no evidence that these cytokines play a critical role in development or normal homeostasis such as metabolism, hematopoiesis, renal and hepatic function, or regulation of blood pressure. During inflammation, injury, immunological challenge, or infection IL-1 and TNF are produced. One concludes from those studies that biological properties of IL-1 and TNF mimic host responses to infection, inflammation, injury, or immunological challenge. In animal models of systemic inflammation (such as in septic shock), specific blockade of *either* IL-1 or TNF results is a reduction in the severity of the inflammation. Moreover, IL-1 and TNF act synergistically in nearly every in vitro and in vivo model of local or systemic inflammation. When *both* cytokines are specifically blocked, the severity of inflammation is reduced further.

Understanding how IL-1 and TNF manifest so many different biological properties can be focused on relatively few mechanistic pathways, mostly those involving changes in constitutive and inducible gene expression or numbers of surface receptors for biologically active molecules. For example, inducible phospholipase A_2(PLA_2) and cyclo-oxygenase, genes controlling increased synthesis of inflammatory leukotrienes (LTs) and prostaglandins (PGs), are highly relevant to understanding the multiple effects of IL-1 and TNF. Another gene is inducible nitric oxide (NO) synthase. Therefore many of the pleiotropic effects of IL-1 and TNF are ameliorated by cyclo-oxygenase or nitric acid synthase inhibitors. Some biological effects of IL-1 and TNF are not mediated by intermediate production of eicosanoids or NO but rather by increased gene expression for other cytokines or regulation of celluar receptors for cytokines. A very prominent cytokine response to IL-1 and TNF is the production of the chemokine family of inflammatory cytokines (Baggiolini et al. 1994). These are highly active molecules which increase the ability of lymphocytes, monocytes, and neutrophils to attach to endothelial cells and migrate into the extravascular space. In septic shock the neutrophil plays an important role in tissue damage, which can be reduced by blocking receptors for IL-8.

2 Biological Effects of IL-1 and TNF Relevant to Septic Shock

2.1 Local and Systemic Effects

A distinction is made between the local effects of IL-1 and TNF and the consequences of their systemic levels. If the function of host defense is the elimination of the invading organism or destruction of foreign tissue, inflammation is the price which is paid for an effective defense. Therefore in systemic inflammation large amounts of IL-1 and TNF are released into the circulation, inducing hypotension and shock which can be lethal in experimental animals. Humans are particularly sensitive to the pyrogenic and hyportensive properties of IL-1 and TNF; a single intravenous injection of 10 ng/kg IL-1 or TNF induces fever (39 °C), whereas hypotension is consistently observed at doses of 100 ng/kg; 300 ng/kg is the maximal dose tolerated because of the severe fall in blood pressure (CHAPMAN et al. 1987; CROWN et al. 1991; SMITH et al. 1992; VAN DER POLL et al. 1990, 1991).

More common than the systemic inflammatory response to IL-1 and TNF are the local effects which result in release of lipid-derived mediators such as platelet-activating factor (PAF) and eiscosanoids. At sites of local injection or production of IL-1 and TNF neutrophil, monocyte, and lymphocyte emigration takes place. This emigration, which is the hallmark of local inflammation, is due to two events: (a) the induction of chemokines and (b) the upregulation of cell adhesion molecules. The chemokine family is best characterized by IL-8. For the purposes of this chapter, the entire family of chemokine peptides are represented by IL-8. IL-1 and TNF are potent inducers of IL-8 synthesis from monocytes, fibroblasts, and endothelial cells. Concentrations of 1 pg/ml IL-1 or TNF induce IL-8 production in fibroblast cultures. In monocytes stimulated with bacterial endotoxin 50% of the IL-8 produced is via an intermediate action of IL-1 (PORAT et al. 1992), and IL-8 produced from endothelial cells stimulated with activated platelets is an IL-1-dependent pathway (KAPLANSKI et al. 1993).

2.2 Synergistic Actions of IL-1 and TNF

The synergistic activities of IL-1 and TNF are as follows:

IL-1 plus TNF
- Hemodynamic shock and lactic acidosis in rabbits
- Radioprotection
- Generation of Shwartzman reaction
- Luteal cell $PGF_{2\alpha}$ synthesis
- PGE_2 synthesis in fibroblasts

Galactosamine-induced hepatotoxicity
Sickness behavior in mice
Circulating nitric oxide and hypoglycemia in malaria
Nerve growth factor synthesis from fibroblasts
Insulin release and islet β-cell death
Insulin resistance
Loss of lean body mass
IL-8 synthesis by mesothelial cells

IL-1 plus bradykinin
Angiogenesis
PGE_2 synthesis in gingival fibroblasts
Arachidonic acid release from synoviocytes
$PGF_{2\alpha}$ synthesis in uterine decidua
IL-6 production from hepatoma cells and fibroblasts

IL-1 plus IL-6
Antigen-induced T-cell IL-2 production
Hepatic synthesis of acid glycoprotein and C3
Hepatic synthesis of LPS binding protein
Endothelial cell synthesis of Mn superoxide dismutase

IL-1 plus fibroblast, platelet-derived, epithelial, or α-transforming growth factor
PGE_2 synthesis in dermal fibroblasts
PGE_2 synthesis in synovial cells
Chemotaxis for fibroblasts
Phospholipase A_2 release from synoviocytes
Degradation of articular cartilage
PGE_2 synthesis in osteoblastic cells

The synergism between IL-1 and TNF is highly consistent and a frequently reported phenomenon. In addition, the synergism between IL-1 and TNF is also observed in vivo whereas the synergism between IL-1 and IL-6, IL-1 and bradykinin, or IL-1 and the various growth factors is mostly on prostanoid synthesis and primarily an in vitro finding. The mechanism for IL-1 synergism in the synthesis for PGE_2 likely involves the ability of one cytokine to release arachidonate and of the ability of IL to stimulate cyclo-oxygenase type 2(COX–2) synthesis. The mechanism for synergism may also involve receptor modulation; however, in the case of IL-1 and TNF synergism receptors for TNF are downregulated by IL-1 (Brakebusch et al. 1994; Holtmann and Wallach 1987). Could the synergism be explained at the level of signal transduction? Although this is an attractive hypothesis, no pathway of IL-1 or TNF signal transduction appears unique to either cytokine at the present time to account for synergism. In fact, since signal mechanisms appear similar, additive rather than synergistic effects should be observed.

The signal transduction of IL-1 appears to be similar to that of TNF. Although the signaling pathway of the IL-1 postreceptor binding remains unclear, several mechanisms have been proposed. These include activation of a GTP-binding protein with no associated increase in adeny1cyclase (O' NEILL et al. 1990), activation of adeny1cyclase (MIZEL 1990; MUNOZ et al. 1990), hydrolysis of three phospholipids by nonphosphatidylinositol phospholipase C (KESTER et al. 1989; ROSOFF et al. 1988), release of ceramide from sphingomyelin following activation of sphingomyelinase (MATHIAS et al. 1993), and release of arachidonic acid from phopholipids via cytosolic PLA_2 and activation of PLAP (CLARK et al. 1991; GRONICH et al. 1994). In addition, some of the kinases that are activated by IL-1 are also activated in cells stimulated with TNF (FRESHNEY et al. 1994; GUESDON et al. 1993, 1994; KRACHT et al. 1994).

2.3 Expression of Various Genes in Cells Exposed to IL-1 and TNF

A fundamental property of IL-1 and TNF in the pathogenesis of septic shock is the ability to induce a variety of genes which affect the vasculature and the local tissue environment. In most cases IL 1 and TNF induce new transcripts in cells which express these genes only during disease. There are several examples, but the most dramatic appear to be other members of the cytokine family and inducible enzymes regulating low molecular weight mediators. Mediators such as PGs, LTs, and NO require cellular enzymes to covert precursors to active molecules. IL-1 and TNF are potent inducers of these enzymes. For PGE2 and LT synthesis IL-1 and TNF stimulate inducible PLA2, which contributes to the synthesis of PAF from lipid precursors. There is also COX-2 gene expression. The genes coding for IL-8 and other members of the chemokine family are not expressed in most cells in health, but picomolar concentrations of IL-1 and TNF trigger gene expression and protein expression in monocytes, fibroblasts, and endothelial cells. IL-1 and TNF induce the expression of their own genes. However, it is important to note that IL-1 and TNF suppress the transcription of the genes coding for albumin and the cytochrome P450 family.

2.4 Effects Mediated by Prostanoids

Many changes induced by IL-1 and TNF are mediated by PGs, particularly PGE2 (Tables 1 and 2). In fact, the use of cyclo-oxygenase inhibitors for a variety of inflammatory conditions is often a therapeutic strategy to reduce IL-1- and TNF-induced PGE2. Humans injected with IL-1 or TNF experience fever, headache, myalgias, and arthralgias, each of which is reduced by coadministration of cyclo-oxygenase inhibitors. One of the more universal activities of IL-1 and TNF is the induction of gene expression for type2 PLA2 and COX-2. IL-1 and TNF induce transcription of COX-2, and neither cytokine increases production of COX-1. Moreover, once triggered, COX-2 production is elevated for several hours, and large amounts of PGE2 are produced in cells stimulated with IL-1 or TNF. It comes as no surprise that inflammation is reduced by administration of cyclo-

Table 1. Cyclo-oxygenase *dependent* activities of IL-1 and TNF

In vivo
Fever
Natriuresis
ACTH and growth hormone release
Suppression of T-cell mitogenesis
Decreased pain threshold
Suppression of appetite and weight loss
Hypotension and shocklike syndrome
Increased melanoma bone marrow metastases
Decreased water intake
In vitro
Suppression of cell proliferation
Increased corticotropin-releasing factor
Formation of osteoclasts
Collagenase and stromelysin
Increased relaxation of arterial vessels
Inhibition of bone mineralization
Induction of leukocyte inhibition factor in fibroblasts

Table 2. Cyclo-oxygenase *independent* activities of IL-1 and TNF

In vivo
Slow-wave sleep
Non-rapid eye movement sleep
Neutrophilia
Lethality in adrenalectomized mice
Decreased mast cell degranulation
Production of colony-stimulating factor
Hypoferremia; hypozincemia
Hepatic acute-phase protein synthesis
Hypoglycemia
IL-6 production
Migration of neutrophils into tissues
In vitro
Gene expression and synthesis of cytokines
Destruction of islect β-cells
Augmentation of T-cell responses
Corticosterone synthesis (direct)
Collagenase/stromelysin gene expression
Induction of NO synthase
Hematopoietic stem cell activation
Enhancement of NO synthase (IL-1-induced PGE2 may augment gene expression and/or synthesis of some IL-1-induced genes, for example other cytokines and NOS via increase cAMP)
Suppression of gastric secretion
Increased progesterone synthesis
Increased nerve growth factor in astrocytes (augmented by addition of cyclo-oxygenase inhibitors)
Increased production of granulocyte-macrophrage colony-stimulating factor, (augmented by addition of cyclo-oxygenase inhibitors)

oxgenase inhibitors, and that many biological activities of IL-1 and TNF are actually due to increased PGE2 production.

There appears to be selectivity in cyclo-oxygenase inhibitors in that some nonsteroidal anti-inflammatory agents are better inhibitors of COX-2 rather than COX-1. Similar to COX-2 induction, IL-1 and TNF preferentially stimulate new transcripts for the inducible form (type 2) of PLA2 which cleaves the fatty acid in the number 2 position of cell membrane phospholipids. In most cases this is arachidonic acid. The release of arachidonic acid is the rate-limiting step in the synthesis of PGs and LTs.

2.5 Effects Mediated by Nitric Oxide

The generation of NO in inflammatory disease appears to be a fundamental event (reviewed in MONCADA et al. 1991). Several studies have demonstrated that IL-1 and TNF initiate transcription and translation of the inducible form of nitric oxide synthase (iNOS). This has been observed in a variety of cells, for example, osteoclaste, murine macrophages, pituitary cells, mast cells, osteoblasts, glial cells, insulin-producing β-cells in the pancreas, smooth muscle cells, chondrocytes, myocytes, and mesangial cells. In mesangial cells IL-1β induced NOS is augmented by elevated levels of cAMP (KUNZ et al. 1994). As with induction of COX-2 and type-2 PLA2, induction of NO likely accounts for a considerable number of biological effects of IL-1 and TNF. In experimental septic shock the fall in mean arterial pressure and the decrease in systemic vascular resistance are thought to be mediated by the induction of NO from smooth muscle cells. LPS injection into animals increases NO in several tissues, and when treated with IL-1 receptor antagonist, there is a 70% decrease in NO (SZABO et al. 1993). In vitro, low concentrations of IL-1 (1pg/ml) are sufficient to induce NO synthase.

In rabbits the blockade of constitutive and inducible NOS with *N*-nitro-L-arginine methyl ester led to IL-1-induced slow-wave sleep, but fever was not inhibited (KAPAS et al. 1994). However, in the rat IL-1-induced fever was inhibited (REIMERS et al. 1994), as were effects of IL-1 on the endocrine pancreas. In the rat, blocking the generation of NO augmented the IL-1 induced release of ACTH (RIVIER and SHEN 1994). In vitro, blocking NO production reduced PAF release from mast cells, inhibition of proteoglycan synthesis by chondrocytes, and apoptosis of the insulin-producing β-islet cell. In a murine macrophage line the induction of COX-2 by LPS was blocked by the NOS inhibitor *N*G-monomethyl-L-arginine and thus has been linked to a requirement for NO generation for PGE2 synthesis (SALVEMI et al. 1993). IL-1β induced PGE2 on fibroblasts was enhanced by the NO donor, nitroprusside, and its enhancement was inhibited by hemoglobin (SALVEMI et al. 1993). IL-1 induced NOS is not inhibited by indomethacin (PALMER et al. 1994).

2.6 Effect on Receptor Expression

Whether at the level of gene or surface expression, an important biological effect of IL-1 and TNF is the regulation of cytokine receptors, receptors for growth factors, and cell adhesion molecules. In general, changes in the binding of ligands to their respective receptors are not due to altered affinities but rather to a change in the number of surface receptors. The expression of surface receptors is usually reflected at the level of receptor gene expression, although there are examples where decreases in surface expression is due to increased receptor shedding. IL-1 and TNF-mediated upregulation of adhesion molecule expression on cultured endothelial cells is a well-established method for to study these adhesins. Low concentrations (1–10 p*M*) of IL-1 alone increases gene and surface receptors expression for ELAM and ICAM-1. In contrast, an increase in IL-2 (p55) receptors by IL-1 requires a costimulant. The p55 TNF receptor is downregulated by IL-1 (Holtmann and Wallach 1987); however, closer examination shows this effect of IL-1 to be due to shedding of the receptor as well as suppression of gene expression (Brakebusch et al. 1994).

2.7 IL-1 and TNF Infusion Mimics Septic Shock

Many of the biological effects of IL-1 and TNF are similar to those observed during a septic event; however, recent studies in humans have confirmed data from animal experiments. IL-1α or IL-1β has been administered to humans in phase I trials. Systemic administration of intravenous IL-1 from 1–10 ng/kg has produced fever, sleepiness, anorexia, generalized myalgias, arthralgias, and headache. However, the most dramatic biological response to IL-1 was observed at doses of 100 ng/kg or higher; in these patients a rapid fall in blood pressure takes place (Smith et al. 1992). Because of these results the dose limiting toxicity for IL-1 of hypotension has been set at 300 ng/kg. In some patients receiving 1 μg/kg stage IV hypotension was reported (Smith et al. 1992). The subcutaneous route is associated with fewer side effects.

TNF infusion into humans is similar to that observed for IL-1 (Chapman et al. 1987). However, TNF also induces a coagulation cascade (van der Poll et al. 1990), which has not been observed in humans injected with IL-1. In addition, low doses of TNF in humans induces neutrophilia whereas higher doses result in a leukopenia (van der Poll et al. 1991).

In the rabbit a single intravenous injection of 10μg/kg recombinant human IL-1β resulted in a shocklike state with hypotension, neutropenia, and thrombocytopenia (Okusawa et al. 1988). This has been confirmed in studies using human IL-1α in baboons (Fischer et al. 1991).

The mechanism for the hypotensive effect of IL-1 appears to be due to the generation of at least three low molecular weight meditators: cyclo-oxygenase products (Okusawa et al. 1988), PAF, and NO (Beasley et al. 1991). The fall in circulating leukocytes and platelets is thought to be due to the stimulation of endothelial adhesion molecules. The effects of IL-1 in inducing a shocklike state are potentiated by coinfusion of TNF. The potentiation of IL-1 and TNF has been observed in both anesthetized (Okusawa et al. 1988) and conscious rabbits (Tredget et al. 1988). In the conscious rabbit coinjection of IL-1 and TNF induced a fall in mean arterial pressure, onset of lactic acidosis, and glucose intolerance. Many effects of IL-1 and TNF are synergistic in a variety of models in vitro and in vivo.

2.8 Effects on the Vascular Wall

In a now "classical" experiment on the local inflammatory response, the roles of IL-1 and TNF were studied for their role in leukocyte infiltration and extravastion of red cells and platelets using intradermal injection of endotoxin. Small doses of endotoxin induce a capillary leak and leukocyte infiltration within hours. The local Swartzman reaction takes place when a systemic injection of endotoxin is given 24 h later. In the hours following the systemic injection of endotoxin there is neutrophil emigration, hemorrhage, and necrosis at the "priming" site of initial the intradermal injection. Using either IL-1 or TNF only, no significant Swartzman reaction was observed as each cytokine was incabable of replacing the "priming" injection of endotoxin. However, when the combination of small doses of IL-1 plus TNF were injected as the priming agent, the full-blown local inflammatory response characteristic of the local Shwartzman reaction was observed (Movat et al. 1987). The proinflammatory effect of IL-1 and TNF was similar to that of acute vasculitis (Movat et al. 1987). Contributing to this event are the effects of IL-1 and TNF on endothelial cells. These include increased production of procoagulant activity, tissue factor, eicosanoids, PAF, and plasmingen activator inhibitor (Dejane et al. 1987; Rossi et al. 1985).

2.9 Comparison with IL-6

The most consistent correlations of clinical severity in inflammatory, autoimmune, or infectious disease with plasma cytokine levels are clearly those with IL-6, not IL-1 or TNF. The best correlation of plasma cytokine levels with mortality from septic shock has been found with IL-6 (Casey et al. 1993). IL-6 levels but not TNFα levels were found to predict fatal outcome in patients with septic shock (Fisher et al. 1993). A recent trial of anti-TNF monoclonal antibodies for the treatment of septic shock observed benefit (significant decrease in 28-day all-cause mortality) only in patients who entered the trial with elevated plasma IL-6 levels (Reinhart et al. 1995). Similar to other intervention studies, there was no correlation with benefit when patients were stratified according to their entry score on the Acute Physiologic and Chronic Health Evaluation.

Therefore, one can conclude that elevated levels of IL-6 in patients with septic shock represent the net effect of *biologically active* IL-1 and TNF. In other words, due to natural inhibitors of IL-1 and TNF (see below) it is the biologically active concentrations of these cytokines which are best correlated with disease severity. It is nearly impossible to measure biologically active plasma IL-1 and TNF levels because plasma contains large concentrations of the IL-1 receptor antagonist (IL-1Ra) and soluble receptors for the two TNF receptors (TNFsRp55 and TNFsRp75; see below). Because of the presence of these inhibitors it is equally impossible to measure the synergistic effect of IL-1 and TNF plasma from patients with septic shock. However, one can use the surrogate marker of IL-6 levels which refect the net, biologically active concentrations of IL-1 and TNF in these patients. In patients with septic shock a 3-day infusion of IL-1Ra resulted in a dramatic, dose-dependent fall in IL-6 levels compared to patients given placebo (FISHER et al. 1994b).

It is important to emphasize, that unlike IL-1 and TNF, there is no evidence that IL-6 is itself an inflammatory cytokine. IL-6 does not induce PGE2 (DINARELLO et al. 1991) but rather suppresses IL-1 inducible cyclo-oxygenase (HAUPTMANN et al. 1991). IL-6 does not cause shock in mice or primates regardless of the amount given, either alone or with TNF. In humans the intravenous administration of IL-6 at 30μg/kg does not produce hypotension whereas at 100 ng/kg IL-1 induces a fall in blood pressure in nearly all patients (SMITH et al. 1992). IL-6 has been given to humans in very high concentrations (100μg/kg) without hypotension; the only sign was headache and fever. Thus human and animal experiments support the concept that IL-6 does not have a causal role in septic shock.

In some models the production of IL-6 appears to be under the strict control of IL-1 and TNF; for example, mice subjected to an inflammatory event induced by intramuscular turpentine fail to produce IL-6 when pretreated with anti-IL-1 receptor antibodies (GERSHENWALD et al. 1990). In baboons injected with *Escherichia coli* anti-TNF antibodies prevent the production of IL-6 in the circulation (FONG et al. 1989). Mice deficient in IL-1β do not produce IL-6 nor serum amyloid A proteins following an inflammatory event due to subcutaneous turpentine injection (ZHENG et al. 1995). In vitro, blocking IL-1 receptors reduces the amount of endotoxin-stimulated IL-6 produced by human monocytes (GRANOWITZ et al. 1992a, b). A reduction in IL-6 levels has also been observed in mice injected with LPS (HENRICSON et al. 1991).

3 Experimental Sepsis Affected by Reducting Production of IL-1 and TNF

3.1 Corticosteroids

Of the drugs that suppress the production of IL-1 and TNF the most studied are corticosteroids. Corticosteroids inhibit the transcription of IL-1, TNF, and nearly all cytokines, and hence they are not specific in this regard. There is also some indication that steroids reduce the secretion of IL-1. In human volunteers injected with corticosteroids just prior to an intravenous injection of endotoxin there are reduced levels of circulating IL-1β, TNF, and IL-6 (Rock et al. 1992; Santos et al. 1993). These reductions in IL-1β, TNF, and IL-6 take place without suppressing IL-1Ra production (Santos et al. 1993). Relevant to these experiments are studies showing that IL-1 and TNF induce endogenous corticosteroids, and that adrenalectomized mice have a profound decrease in their resistance to IL-1 or TNF-mediated lethality (reviewed in Fantuzzi and Ghezzi 1993). Therefore some studies suggest that IL-1 and TNF-induced endogenous corticosteroid production acts as an intrinsic anti-inflammatory negative feedback mechanism.

3.2 Inhibition of Lipoxygenase

Inhibitors of 13-lipoxygenase (Schade et al. 1987) but not 5-lipoxygenase (Sirko et al. 1991) reduce TNFα and IL-1β transcription. Earlier studies had implicated as the lipoxygenase product triggering IL-1 and TNF synthesis. However, using specific inhibitors of 5-lipoxygenase, the importance of LTB4 in the production of cytokines is unlikely. Nevertheless, the role of lipoxygenase products in stimulating IL-1 and TNF production is supported by several controlled studies in humans consuming dietary supplements of eicosapentaenoic (ω-3) fatty acids or a diet rich in these fatty acids (see below). When ω-3 fatty acids are incorporated into cell membranes, the cyclo-oxygenase and lipoxygenase products following phopholipase-mediated hydrolysis of membrane phospholipids are not PGE2 and LTB4 but rather PGE3 and LTB5. This change alters signal transduction pathway induced by exogenous stimulants and results in an attenuation in the synthesis of proinflammatory cytokines.

3.3 Inhibition by Dietary ω-3 Fatty Acids

Several clinical trials have demonstrated a beneficial effect of dietary supplementation of ω-3 fatty acids (reviewed in Meydani and Dinarello 1993). Controlled studies have consistently shown a 50%–60% decrease in IL-1, IL-6, and TNF production in peripheral blood mononuclear cells (PBMC) of subjects ingesting ω-3 fatty acid supplements compared to PBMC taken before this dietary intervention (Endres et al. 1989b; Meydani et al. 1990) One study reported a selective decrease in cell content of IL-1β (Molvig et al. 1991). The phenomenon can also be demonstrated by measuring cytokine production in whole blood (Cooper et al. 1993). The source of the ω-3 fatty acid does not have to be high-dose dietary supplementation. In fact, a significant suppression of in vitro IL-1 and TNF production was observed in volunteers consuming low fat diets using fish as the primary source of protein (Meydani et al. 1993). When the major source of protein was changed to nonfish foods (poultry), there was no reduction in cytokine production (MEYDANI et al. 1993).

3.4 Effect of Cyclo-oxygenase Inhibitors on the Production of IL-1 and TNF

Despite several studies there is no clear answer as to whether PGs suppress IL-1 production in cultured PBMC in vitro. This is probably due to the type of stimulant used and to the contribution of endogenous IL-1 production (IL-1 induced IL-1) to the total IL-1 synthesized (GRANOWITZ et al. 1992a). In general, adding cyclo-oxygenase inhibitors to LPS-stimulated PBMC can suppress, augment, or have no effect on IL-1 production. On the other hand, under the same culture conditions, LPS-induced TNF gene expression and synthesis is exquisitively sensitive to suppression by PGE2 and PGI2. In humans injected with LPS and pretreated with oral cyclo-oxygenase inhibitors the circulating levels of TNF and IL-6 are higher than in controls not given cyclo-oxygenase inhibitors (SPINAS et al. 1991). This observation is consistent with the mechanism that PGI2/PGE2-induced TNF suppression is via elevation in cAMP (ENDRES et al. 1991) and explains why drugs such as pentoxiphylline suppress TNF production. IL-1 induces cAMP via the intermediate induction of PGI2 (BEASLEY and McGUIGGIN 1995). An increase in cAMP can also be accomplished by triggering the H_2 receptor on moncytes with histamine and without the intermediate production of PGE2 (VANNIER et al. 1991b).

In the absence of an external stimulus, raising intracellular levels of cAMP has a slight or no effect on IL-1β gene expressin (VANIER AND DINARELLO 1993). However, the secretion of meture IL-1β is suppressed by elevated PGE2; therefore cyclo-oxygenase inhibitors or IFNγ increases the secretion of IL-1β (SCHINDLER et al. 1990a). It is presently unclear whether this effect is due to decreased myristoylation (STEVENSON et al. 1992b, 1993), decreased IL-1β converting enzyme (ICE) activity or a reduction in the secretory ("channel") pathway proposed for IL-1β (SINGER et al. 1993). In contrast to PBMC stimulated with LPS, when these cells are stimulated with IL-1, cyclo-oxgenase inhibitors suppress IL-1 and IL-6 production. To confirm this observation the addition of either exogenous PGE2 or histamine to cell cultures increase intracellular cAMP and results in enhanced IL-1-induced IL-1, IL-6, and IL-8 production (VANNIER and DINARELLO 1993, 1994; VANNIER et al. 1991a). In terms of noninfectious diseases cyclo-oxygenase inhibitors likely reduce IL-1-induced cytokine production. One can conclude that IL-1 induced cytokine synthesis is positively influenced by prostanoid-induced cAMP, but that LPS-induced cytokines are either unaffected or slightly negatively influenced. The effect of prostanoids on TNF production is, however, clearly negative.

3.5 Inhibition of IL-1 and TNF Processing Enzymes

Studies on the production of IL-1β and TNFα have established that synthesis of the precursors of each cytokine and release into the extracellular compartment are distinct events. IL-1β is unique in that the precursor lacking a leader sequence is barely active and remains in the cytosol until cleavage and release. TNFα has a weak leader sequence, appears to be associated with the Golgi, and exists in a cell membrane form before being cleaved and released (KRIEGLER et al. 1988). The precursors for IL-1β (STEVENSON et al. 1993) and TNFα (STEVENSON et al. 1992a) undergo myristoylation on lysines, which is thought to contribute to membrane localization. Although the primary N-terminal amino acids for extracellular IL-1β and TNFα have been known for several years, the way in which the respective precursors are cleaved and transported out of the cell was poorly understood.

ICE is a constitutively produced intracellular cysteine protease which appears to be the sole enzyme for cleaving precursor IL-1β between aspartic acid (116) and alanine (117). ICE is stored in cells in an inactive form but becomes enzymatically active by the same cell stimuli which induce the synthesis of IL-1β. Substrate inhibitors of ICE have been used to prevent cleavage of the 31-KDa precursor IL-1β and release of mature 17-Kda in human blood and animal monocytes (CHIN and KOSTURA 1993). Reduction in inflammatory disease activity also takes place when ICE is inhibited.

Serine proteases cleave the 26-Kda TNFα precursor between alanine (76) and valine (77), yielding the 17-Kda mature TNFα (MCGEEHAN et al. 1994; MOHLER et al. 1994). Unlike ICE, a putative, specific TNFα-converting enzyme is presently unknown, although it appears to be in the general class of metalloproteinases with a zinc binding motif of HEXGH (GEARING et al. 1994; MCGEEHAN et al. 1994). In vitro, metalloproteinase inhibitors and zinc chelators suppress the processing of TNFα from human blood monocytes and murine macrophages but do not affect the release of lymphotoxin-α from T-lymphocytes or the release of other cytokines (GEARING et al. 1994; MCGEEHAN et al. 1994; MOHLER et al. 1994). These metalloproteinsase inhibitors do not affect the production of IL-1β or IL-6 in whole human blood incubated with LPS (GEARING et al. 1994). Even membrane-associated cytokines such as macrophage colony-stimulating factor and transforming growth factor (TGF)α were unaffected. The proteinase inhibitors used in these studies are not specific since they can cleave other proteins. However, when administered to rats or mice, these metalloproteinase inhibitors reduce circulating levels of LPS-induced TNFα (GEARING et al. 1994; MOHLER et al. 1994).

Other examples of cell-associated cytokines requiring a proteolytic cleavage step include macrophage colony-stimulating factor, TGFα and endothelin-1. Of relevance to inflammation, the precursor of IL-1α is vulnerable to cleavage by a calcium-dependent, membrane-associated serine protease (calpain; KOBAYASHI et al. 1990, 1991; WATANABE and KOBAYASHI 1994). However, unlike TNFα and IL-1β, IL-1α is primarily an intracellular molecule; precursor IL-1α is active as a membrane form and requires cell-cell contact for biological activity (KAPLANSKI et al. 1994). Processing and release of IL-1α requires the special circumstance of activating calpain using calcium ionophores or stimulating cells with very large numbers of bacteria. Ordinarily LPS or cytokine stimulation of human moncytes is accompained by the spontaneous release of IL-1β and TNFα but not of IL-1α (ENDRES et al. 1989a; LONNEMANN et al. 1989; SCHINDLER et al. 1990b).

Do these cytokine cleaving proteases play a critical role in the progression of inflammation? In ICE-deficient mice there is prolonged survival following a lethal challenge of endotoxin and decrease in vitro release of IL-1α (LI et al. 1995). In another strain of mice deficient in ICE a reduction in IL-1α, TNF, and IL-6 production was observed (KUIDA et al. 1995). There was a reduction in the near 100% lethality of mice pretreated with one of the TNFα-processing inhibitors before receiving galactosamine and challenged with LPS (MOHLER et al. 1994). Figure 1 summarizes the processing enzymes for IL-1β, IL-1α, and TNFα.

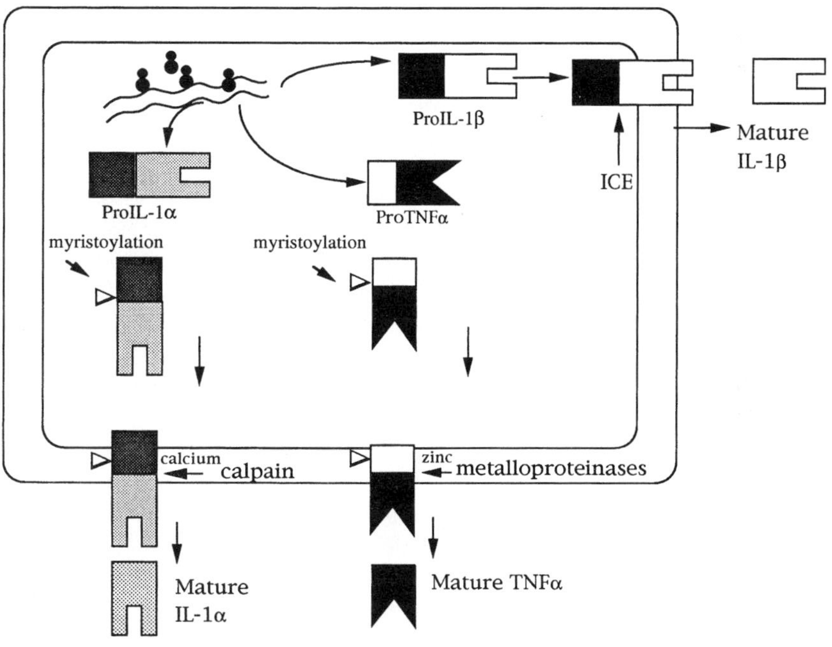

Fig. 1. Processing of IL-1β, IL-1α, and TNFα by proteases. The 31-kDa precusor for IL-1β (*proIL-1β*) is initially translated and found in the cytosol. It is transported to the vicinity of the cell membrane where the intracellular IL-1β converting enzyme (*ICE*) cleaves proIL-1β and the mature (17-kDa) is secreted into the extracellular space. ICE must be activated before it can cleave proIL-1β. Precursor IL-1α (*proIL-1α*) is also found in the cytosol. It is myristoylated and inserted into the cell membrane. The cell membrane associated calpains are activated by calcium and cleave proIL-1α. The mature (17-kDa) IL-1α is then secreted into the extracellular space. The precursor for TNFα is also myristoylated and found in the cell membrane. Zinc-dependent metalloproteinases cleave membrane-bound TNFα, and mature 17-kDa TNFα is released

3.6 Cytokine-Suppressing Anti-inflammatory Drugs

Recent studies have taken advantage of pyridinyl-imidazol compounds which block the synthesis of IL-1β and TNFα without affecting transcription or their steady-state levels of mRNA. mRNA levels for either IL-1 or TNFα in PBMC stimulated with LPS in the presence of these compounds are indistinguishable from those in PBMC stimulated with C5a, hypoxia, or adherence. In other words, there is ample cytokine mRNA but no cytokine protein. These are interesting compounds which have entered clinical medicine. Some are cyclo-oxygenase/lipoxygenase inhibitors because, by classification, they block these enzymes. Hence they are often called "dual inhibitors." However, their mechanism of action in suppressing IL-1 and TNF have never been linked to their ability to suppress cyclo-oxygenase/lipoxygenase (Sirko et al. 1991). These and other imidazol-like drugs have recently been called "cytokine-suppressing anti-inflammatory drugs" or CSAIDs (Young et al. 1994).

Two agents have received careful study for their ability not to affect transcription but rather to inhibit translation (Olivera et al. 1993; RordorfAdam et al. 1994). Recent studies suggest that the mechanism by which CSAIDs reduce IL-1 and TNF translations their ability to bind to and inactivate two related mitogen-activating protein (MAP) kinases (Lee et al. 1994). As with most MAP kinases, the novel kinases are serine-threonine kinases. These kinases phosphorylate proteins are required for translation of cytokine mRNAs into their respective proteins (Lee et al. 1994). These MAP kinases also have the same nucleotide sequences as those of the IL-1 and TNF signal-associated MAP kinase p38 (Galcheva-Gargova et al. 1994; Han et al. 1994). The p38 MAP kinase is a homologue of the yeast *HOG-1* gene. The cytokine synthesis inhibiting drugs bind and inactivate these MAP kinases in cells stimulated with LPS or hyperosmolarity (Galcheva-Gargova et al. 1994; Han et al. 1994). A *HOG-1* gene related p38 MAP kinase is part of the IL-1 and TNF signal transduction phosphorylation cascade (Freshney et al. 1994; Kracht et al. 1994). Recently it has been shown that the initiation factor eIF-4E requires a MAP-like phosphorylation step in order to dissociate from PHAS-I, a translational regulatory molecule (Lin et al. 1994). Although speculative, an IL-1, LPS, or TNF-directed MAP kinase phosphorylation of PHAS-I may be related to the MAP kinases inhibited by CSAIDs.

4 The Anti-inflammatory Cytokine Network in Sepsis

In most situations cytokine responses are self-limiting. In some patients sepsis is also a limited event, even without the intervention of antibiotics. Several mediators of the inflammatory response contribute to the downregulation of cytokine production and activity. For example, PGE2 suppresses a variety of immune responses such as IL-2 production. On the other hand, in some infectious diseases inflammation persists because the infectious agent is still present, and cytokine production persists. Proinflammatory cytokines, which are produced in response to an infection, in turn induce the synthesis of other cytokines which have the ability to suppress the inducing cytokine. For example, IL-1 stimulates a cascade of inflammatory mediators as well as a cascade of cytokines which act to suppress further production of IL-1. This is simple negative feedback, a well-established biological process. Only recently has the extent and complexity of the "anti-inflammatory cytokine" network been investigated. The evidence that "anti-inflammatory cytokines" do in fact play a role in downregulating inflammation comes from (a) experiments in which neutralizing antibodies to certain cytokines worsen the inflammation and (b) gene deletion studies in mice reveal a role for a particular cytokine as an anti-inflammatory agent. For example, antibodies to IL-1Ra worsen colitis (FERRETTI et al. 1994) and mice deficient in IL-10 develop spontaneous inflammatory bowel disease (MOORE et al. 1993).

4.1 IL-4, IL-10, IL-13, and TGFβ

IL-4, IL-10, IL-13, and TGFβ each suppress gene expression and synthesis of IL-1, TNF, and other cytokines. In vitro, these cytokines can reduce endotoxin-induced gene expression and synthesis of IL-1 and TNF as much 90%, and when given to mice or rats, can reduce lethal endotoxemia. As such they are potentially useful in some clinical situations. IL-10 appears to be particularly useful because, unlike IL-4 and TGFβ, IL-10 has no clinical side effects. A randomized, double-blind, placebo-controlled trial (phase I) in healthy human volunteers demonstrated the absence of clinical toxicity and also studied the effect of a single intravenous injection of IL-10 on cytokine production (CHERNOFF et al. 1995). Blood was removed before and 3, 6, 24, and 48 h after the injection and incubated in vitro with endotoxin, and the amounts of IL-1β, TNFα, IL-6, IL-8, IL-1Ra, and TNF soluble receptor p55 (TNFsRp55) were measured. At doses of 10 and 25 μg/kg there was a 90% reduction in IL-1β, TNFα, and IL-6 production in blood taken 3 and 6 h after the injection; at 25μg/kg a 50% reduction IL-1β, TNFα, and IL-6 production was present after 24 and 48 h. In contrast, there was no suppression of IL-1Ra or TNFsRp55.

A 40%–60% reduction in circulating lymphocytes expressing CD4, CD8, and CD3 was observed after infusion of IL-10 (CHERNOFF et al. 1995). The proliferative response to the mitogen phytohemagglutinin was suppressed in PBMC from volunteers given 10 or 25 μg/kg IL-10 and was not reversed by using higher concentrations of the mitogen. It is unclear whether the suppression of lymphocyte responses was due to low numbers of CD3 cells or to an intrinsic reduction in response to a proliferative signal. Although CD2 is the major receptor for mitogenic responses, CD2-bearing lymphocytes did not change following IL-10 (Chernoff et al. 1995). These human studies confirm the in vitro effects of IL-10 and suggest that IL-10 may be useful in suppressing inflammatory cytokine production in selected diseases.

IL-4 and IL-13 also suppress LPS -induced IL-1 and TNF gene expression and synthesis. In addition, they increase IL-1Ra production (Vannier et al. 1992). IL–4 and IL-13 share the same receptor complex on monocytes, and hence similar biological effects for both cytokines are often observed. There are, however, few if any receptors for IL-13 on T-lymphocytes, and hence the immunological suppressive effects of IL-4 and IL-10 are not observed for IL-13. Similarly to IL-4 –IL-10, and IL-13, TGFβ suppress gene expression and synthesis of IL-1 and TNF and also increase IL-1Ra production (Chantry et al. 1989). However, TGFβ, which has profound immunosuppressive effects, is a growth factor for normal and neoplastic cells.

4.2 IL-6 and Other Ligands for the gp130 Signal Transducer

In mice LPS induces higher circulating levels of TNF in mice deficient in producing IL-6 than in wild-strain mice (Fattori et al. 1994). This experiment suggests that IL-6 production during systemic inflammation downregulates TNF production, and that IL-6 functions as an anti-inflammatory cytokine. Other evidence supports the view that IL-6 may be a naturally occurring anti-inflammatory cytokine. Infusions of IL-6 into humans results in high levels of IL-1Ra and high levels of TNFsRp55 (Tilg et al. 1994a). In vitro, IL-6 suppresses LPS - and TNF-induced IL-1 production (Aderka et al. 1989; Schindler et al. 1990b). The ability of IL-6 to function as an anti-inflammatory molecule is consistent with it being part of the cytokine family, which use the gp130 signal apparatus. IL-6 shares its signaling event with leukemia-inhibitory factor (LIF), ciliary neurotropic factor (CNTF), IL-11, oncostatin M, and cardiotropin. CNTF suppress IL-1 stimulated IL-8 and PGE2 synthesis (Shapiro et al. 1994). In vivo, IL-6 stimulates the release of corticosteroids and synthesis of the full spectrum of acute-phase proteins, including many antiproteases. One interpretation of the biological significance of IL-6 is the anti-inflammatory property of these anti-proteases (Tilg et al. 1994a).

A specific receptor for IL-6 binds the ligand and the IL-6/IL-6R complex triggers homodimerization of gp130. Recent studies indicate that in various animal models of sepsis and inflammation the adminstration of IL-11 or LIF reduces the severity of disease (Waring et al. 1995). Similarly to IL-6, CNTF binds to its specific soluble receptor and triggers dimerization of gp130 (with the LIF receptor). When incubated with human blood monocytes, the CNTF/soluble receptor complex supresses LPS-induced PGE2 and IL-1α synthesis (Shapiro et al. 1994). In addition, the CNTF/soluble receptor complex suppresses IL-1α-induced IL-8 and PGE2 synthesis (Shapiro et al. 1994). In general, inhibition is 50%, and it is unclear whether this is at the transcriptional or posttranscriptional level. The anti-inflammatory effects of LIF in some cells are similar to those of IL-6 and CNTF since they use the same gp130 signal transduction pathways. However, for LIF the signaling through gp130 does not require a soluble receptor since dimerization occurs between gp130 and the LIFβ receptor.

The above studies suggest that signaling of gp130 may provide an insight into the mechanism of inhibiting IL-1 mediated events. Following dimerization of gp130, tyrosine phosphorylation, and translocation of p91 a well-known nuclear factor (NF) associated with the induction of acute-phase proteins in the liver, appears . p91 is one of three NFs making a complex which initiates gene expression. Several acute phase proteins preferentially induce IL-1Ra over IL-1 in vitro (TILG et al. 1993). In hepatocytes the same signal that induces hepatic acute-phase protein gene expression also inhibits transcription of albumin and cytochrome p450. Therefore this pathway, although studied for increased gene expression of hepatic proteins, also leads to events which decrease gene expression of IL-1 inducible genes. Therefore a concept has developed that IL-6 and related gp130 signaling cytokines lead to an anti-inflammatory state to counter the proinflammatory state initiated by IL-1 and TNF (TILG et al. 1993).

4.3 Interferons

IFNγ is not part of the circulating cytokine profile in humans injected with LPS. IFNγ has been shown to increase receptors for TNF and thus can act to enhance the biological effects of TNF. The role of IFNγ in septic shock has been reviewed by BILLIAU and VANDEKERCKHOVE (1991). No clear role for IFNγ emerges from these studies. However, in some clinical situations IFNs are thought to exert their effects by acting as anti-inflammatory agents. Early studies showed that INFγ suppresses LPS or IL-1 induced PGE2 production in human monocytes. Although it is well-known that IFNγ increases IL-1 and TNF production induced by LPS (UCLA et al. 1990), using IL-1 as a stimulant, IFNγ suppresses IL-1 induced IL-1 (GHEZZI and DINARELLO 1988)> Subsequently, it was shown that IFNα also suppresses IL-1-induced IL-1 synthesis due to reduced transcription (SCHINDLER et al. 1990a). Similarly to CNTF (SHAPIRO et al. 1994), suppression of IL-1 induced IL-8 transcription by IFNγ has been reported to be mediated by NFκB (OLIVEIRA et al. 1994). Of considerable importance is the finding that tyrosine phosphorylations following triggering of IFN receptors result in a kinase cascade similar to those described for gp130. Thus, depending on the signal, the phosphorylation of p91 can lead to suppression of IL-1 induced gene expression.

5 Naturally Occurring Inhibitors of IL-1

5.1 IL-1 Receptor Antagonist

IL-1Ra is produced primarily from macrophagic cells as a 22-kDa glycosylated protein. Unlike IL-1α and IL-1β, which lack a leader peptide, IL-1Ra possesses a leader sequence and is synthesized, processed, and secreted from the cell (reviewed in AREND 1993). The three-dimensional structure of each member of the IL-1 family (IL-1β, IL-1α, and IL-1Ra) is comprised of all β strands, and each member binds to the same IL-1 type I receptors (IL-1RI). However, IL-1Ra binds to IL-1RI with virtually the same affinity as IL-1α or IL-1β but does not trigger a response (DRIPPS et al. 1991). The cytokine is thus the naturally occurring inhibitor or IL-1. IL-1Ra has nearly the same affinity (approx. 200 pM) for the IL-1RI as that of human IL-1α and IL-1β (EISENBERG et al. 1990; HANNUM et al. 1990; SECKINGER et al. 1987).

In animal studies the administration of IL-1Ra reveals that IL-1 plays an important role in the pathogenesis of inflammatory and immunologically mediated disease, including animal models of septic shock. Many studies report a reduction in severity or mortality of at least 50%, but in many the amelioration of pathological changes is complete. One consistent observation is the reduction in the number of infiltrating neutrophils associated with local inflammation, and this effect of IL-1Ra may be due to preventing IL-1 induced synthesis of cell adhesion molecules, PAF, IL-8, and related chemokines. Only a few IL-1RI need be occupied to trigger a biological response, and therefore it is necessary to sustain a high level of IL-1Ra to block unoccupied receptors. When exogenous IL-1Ra is injected into animals, high plasma levels (10–20 μg/ml) are needed before one observes a reduction in disease. In humans similar levels of IL-1Ra are needed to block the hematological response to LPS (GRANOWITZ et al. 1993).

5.2 IL-1Ra Production in Sepsis

It is not unusual to measure high and more sustained levels of IL-1Ra than IL-1β in patients with septic shock (FISCHER et al. 1992). For example, in health volunteers injected intravenously with a low dose of *E. coli* endotoxin circulating IL-1Ra levels are at a 100-fold molar excess (peak level of 6000–7000 pg/ml) to those of IL-1β (70–80 pg/ml) and are significantly elevated above the baseline levels for over 24 h (GRANOWITZ et al. 1991). In patients with septic shock, juvenile rheumatoid arthritis, or inflammatory bowel disease a similar ratio can be observed, and elevated IL-1Ra levels may be correlated with the severity of disease (HYAMS et al. 1994). In patients with thermal burns the levels of IL-1Ra are correlated with the burn surface area, and the highest levels of IL-1Ra have been measured in non-survivors (MANDRUP-POULSEN et al. 1995). The intravenous injection of 30 ng/kg of IL-1α into humans induces 25–30 ng/ml of IL-1Ra (TILG et al. 1994b), which is fourfold higher than that induced by LPS. Injection of IL-1β into humans results in an 86-fold increase in plasma IL-1Ra after 1 h (Bargetzi et al. 1993). In humans, endogenous TNF production during endotoxemia contributes to IL-1Ra production (van der Poll et al. 1994).

In patients with acute Lyme arthritis the duration of joint inflammation is shortest in those patients with the highest joint fluid levels of IL-1Ra whereas it is prolonged in those with low levels of IL-1Ra (Miller et al. 1993). Is the amount of IL-1Ra produced in disease sufficient to dampen the response to IL-1? Using specific, neutralizing antibodies to mouse IL-1Ra, an increase in the formation of Schistosoma egg granulomata was observed when endogenous IL-1Ra was neutralized (Chensue et al. 1993). In rabbits with immune complex colitis the infusion of a neutralizing antibody to rabbit IL-1Ra resulted in exacerbation and prolongation of the colitis (Ferretti et al. 1994). At the present, the phenotype of an IL-1Ra deficient mouse is unknown, but neutralizing endogenous IL-1Ra appears to worsen inflammations.

5.3 IL-1Ra in Experimental Endotoxemia in Humans

IL-1Ra given intravenously to healthy volunteers is without side effects or changes in biochemical, hematological and endocrinological parameters, even when peak blood levels reach 30 μg/ml and are sustained above 10 μg/ml for several hours (Granowitz et al. 1992b). To evaluate the effect of IL-1 receptor blockade on clinical disease under controlled experimental conditions, healthy volunteers were challenged with intravenous endotoxin and administered an infusion of 10 mg/kg IL-1Ra at the same time. There was no effect on endotoxin-induced fever, although blood levels of IL-1Ra were not significantly elevated until 1 h after the bolus injection of endotoxin. In animal studies peripheral endotoxin induces fever by triggering IL-1 induction of IL-6 synthesis in the central nervous system (LeMay et al. 1990). Since IL-1Ra does not cross the blood-brain barrier, this may account for the inability of IL-1Ra to diminish endotoxin fever (Dinarello et al. 1992). However, there was a 50% reduction in the endotoxin-induced neutrophilia and a reduction in the circulating levels of granulocyte colony-stimulating factor compared to subjects injected with endotoxin plus saline (Granowitz et al. 1993).

Endotoxin injection suppresses the mitogen-induced proliferative response of PBMC in vitro. However, in volunteers injected with endotoxin plus IL-1Ra there was no suppression of the response (Granowitz et al. 1993). Mitogen- and antigen-induced proliferation is a well-established parameter of immunocompetence and is associated with decreased production of IL-2. Similarly to experimental endotoxin injection, this suppression is observed in patients with multiple trauma, sepsis, and cardiopulmonary bypass. In experimental endotoxemia and the above clinical conditions, treatment with cyclo-oxygenase inhibitors restores these cell-mediated immune responses (Markewitz et al. 1993). This effect of cyclo-oxygenase inhibitors is consistent with the well-known suppressive effects of PGE2 on IL-2 production and T-cell proliferation. Since IL-1 is a potent inducer of COX-2, it is not surprising that blocking IL-1 receptors during endotoxemia reduces IL-1 induced PGE2 production during endotoxemia. Thus, the studies establish that under conditions of low-dose endotoxemia it is possible to block IL-1 mediated responses with IL-1Ra. Those host response parameters which are unaffected by IL-1Ra are likely due to other cytokines such as TNF or IL-6 or the combination of these cytokines with IL-1.

5.4 Clinical Trials of IL-1Ra in Septic Shock

IL-1Ra has been in given to patients with septic shock. The initial (phase II) trial was a randomized, placebo-controlled, open-label study in 99 patients. Patients received either placebo or a loading bolus of 100 mg followed by a 3-day infusion of 17, 67, or 133 mg/h IL-1Ra (Fisher et al. 1994b). A dose-dependent improvement in 28-day mortality was observed; mortality was reduced from 44% in the placebo group to 16% in the group receiving the highest dose of IL-1Ra ($p=0.015$). In this study, there was a dose-related fall in the circulating levels of IL-6 24 h after the initiation of IL-1Ra infusion. This fall is consistent with the well-established control of circulating IL-6 levels by IL-1 (Fischer et al. 1991; Gershenwald et al. 1990) and the correlation of disease severity and outcome with IL-6 levels (Casey et al. 1993). The mean plasma level of IL-1Ra was 25–28 μg/ml in the high-dose group, and this order of magnitude of curculating IL-1Ra concentration is measured in animals which benefit from IL-1Ra during experimental shock (Aiura et al. 1993).

A large phase III trial in 893 patients revealed a trend but without a statistically significant reduction in 28-day mortality (Fisher et al. 1994b). However, a retrospective analysis of 563 patients with a predicted risk of mortality of 24% or greater (Knaus et al. 1993) revealed a significant reduction in 28-day mortality (45% in the placebo group and 35% in patients receiving 2 mg/kg per hour for 72 h, $p=0.005$; Fisher et al. 1994b). Similar improvement was observed when patients were scored on the basis on organ failure at entry. Circulating levels of thromboxane B2, PGI2, LTC4, LTD4, and LTE4 were attenuated ($p<0.05$) at 72h in patients receiving the high dose of IL-1Ra whereas in patients receiving the placebo these eicosanoids were increased at 72 h (Friedman et al. 1994). A second phase III trial using 10 g IL-1Ra infused over 3 days was undertaken but terminated during an interim analysis because a reduction in overall 28-day mortality would not likely reach statistical significance. Similarly to other trials in septic shock using strategies of cytokine or endotoxin blockade, the efficacy of IL-1Ra in reducing 28-day all-cause mortality is not easily demonstrated despite the impressive results in animal models of septic shock and death. Patient heterogeneity versus animal homogeneity is thought to contribute to a failure to bridge the gap between animal and clincal data in sepsis.

5.5 Soluble IL-1 Receptors

The extracellular domains of the type I and type II receptors are found in the circulation of healthy humans and are further elevated under conditions of infection and inflammation. The concentration of soluble type I receptor (IL-1sRI) in health is approximately 50–100 pM (AREND et al. 1994; GIRI et al. 1994) whereas the concentration of the soluble form of the type II receptor (IL-1sRII) in health is approximately 100–200 p*M* (AREND et al. 1994; ORENCOLE et al. 1994). Therefore there is a rather high level of IL-1 soluble receptors present in health, and their circulating concentrations are certainly at two-threefold molar excesses of the concentrations of circulating IL-1 produced in severe disease. The tissue levels (interstitial fluid) of these receptors are unknown, and therefore the degree to which they reduce the response to locally produced IL-1 is presently unclear.

IL-1sRI has been used in several models of inflammatory and autoimmune disease. Administration of murine IL-1sRI to mice has increased the survival of heterotopic heart allografts and reduced the hyperplasic lymph node response to allogeneic cells (FANSLOW et al. 1990). In a rat model of antigen-induced arthritis the local instillation of the murine IL-1sRI reduced joint swelling and tissue destruction (DOWER et al. 1994). When a dose of soluble receptor (1 μg) was instilled into the contralateral, unaffected joint, a reduction in the degree of tissue damage was observed in the affected joint. These data suggest that the amount of IL-1sRI given in the normal, contralateral joint acted systemically. In a model of experimental autoimmune encephalitits the IL-1sRI reduced the severity of this disease (JACOBS et al. 1991). Administration of IL-1sRI to animals has also been reported to reduce the physiological response to LPS, acute lung injury, and delayed-type hypersensitivity (DOWER et al. 1994). To date there are no data on the efficacy of in animal models of disease.

5.6 Soluble IL-1 Receptor Type I in Human Studies

Recombinant human IL-1sRI has been administered intravenously to healthy humans in a phase I trial without side effects or changes in physiological, hematological, or endocrinological parameters. Thus, similarly to infusions of IL-1Ra, IL-1sRI appears safe and reinforces the conclusion that IL-1 does not have a role in homeostasis in humans. Administration of IL-1sRI in humans has been effective in reducing the delayed hypersensitivity skin reaction to recall antigens. Patients with known antigen hypersensitivity received an intradermal injection of a specific antigen and also an injection of either placebo or IL-1sRI near the site. In addition, IL-1sRI was injected locally at a contralateral site. As the amount of IL-1sRI increased, there was a progressive decrease in the inflammatory lesion due to the allergen. However, there was also a decrease in the lesion size as the amount of IL-1sRI injected at the contralateral site was increased to 10–100 μg (DOWER et al. 1994). These findings are similar to observations in rats receiving a contralateral instillation of IL-1sRI during antigen-induced arthritis.

6 Reducing the Activity of TNF

6.1 Soluble TNF Receptors

Unlike IL-1, a naturally occurring receptor antagonist to TNF has not been found. However, soluble receptors to TNF are present in the circulation of healthy humans and may act as naturally occurring inhibitors of TNF activity. The situation is similar to that of soluble IL-1 receptors. There are two cell surface TNF receptors: p55 and p75 (ENGELMANN et al. 1989, 1990a,b; OLSSON et al. 1993). The extracellular domains of each TNF receptor are shed from the cell surface by a serine protease associated with cell activation (BRAKEBUSCH et al. 1994) and are found in the circulation of healthy humans. The concentration of the p75 is approximately 300 p*M* and is threefold greater than that of the p55 form (CHIKANZA et al. 1993; GIRARDIN et al. 1992). However, increases in the circulating levels of soluble receptors to TNF during disease states appears greater than that for suluble IL-1 receptors. For example, endotoxemia induces the release of both TNF receptors into the circulation and the increase is several-fold that of the concentration in healthy subjects (SHAPIRO et al. 1993; VAN ZEE et al. 1992). TNF itself induces the release of its soluble receptors (JANSEN et al. 1995). Soluble TNF receptors are also elevated in patients with cancer, and the levels are correlated with the tumor burden or extent of the metastases (ADERKA et al 1991). Other studies have documented the presence of soluble TNF receptors in the circulation or joint fluid in a variety of autoimmune and inflammatory diseases.

The soluble receptors for IL-1 inhibit the action of IL-1 in a dose dependent fashion. In contrast, soluble TNF receptors can act as "carriers" of TNF in certain experimental models. This phenomenon was first shown by adding increasing amounts of soluble TNF receptors to cells exposed to TNF. The biological acitivity of TNF was enhanced at low molar ratios of receptor to ligand (ADERKA et al. 1992). At higher molar ratios of soluble receptor to TNF the activity was decreased, and there was dose-dependent inhibition of TNF activity (ADERKA et al. 1992). Therefore at low molar ratios, the soluble TNF receptors protect the TNF from degradation and destabilization. One likely mechanism for the stabilization of TNF by the soluble receptors is to maintain the trimer structure of TNF since monomeric TNF is biologically inactive. The amount of natural inhibition or natural "stabilization" of TNF by the soluble receptors during inflammation is unclear. Mice deficient in the the p55 TNF receptor do not manifest increased susceptibility to infection or inflammation. However, the p75 receptor is thought to function as the natural carrier of TNF compared to the p55 receptor.

In several animal models of sepsis and inflammatory disease the administration of recombinant forms of soluble p55 TNF receptor has reduced inflammation or prolonged survival (BERTINI et al. 1993; LESSLAUER et al. 1991a,b; PORAT et al. 1995; VAN ZEE et al. 1992). The subject has recently been reviewed (Olsson et al. 1993). Although a chimeric receptor to TNF p75 has been observed in human sepsis, the survival at higher doses was without benefit (reviewed in Opal 1996). This may be due to prolongation of the half-life of TNF due to trimer stabilization. In general, clinical trials for soluble forms of the p55 TNF receptor are presently underway for sepsis, inflammatory bowel disease, and rheumatoid arthritis. The use of soluble forms of the p55 TNF receptor in these clinical situations is based upon the beneficial effects of monoclonal antibodies to TNF in several trials. Therefore it is anticipated that soluble forms of the p55 TNF receptor will be used to treat acute and chronic graft rejection, sepsis, graft versus host disease, and a variety of inflammatory diseases.

6.2 Neutralizing Antibodies to TNF

From a historical viewpoint, the first experiment which implicated the importance of endogenous cytokines in the pathogenesis of septic shock was the demonstration that neutralizing antibodies to TNF (also called cachetin) reduce the lethality of LPS in mice (Beutler et al. 1985). This study was then expanded to primates and employed live *E. coli* organisms (Tracey et al. 1987) and in rabbits using LPS (Mathison et al. 1988). A protective role was again observed. Furthermore, anti-TNF antibodies had a dramatic effect in reducing the circulating levels of IL-1 and IL-6 (Fong et al. 1989). Following these studies many reports confirmed a role for TNF in the lethality of endotoxin shock in animals. Only a few studies, namely those using cecal ligation as a model, have shown that anti-TNF antibodies do not affect outcome. Nevertheless, animal studies formed the basis for using anti-TNF antibodies in humans with septic shock. Clinical trials have yielded mixed results, not dissimilar to those reported in patients treated with IL-1Ra.

A large, randomized, placebo-controlled, double-blind multicenter study of a murine anti-human TNFα monoclonal antibody in 971 patients with the sepsis syndrome (Abraham et al. 1995). There were two doses of anti-human TNFα, a single infusion of 7.5 mg/kg or one of 15 mg/kg. There was no overall benefit in 28-day all-cause mortality in patients receiving the antibody. However, in a subset of 478 patients who had septic shock upon entry into the study there was reduction in 3-day all-cause mortality compared to matched placebo controls (44% reduction at 15 mg/kg, p=0.01; 48% reduction at 7.5 mg/kg, p=0.004). At 28 days there were no differences in mortality in patients treated with the different doses of the antibody. Similar results have been reported in smaller studies (Fisher et al. 1993). Short-term benefits on left ventricular function have been observed in patients with septic shock treated with anti-TNF (Vincent et al. 1992).

Why do these trials fail to show overall efficacy in reducing the mortality in these patients? The topic of heterogeneity of patients as well as that of disease causation has been used to explain these results (Opal 1995). Is there anything wrong with the concept that blocking (or reducing) TNF activity or production should reduce the mortality of septic shock, as has been observed in the vast majority of animal studies? The best test for efficacy an anti-TNF in terms of patient heterogeneity are studies in patients with rheumatoid arthritis. This patient group is not as heterogeneous as the group with septic shock. In a multicenter, placebo-controlled study monclonal anti-TNF reduced the both severity and the biochemical markers of rheumatoid arthritis (ELLIOTT et al. 1994).

Therefore one may conclude that for patients with septic shock to show a benefit with anti-TNF treatment a better set of entry criteria need to be selected. Using plasma IL-6 levels as the surrogate marker of biologically active TNF (and IL-1) may provide a better selection criterion (REINHART et al. 1995). An alternate approach is to use patients in which the underlining disease does not contribute to 28-day mortality other than that due to the sepsis episode. This approach, however, will require a far greater amount of time to complete a large pivotal study.

Acknowledgement. These studies were supported by NIH grant AI 15614.

References

Abraham E, Wunderink R, Silverman H, Perl TM, Nasraway S, Levy H, Bone R, Wenzel RP, Balk R, Allred R, Pennington JE, Wherry JC (1995) Efficacy and safety of monoclonal antibody to human tumor necrosis factor-α in patients with sepsis syndrome. JAMA 273: 934–941

Aderka D, Le JM, Vilcek J (1989) IL-6 inhibits lipopolysaccharide-induced tumor necrosis factor production in cultured human monocytes, U937 cells, and in mice. J Immunol 143: 3517–3523

Aderka D, Engelmann H, Hornik V, Skorick Y, Levo Y, Wallach D, Kushtai G (1991) Increased serum levels of soluble receptors for tumor necrosis in cancer patients. Cancer Res 51: 5602–5607

Aderka D, Engelmann H, Maor Y, Brakebusch C, Wallach D (1992) stabilization of the bioactivity of tumor necrosis factor by its soluble receptors. J Exp Med 175: 323–329

Aiura K, Gelfand JA, Wakabayashi G, Burke JF, Thompson RC, Dinarello CA (1993) Interleukin-1 (IL-1) receptor antagonist prevents *Staphylococcus epidermidis*-induced hypotension and reduces circulating levels of tumor necrosis factor and IL-1β in rabbits. Infect Immun 61: 3342–3350

Arend WP (1993) Interleukin-1 receptor antagonist. Adv Immunol 54: 167–227

Arend WP, Malyak M, Smith MF, Whisenand TD, Slack JL, Sims JE, Giri JG, Dower SK (1994) Binding of IL-1α, IL-1β, and IL-1 receptor antagonist by soluble IL-1 receptors and levels of soluble IL-1 receptors in synovial fluids. J Immunol 153: 4766–4774

Baggiolini M, Dewald B, Moser B (1994) Interleukin-8 and related chemotactic cytokines-CXC and CC chemokines. Adv Immunol 55: 97–179

Bargetzi MJ, Lantz M, Smith CG, Torti FM, Olsson I, Eisenberg SP, Starnes HF (1993) Interleukin-1 beta induces interleukin-1 receptor antagonist and tumor necrosis factor binding proteins. Cancer Res 53: 4010–4013

Beasley D, McGuiggin M (1995) Interleuin-1 induces prostacyclin-dependent increases in cyclic AMP production and does not affect cyclic GMP production in human vascular smooth muscle cells. Cytokine (in press)

Beasley D, Schwartz JH, Brenner BM (1991) Interleukin 1 induces prolonged L-arginine-dependent cyclic guanosine monophosphate and nitrite production in rat vascular smooth muscle cells. J Clin Invest 87: 602–608

Bertini R, Delgado R, Faggioni R, Gascon MP, Ythier A, Ghezzi P (1993) Urinary TNF-binding protein (TNF soluble receptor) protects mice against the lethal effect of TNF and endotoxic shock. Eur Cytokine Netw 4: 39–42

Beutler B, Milsark IW, Cerami A (1985) Passive immunization against cachetin/tumor necrosis factor protects mice from lethal effect of endotoxin. Science 229: 869–871

Billiau A, Vandekerckhove F (1991) Cytokines and their interactions with other inflammatory mediators in the pathogenesis of sepsis and septic shock. Eup J Clin Invest 21: 559–573
Brakebusch C, Varfolomeev EE, Batkin M, Wallach D (1994) Structural requirements for inducible shedding of the p55 tumor necrosis factor receptor. J Biol Chem 269: 32488–32496
Casey LC, Balk RA, Bone RC (1993) Plasma cytokines and endotoxin levels correlate with survival in patients with the sepsis syndrome. Ann Intern Med 119: 771–778
Chantry D, Turner M, Abney E, Feldmann M (1989) Modulation of cytokine production by transforming growth factor-beta. J Immunol 142: 4295–4300
Chapman PB, Lester TJ, Casper ES, Gabrilove JL, Wong GY, Kempin SJ, Gold PJ, Welt S, Warren RS, Starnes HF (1987) Clinical pharmacology of recombinant human tumor necrosis factor in patients with advanced cancer. J Clin Oncol 5: 1942–1951
Chensue SW, Bienkowski M, Eessalu TE, Warmington KS, Hershey SD, Lukas NW, Kunkel SL (1993) Endogenous IL-1 receptor antagonist protein (IRAP) regulates schistosome egg granuloma formation and the regional lymphoid response. J Immunol 151: 3654–3662
Chernoff AE, Granowitz EV, Shapiro L, Vannier E, Lonnemann G, Angel JB, Kennedy JS, Rabson AR, Wolff SM, Dinarello CA (1995) A randomized, controlled trial of interleukin-10 in humans: inhibition of inflammatory cytokine production and immune responses. J Immunol (in press)
Chikanza IC, Roux-Lombard P, Dayer JM, Panayi GS (1993) Tumour necrosis factor soluble receptors behave as acute phase reactants following surgery in patients with rheumatoid arthritis, chronic osteomyelitis and osteorthritis. Clin Exp Immunol 92: 19–22
Chin J, Kostura MJ (1993) Dissociation of IL-1b synthesis and secretion in human blood monocytes stimulated with bacterial cell waste products. J Immunol 151: 5574–5585
Clark MA, Özgür LE, Conway TM, Dispoto J, Crooke ST, Bomalaski JS (1991) Cloning of a phospholipase A2-activating protein. Proc Natl Acad Sci USA 88: 5418–5422
Cooper AL, Gibbons L, Horan MA, Little RA, Rothwell NJ (1993) Effect of dietary fish oil supplementation on fever and cytokine production in human volunteers. Clin Nutr 12: 321–328
Crown J, Jakubowski A, Kemeny N, Gordon M, Gasparetto C, Wong G, Toner G, Meisenberg B, Botet J, Applewhite J, Sinha S, Moore M, Kelsen D, Buhles W, Gabrilove J (1991) A phase I trial of recombinant human interleukin-1β alone and in combination with myelosuppressive doses of 5-fluoruracil in patients with gastrointestinal cancer. Blood 78: 1420–1427
Dejana E, Breviario F, Erroi A, Bussolino F, Mussoni L, Gramse M, Pintucci G, Casali B, Dinarello CA, Van Damme J, Mantovani A (1987) Modulation of endothelial cell functions by different molecular species of interleukin 1. Blood 69: 695–699
Dinarello CA, Cannon JG, Mancilla J, Bishai I, Lees J, Coceani F (1991) Interleukin-6 as an endogenous pyrogen: induction of prostaglandin E2 in brain but not in peripheal blood mononuclear cells. Brain Res 562: 199–206
Dinarello CA, Zhang XX, Wen HD, Wolff SM, Ikejima T (1992) The effect of interleukin-1 receptor antagonist on IL-1, LPS, *Staphylococcus epidermidis* and tumor necrosis factor fever. In: Bartfai T, Ottoson D (eds) Neuro-immunology of fever. Pergamon Oxford, pp11–18
Dower SK, Fanslow W, Jacobs C, Waugh S, Sims JE, Widmer MB (1994) Interleukin-1 antagonists. Ther Immunol 1: 113–122
Dripps DJ, Brandhuber BJ, Thompson RC, Eisenberg SP (1991) Effect of IL-1ra on IL-1 signal transduction. J Biol Chem 266: 10331–10336
Eisenberg SP, Evans RJ, Arend WP, Verderber E, Brewer MT, Hannum CH, Thompson RC (1990) Primary structure and functional expression from complementary DNA of a human interleukin-1 receptor antagonist. Nature 343: 341–346
Elliott MJ, Maini RN, Feldmann M, Kalden JR, Antoni C, Smolen JS, Leeb B, Breedveld FC, Macfarlane JD, Long-Fox A, Charles P, Bijl H, Woody JN (1994) Randomised double-blind comparison of chimeric monoclonal antibody to tumour necrosis factor alpha (cA2) versus placebo in rheumatoid arthritis. Lancet 344: 1105–1110
Endres S, Cannon JG, Ghorbani R, Dempsey RA, Sisson SD, Lonnemann G, van der Meer JWM, Wolff SM, Dinarello CA (1989a) In vitro production of IL-1β, IL-1α, TNF, and IL-2 in healthy subjects: distribution, effect of cyclooxygenase inhibition and evidence of independent gene regulation Eur J Immunol 19: 2327–2333
Endres S, Ghorbani R, Kelley VE, Georgilis K, Lonnemann G, van der Meer JW, Cannon JG, Rogers TS, Klempner MS, Weber PC, Wolff SM, Dinarello CA (1989b) The effect of dietary supplementation with n-3 polyunsaturated fatty acids on the synthesis of interleukin-1 and tumor necrosis factor by mononuclear cells. N Engl J Med 320: 265–271

Endres S, Fuelle H-J, Sinha B, Stoll D, Dinarello CA, Gerzer R, Weber PC (1991) Cyclic nucleotides differentially regulate the synthesis of tumour necrosis factor-α and interleukin-1β by human mononuclear cells. Immunology 72: 56–60

Engelmann H, Aderka D, Rubinstein M, Rotman D, Wallach D (1989) A tumor necrosis factor-binding protein purified to homogeneity from human urine protects cells from tumor necrosis factor toxicity. J Biol Chem 264: 11974–11980

Engelmann H, Holtmann H, Brakebusch C, Avni YS, Sarov I, Nophar Y, Hadas E, Leitner O, Wallach D (1990a) Antibodies to a soluble form of a tumor necrosis factor (TNF) receptor have TNF-like activity. J Biol Chem 265: 14497–14504

Engelmann H, Novick D, Wallach D (1990b) Two tumor necrosis factor-binding proteins purified from human urine. Evidence for immunological cross-reactivity with cell surface tumor necrosis factor receptors. J Biol Chem 265: 1531–1536

Fanslow WC, Sims JE, Sassenfeld H, Morrissey PJ, Gillis S, Dower SK, Widmer MB (1990) Regulation of alloreactivity in vivo by a soluble form of the interleukin-1 receptor. Science 248: 739–742

Fantuzzi G, Ghezzi P (1993) Glucocorticoids as cytokine inhibitors: role in neuroendocrine therapy of inflammatory diseases. Mediator Inflamm 2: 263–270

Fattori E, Cappelletti M, Cost P, Sellitto C, Cantoni L, Carelli M, Faggioni R, Fantuzzi G, Ghezzi P, Poli V (1994) Defective inflammatory response in interleukin 6-deficient mice. J Exp Med 180: 1243–1250

Ferretti M, Casini-Raggi V, Pizarro TT, Eisenberg SP, Nast CC, Cominelli F (1994) Neutralization of endogenous IL-1 receptor antagonist exacerbates and prolongs inflammation in rabbit immune colitis. J Clin Invest 94: 449–453

Fischer E, Marano MA, Barber AE, Hudson AA, Lee K, Rock CS, Hawes AS, Thompson RC, Hayes TV, Benjamin WR, Lowry SF, Moldawer LL (1991) A comparison between the effects of interleukin-1α administration and sublethal endotoxemia in primates. Am J Physiol 261: R442–R449

Fischer E, van Zee KJ, Marano MA, Rock CS, Kenney JS, Poutsiaka DD, Dinarello CA, Lowry SF, Moldawer LL (1992) Interleukin-1 receptor antagonist circulates in experimental inflammation and in human disease. Blood 79: 2196–2200

Fisher CJJ, Opal SM, Dhainaut J-F, Stephens S, Zimmerman JL, Nightingale P, Harris SJ, Schein RMH, Panacek EA, Vincent J-L, Foulke GE, Warren EL (1993) Influence of an anti-tumor necrosis factor monoclonal antibody on cytokine levels in patients with sepsis. Crit Care Med 21: 318–327

Fisher CJJ, Dhainaut JF, Opal SM, Pribble JP, Balk RA, Slotman GJ, Iberti TJ, Rackow EC, Shapiro MJ, Greenman RL (1994a) Recombinant human interleukin-1 receptor antagonist in the treatment of patients with sepsis syndrome. Results from a randomized, double blind, placebo-controlled trial. JAMA 271: 1836–1843

Fisher CJJ, Slotman GJ, Opal SM, Pribble J, Bone RC, Emmanuel G, Ng D, Bloedow DC, Catalano MA (1994b) Initial evaluation of human recombinant interleukin-1 receptor antagonist in the treatment of sepsis syndrome: a randomized, open-label, placebo-controlled multicenter trial. Crit Care Med 22: 12–21

Fong Y, Tracey KJ, Moldawer LL, Hesse DG, Manogue KB, Kenney JS, Lee AT, Kuo GC, Allison AC, Lowry SF, Cerami A (1989) Antibodies to cachectin/tumor necrosis factor reduce interleukin 1β and interleukin 6 appearance during lethal bacteremia. J Exp Med 170: 1627–1633

Freshney NW, Rawlinson L, Guesdon F, Jones E, Cowley S, Hsuan J, Saklatvala J (1994) Interleukin-1 activates a novel protein cascade that results in the phosphorylation of hsp27. Cell 78: 1039–1049

Friedman B, Mure A, Brathwaite C, Quinn J, Shapiro E, Melhorn L, Slotman G (1994) The effects of human recombinant interleukin-1 receptor antagonist (IL-1ra) on plasma levels of protaglandins and leukotrienes in septic patients. Crit Care Med 22: A193

Galcheva-Gargova Z, Dèrijard B, Wu I-H, Davis RJ (1994) An osmosensing signal transduction pathway in mammalian cells. Science 265: 806–809

Gearing AJH, Beckett P, Christodoulous M, Churchill M, Clements J, Davidson AH, Drummond AH, Galloway WA, Gilbert R, Gordon JL, Leber TM, Mangian M, Miller K, Nayee P, Owen K, Patel S, Thomas W, Wells G, Wood LM, Woodley K (1994) Processing of tumour necrosis factor-a precursor by metalloproteinases. Nature 370: 555-567

Gershenwald JE, Fong YM, Fahey TJ, Calvano SE, Chizzonite R, Kilian PL, Lowry SF, Moldawer LL (1990) Interleukin 1 receptor blockade attenuates the host inflammatory response. Proc Natl Acad Sci USA 87: 4966–4970

Ghezzi P, Dinarello CA (1988) IL-1 induces IL-1. III. Specific inhibition of IL-1 production by IFN-γ. J Immunol 140: 4238–4244

Girardin E, Roux-Lombard P, Grau GE, Suter P, Gallati H, Group JS, Dayer JM (1994) Imbalance between tumor necrosis factor α and soluble TNF receptor concentrations in severe meningococcemia. Immunology (in press)

Giri JG, Wells J, Dower SK, Mc Call CE, Guzman RN, Slack J, Bird TA, Shanebeck K, Grabstein KH, Sims JE, Alderson MR (1994) Elevated levels of shed type II IL-1 receptor in sepsis. J Immunol 153: 5802–5813

Granowitz EV, Santos A, Poutsiaka DD, Cannon JG, Wilmore DA, Wolff SM, Dinarello CA (1991) Circulating interleukin-1 receptor antagonist levels during experimental endotoxemia in humans. Lancet 338: 1423–1424

Granowitz EV, Clark BD, Vannier E, Callahan MV, Dinarello CA (1992a) Effect of interleukin-1 (IL-1) blockade on cytokine synthesis. I. IL-1 receptor antagonist inhibits IL-1-induced cytokine synthesis and blocks the binding of IL-1 to its type II receptor on human monocytes. Blood 79: 2356–2363

Granowitz EV, Porat R, Mier JW, Pribble JP, Stiles DM, Bloedow DC, Catalano MA, Wolff SM, Dinarello CA (1992b) Pharmacokinetics, safety, and immunomodulatory effects of human recombinant interleukin-1 receptor antagonist in healthy humans. Cytokine 4: 353–360

Granowitz EV, Vannier E, Poutsiaka DD, Dinarello CA (1992c) Effect of interleukin-1 (IL-1) blockade on cytokine synthesis. II. Interleukin-1 receptor antagonist inhibits lipopolysaccharide-induced cytokine synthesis by human monocytes. Blood 79: 2364–2369

Granowitz EV, Porat R, Mier JW, Orencole SF, Callahan MV, Cannon JG, Lynch EA, Ye K, Poutsiaka DD, Vannier E, Shapiro L, Pribble JP, Stiles DM, Catalano MA, Wolff SM, Dinarello CA (1993) Hematological and immunomodulatory effects of an interleukin-1 receptor antagonist coinfusion during low-dose endotoxemia in healthy humans. Blood 82: 2985–2990

Gronich J, Konieczkowski M, Gelb MH, Nemenoff RA, Sedor JR (1994) Interleukin-1α causes a rapid activation of cytosolic phospholipase A2 by phosphorylation in rat mesangial cells. J Clin Invest 93: 1224–1233

Guesdon F, Freshney N, Waller RJ, Rawlinson L, Saklatvala J (1993) Interleukin 1 and tumor necrosis factor stimulate two novel protein kinases that phosphorylate the heat shock protein hsp27 and betacasein. J Biol Chem 268: 4236–4243

Guesdon F, Waller RJ, Saklatvala J (1994) Specific activation of β-casein kinase by the inflammatory cytokines interleukin-1 and tumour necrosis factor. Biochem J 304: 761–768

Han J, Lee J-D, Bibbs L, Ulevitch RJ (1994) A MAP kinase targeted by endotoxin and hyperosmolarity in mammalian cells. Science 265: 808–811

Hannum CH, Wilcox CJ, Arend WP, Joslin FG, Dripps DJ, Heimdal PL, Armes LG, Sommer A, Eisenberg SP, Thompson RC (1990) Interleukin-1 receptor antagonist activity of a human interleukin-1 inhibitor. Nature 343: 336–340

Hauptmann B, Van Damme J, Dayer JM (1991) Modulation of IL-1 inflammatory and immunomodulatory properties by IL-6. Eur Cytokine Netw 2: 39–46

Henricson BE, Neta R, Vogel SN (1991) An interleukin-1 receptor antagonist blocks lipopolysaccharide-induced colony stimulating factor production and early endotoxin tolerance. Infect Immun 59: 1188–1192

Holtmann H, Wallach D (1987) Down regulation of the receptors for tumor necrosis factor by interleukin 1 and 4 beta-phorbol-12-myristate-13-acetate. J Immunol 139: 1161–1167

Hyams JS, Fitzgerald JE, Wyzga N, Treem WR, Justinich CJ, Kreutzer DL (1994) Characterization of circulating interleukin-1 receptor antagonist expression in children with inflammatory bowel disease. Dig Dis Sci 39: 1893–1899

Jacobs CA, Baker PE, Roux ER, Picha KS, Toivola B, Waugh S, Kennedy MK (1991) Experimental autoimmune encephalomyelitis is exacerbated by IL-1α and suppressed by soluble IL-1 receptor. J Immunol 146: 2983–2989

Jansen J, van der Poll T, Levi M, van Deventer SJH (1995) Inhibition of the release of soluble tumor necrosis factor receptors in experimental endotoxemia by anti-tumor necrosis factor antibody. J Clin Immunol (in press)

Kapas L, Shibaya M, Kimura M, Krueger JM (1994) Inhibition of nitric oxide synthesis suppresses sleep in rabbits. Am J Physiol 266: R151–R157

Kaplanski G, Porat R, Aiura K, Erban JK, Gelfand JA, Dinarello CA (1993) Activated platelets induce endothelial secretion of interleukin-8 in normal vitro via an interleukin-1-mediated event. Blood 81: 2492–2495

Kaplanski G, Farnarier C, Kaplanski S, Porat R, Shapiro L, Bongrand P, Dinarello CA (1994) Interleukin-1 induces interleukin-8 from endothelial cells by a juxacrine mechanism. Blood 84: 4242–4248

Kester M, Siomonson MS, Mene P, Sedor JR (1989) Interleukin-1 generate transmembrane signals from phospholipids through novel pathways in cultured rat mesangial cells. J Clin Invest 83: 718–723

Knaus WA, Harrell FE, Fisher CJ, Wagner DP, Opal SM, Sadoff JC, Draper EA, Walawander CA, Conboy K, Grasela TH (1993) The clinical evaluation of new drugs for sepsis. JAMA 270: 1233–1241

Kobayashi Y, Yamamoto K, Saido T, Kawasaki H, Oppenheim JJ, Matsushima K (1990) Identification of calcium-activated neutral protease as a processing enzyme of human interleukin 1 alpha. Proc Natl Acad Sci USA 87: 5548–5552

Kobayashi Y, Oppenheim JJ, Matsushima K (1991) Human pre-interleukin-1α and 1β: structural features revealed by limited proteolysis. Chem Pharm Bull (Tokyo) 39: 1513–1517

Kracht M, Truong O, Totty NF, Shiroo M, Saklatvala J (1994) Interleukin-1α activates two forms of p54α mitogen-activated protein kinase in rabbit liver. J Exp Med 180: 2017–2027

Kriegler M, Perez C, DeFay K, Albert I, Lu SD (1988) A novel form of TNF/cachectin is a cell surface cytotoxic transmembrane protein: ramifications for the complex physiology of TNF. Cell 53: 45–53

Kuida K, Lippke JA, Ku G, Harding MW, Livingston DJ, Su MS-S, Flavell RA (1995) Altered cytokine export and apoptosis in mice deficient in interleukin-1β converting enzyme. Science (in press)

Kunz D, Mühl H, Walker G, Pfeilschifter J (1994) Two distinct signaling pathways trigger the expression of inducible nitric oxide synthase in rat mesangial cells. Proc Natl Acad Sci USA 91: 5387–5391

Lee JC, Laydon JT, McDonnell PC, Gallagher TF, Kumar S, Green D, McNulty D, Blumenthal MJ, Heys JR, Landvatter SW, Strickler JE, McLaughlin MM, Slemens IR, Fisher SM, Livi GP, White JR, Adams JL, Young PR (1994) A protein kinase involved in the regulation of inflammatory cytokine biosynthesis. Nature 372: 739–747

LeMay LG, Otterness IG, Vander AJ, Kluger MJ (1990) In vivo evidence that the rise in plasma IL-6 following injection of a fever–inducing dose of LPS is mediated by IL-1 beta. Cytokine 2: 199–204

Lesslauer W, Tabuchi H, Gentz M, Brockhaus M, Schlaeger EJ, Grau G, Piquet H, Pointaire P, Vassali P, Loetscher HR (1991a) Recombinant soluble TNF receptor proteins inhibit LPS-induced lethality in mice (Abstr). Cytokine 3: 497

Lesslauer W, Tabuchi H, Gentz R, Brockhaus M (1991b) Recombinant soluble tumor necrosis factor receptor proteins protect mice from lipopolysaccharide-induced lethality. Eur J Immunol 21: 2883–2886

Li P, Allen H, Banerjee S, Franklin S, Herzog L, Johnston C, McDowell J, Paskind M, Rodman L, Salfeld J, Towne E, Tracey D, Wardwell S, Wei F-Y, Wong W, Kamen R, Seshadri T (1995) Mice deficient in interleukin-1 converting enzyme (ICE) are defective in production of mature interleukin-1β and resistant to endotoxic shock. Cell 80: 401–411

Lin T-A, Kong X, Haystead TAJ, Pause A, Belsham G, Sononberg N, Lawrence JC (1994) PHAS-I as a link between mitogen activated protein kinase and translation initiation. Science 266: 653–655

Lonnemann G, Endres S, van der Meer JW, Cannon JG, Koch KM, Dinarello CA (1989) Differences in the synthesis and kinetics of release of interleukin 1 alpha, interleukin 1 beta and tumor necrosis factor from human mononuclear cells. Eur J Immunol 19: 1531–1536

Mandrup-Poulsen T, Wogensen L, Jensen M, Svensson P, Nilsson P, Emdal T, Mølvig J, Dinarello CA, Nerup J (1995) Circulating interleukin-1 receptor antagonist concentrations are increased in adult patients with thermal injury. Crit Care Med 23: 26–33

Markewitz A, Faist E, Lang S, Endres S, Fuchs B, Reichart B (1993) Successful restoration of cell-mediated immune response after cardiopulmonary bypass by immunomodulation. J Thorac Cardiovasc Surg 105: 15–24

Mathias S, Younes A, Kan C-C, Orlow I, Joseph C, Kolesnick RN (1993) Activation of the sphingomyelin signaling pathway in intact EL4 cells and in a cell-free system by IL-1β. Science 259: 519–522

Mathison JC, Wolfson E, Ulevitch RJ (1988) Participation of tumor necrosis factor in the mediation of gram negative bacterial lipopolysaccharide-induced injury in rabbits. J Clin Invest 81: 1925–1937

McGeehan GM, Becherer JD, Bast R Jr, Boyer CM, Champion B, Connolly KM, Conway JG, Furdon P, Karp S, Kidao S (1994) Regulation of tumour necrosis factor-alpha processing by a metalloproteinase inhibitor. Nature 370: 558–561

Meydani SN, Dinarello CA (1993) Influence of dietary fatty acids on cytokine production and its clinical implications. Nutr Clin Pract 8: 65–72

Meydani SN, Endres S, Woods MM, Goldin BR, Soo C, Morrill-Labrode A, Dinarello CA, Gorbach SL (1990) Oral (n-3) fatty acid supplementation suppresses cytokine production and lymphocyte proliferation: comparison between young and older women. J Nutr 121: 547–555

Meydani SN, Lichtenstein AH, Cornwall S, Meydani M, Goldin BR, Rasmussen H, Dinarello CA, Schaefer EJ (1993) Immunological effects of National Cholesterol Education Panel (NCEP) step-2 diets with and without fish-derived n-3 fatty acid enrichment. J Clin Invest 92: 105–113

Miller LC, Lynch EA, Isa S, Logan JW, Dinarello CA, Steere AC (1993) Balance of synovial fluid IL-1β and IL-1 receptor antagonist and recovery from Lyme arthritis. Lancet 341: 146–148

Mizel SB (1990) Cyclic AMP and interleukin-1 signal transduction. Immunol Today 11: 390–391

Mohler KM, Sleath PR, Fitzner JN, Cerretti DP, Alderson M, Kerwar SS, Torrance DS, Otten-Evans C, Greenstreet T, Weerawarna K, Kronhein SR, Petersen M, Gerhart M, Kozlosky CL, March CJ, Black RA (1994) Protection against a lethal dose of endotoxin by an inhibitor of tumour necrosis factor processing. Nature 370: 218–220

Molvig J, Pociot F, Svingen BA, Mandrup-Poulsen T, Nerup J (1991) Dietary supplementation with n-3 fatty acids decreases mononuclear proliferation and interleukin-1β content but not monokine secretions in insulin-dependent diabetic individuals. Scand J Immunol 34: 399–410

Moncada S, Palmer RMJ, Higgs EA (1991) Nitric oxide: physiology, pathophysiology, and pharmacology. Pharmacol Rev 43: 109–142

Moore KW, O'Garra A, de Waal Malefyt R, Vieira P, Mosmann TR (1993) Interleukin-10. Annu Rev Immunol 11: 165–190

Movat HZ, Burrowes CE, Cybulsky MI, Dinarello CA (1987) Acute inflammation and a Shwartzman-like reaction induced by interleukin-1 and tumor necrosis factor. Synergistic action of the cytokines in the induction of inflammation and microvascular injury. Am J Pathol 129: 463–476

Munoz E, Beutner U, Zubiaga A, Huber BT (1990) IL-1 activates two separate signal transduction pathways in T helper type II cells. J Immunol 144: 964–969

O'Neill LAJ, Bird TA, Saklatvala J (1990) Interleukin-1 signal transduction. Immunol Today 11: 392–394

Okusawa S, Gelfand JA, Ikejima T, Connolly RJ, Dinarello CA (1988) Interleukin 1 induces a shock-like state in rabbits. Synergism with tumor necrosis factor and the effect of cyclooxygenase inhibition. J Clin Invest 81: 1162–1172

Oliveira IC, Mukaida N, Matsushima K, Vilcek J (1994) Transcriptional inhibition of the interleukin-8 gene by interferon is mediated by the NFκB site. Mol Cell Biol 14: 5300–5308

Olivera DL, Laydon JT, Hillegass L, Badger AM, Lee JC (1993) Effects of pyridinyl imidazole compounds on murine TNFα production. Agents Actions 39: C55–C57

Olsson I, Gatanaga T, Gullberg U, Lantz M, Granger GA (1993) Tumour necrosis factor (TNF) binding proteins (soluble TNF receptor forms) with possible roles in inflammation and malignancy. Eur Cytokine Netw 4: 169–180

Opal SM (1996) Clinical trials of novel therapeutic agents: why did they fail? In: Vincent J-L (ed) Yearbook of intensive care and emergency medicine. Springer, Berlin, Heidelberg, New York, 425–436

Orencole SF, Vannier E, Dinarello CA (1994) Detection of soluble IL-1 receptor type Ii (IL-1sRII) in sera and plasma from healthy volunteers (Abstr). Cytokine 6: 554

Palmer RM, Hickery MS, Charles IG, Moncada S, Bayless MT (1994) Induction of nitric oxide synthase in human chrondrocytes. Biochem Biophys Res commun 193: 398–405

Porat R, Poutsiaka DD, Miller LC, Granowitz EV, Dinarello CA (1992) Interleukin-1 (IL-1) receptor blockade reduces endotoxin and *Borrelia burgdorferi*-stimulated IL-8 synthesis in human mononuclear cells. FASEB J 6: 2482-2486

Porat R, Paddock HN, Schwaitzberg SD, Connolly RJ, Wilkens T, Dasch JR, Gascon M-P, Hutchison JS, Ythier A, Wallach D, Dinarello CA (1995) Glycosylated recombinant human TNF binding protein-1 reduces mortality, shock and production of TNF in rabbit *Escherichia coli* sepsis. Crit Care Med (in press)

Reimers JI, Bjerre U, Mandrup-Poulsen T, Nerup J (1994) Interleukin-1β induces diabetes and fever in normal rats by nitric oxide via induction of different nitric oxide synthases. Cytokine 6: 512–520

Reinhart K, Wiegand-Löhnert C, Grimminger F, Kaul M, Withington D, Treatcher J, Eckart J, Willats S, Guerrero-Sanz J, Krausch D, Stockenhuber F, Eiselstein J, Kempeni J (1995) Treatment of severe sepsis with antiTNF monoclonal antibody MAK 195F: dose-dependent reduction of mortality in patients with elevated IL-6 serum levels (Abstr). Clin Intensive Care 6: 82

River C, Shen GH (1994) In the rat, endogenous nitric oxide modulates the response of the hypothalamic-pituitary-adrenal axis to interleukin-1β, vasopressin and oxytocin. J Neurosci 14: 1985–1993

Rock CS, Coyle SM, Keogh CV, Lazarus DD, Hawes AS, Leskiw M, Moldawer LL, Stein TP, Lowry SF (1992) Influence of hypercortisolemia on the acute-phase protein response to endotoxin in humans. Surgery 112: 467–74

Rordorf-Adam C, Geiger T, Henn R, Arnold J, Solf I, Wiesenberg PG, Ferrini PG, Vosbeck K (1994) CGP 47969A: a novel inhibitor of the synthesis of inflammatory cytokines. Agents Actions 41: 29–35

Rosoff PM, Savage N, Dinarello CA (1988) Interleukin-1 stimulates diacylglycerol production in T lymphocytes by a novel mechanism. Cell 54: 73–81

Rossi V, Breviario F, Ghezzi P, Dejana E, Mantovani A (1985) Prostacyclin synthesis induced in vascular cells by interleukin-1. Science 229: 174–176

Salvemi D, Misko TP, Masferrer JL, Seibert K, Currie MG, Needleman P (1993) Nitric oxide activates cyclooxygenase enzymes. Proc Natl Acad Sci USA 90: 7240–7244

Santos AA, Scheltinga MR, Lynch E, Brown EF, Lawton P, Chambers E, Browning J, Dinarello CA, Wolff SM, Wilmore DW (1993) Elaboration of interleukin-1 receptor antagonist is not attenuated by glucocorticoids after endotoxemia. Arch Surg 128: 138–144

Schade UF, Burmeister I, Engel R (1987) Increased 13-hydroxyoctadecadienoic acid content in lipopolysacharide stimulated macrophages. Biochem Biophys Res Commun 147: 695–700

Schindler R, Ghezzi P, Dinarello CA (1990a) IL-1 induces IL-1. IV. IFN-γ suppresses IL-1 but not lipopolysaccharide-induced transcription of IL-1. J Immunol 144: 2216–2222

Schindler R, Mancilla J, Endres S, Ghorbani R, Clark SC, Dinarello CA (1990b) Correlations and interactions in the production of interleukin-6 (IL-6), IL-1, and tumor necrosis factor (TNF) in human blood mononuclear cells: IL-6 suppresses IL-1 and TNF. Blood 75: 40–47

Seckinger P, Lowenthal JW, Williamson K, Dayer JM, MacDonald HR (1987) A urine inhibitor of interleukin-1 activity that blocks ligand binding. J Immunol 139: 1546–1549

Shapiro L, Clark BD, Orencole SF, Poutsiaka DD, Granowitz EV, Dinarello CA (1993) Detection of tumor necrosis factor soluble receptor p55 in blood samples from healthy and endotoxemic humans. J Infect Dis 167: 1344–1350

Shapiro L, Panayotatos N, Meydani SN, Wu D, Dinarello CA (1994) Ciliary neurotrophic factor combined with soluble receptor inhibits synthesis of pro-inflammatory cytokines and prostaglandin-E2 in vitro. Exp Cell Res 215: 51–56

Singer II, Scott S, Chin J, Kostura MJ, Miller DK, Chapman K, Bayne EK (1993) Interleukin-1β converting enzyme is localized on the external cell-surface membranes and in the cytoplasmic ground substance of activated human monocytes by immuno-electronmicroscopy (Abstr). Lymphokine Cytokine Res 12: 340

Sirko S, Schindler R, Doyle MJ, Weisman SM, Dinarello CA (1991) Transcription, translation and secretion of interleukin-1 and tumor necrosis factor: effects of tebufelone, a dual cyclooxygenase/5-lipoxygenase inhibitor. Eur J Immunol 21: 242–250

Smith JW, Urba WJ, Curti BD, Elwood LJ, Steis RG, Janik JE, Sharfman WH, Miller LL, Fenton RG, Conlon KC, Rossio J, Kopp W, Shimuzut M, Oppenheim JJ, Longo D (1992) The toxic and hematologic effects of interleukin-1 alpha administered in a phase I trial to patients with advanced malignancies. J Clin Oncol 10: 1141–1152

Spinas GA, Bloesch D, Keller U, Zimmerli W, Cammisuli S (1991) Pretreatment with ibuprofen augments circulating tumor necrosis factor, interleukin-6 and elastase during acute endotoxemia. J Infect Dis 163: 89–95

Stevenson FT, Bursten SL, Locksley RM, Lovett DH (1992a) Myristyl acylation of the tumor necrosis factor-a precursor on specific lysines. J Exp Med 176: 1053–1062

Stevenson FT, Torrano F, Locksley RM, Lovett DH (1992b) Interleukin-1: the patterns of translation and intracellular distribution support alternative secretory mechanisms. J Cell Physiol 152: 223–231

Stevenson FT, Bursten SL, Fanton C, Locksley RM, Lovett DH (1993) The 31-kDa precursor of interleukin-1α is myristoylated on specific lysines within the 16-kDa N-terminal propiece. Proc Natl Acad Sci USA 90: 7245-7249

Szabo C, Wu C-C, Gross SS, Thiemermann C, Vane JR (1993) Interleukin-1 contributes to the induction of nitric oxide synthase by endotoxin in vivo. Eur J Pharmacol 150: 157–160

Tilg H, Vannier E, Vachino G, Dinarello CA, Mier JW (1993) Antiinflammatory properties of hepatic acute phase proteins: preferential induction of interleukin-1 (IL-1) receptor antagonist over interleukin-1β synthesis by human peripheral blood mononuclear cells. J Exp Med 178: 1629–1636

Tilg H, Trehu E, Atkins MB, Dinarello CA, Mier JW (1994a) Interleukin-6 (IL-6) as an anti-inflammatory cytokine: induction of circulating IL-1 receptor antagonist and soluble tumor necrosis factor receptor p55. Blood 83: 113–118

Tilg H, Trehu E, Shapiro L, Pape D, Atkins MB, Dinarello CA, Mier JW (1994b) Induction of circulating soluble tumour necrosis factor receptor and interleukin 1 receptor antagonist following interleukin-1α infusion in humans. Cytokine 6: 215–219

Tracey K, Fong Y, Hesse DG, Manogue KR, Lee AT, Kuo GC, Lowry SF, Cerami A (1987) Anti-cachectin/TNF monoclonal antibodies prevent septic shock during lethal bacteremia. Nature 330: 662–664

Tredget EE, Yu YM, Zhong S, Burini R, Okusawa S, Gelfand JA, Dinarello CA, Young VR, Burke JF (1988) Role of interleukin 1 and tumor necrosis factor on energy metabolism in rabbits. Am J Physiol 255: E760–E768

Ucla C, Roux-Lombard P, Fey S, Dayer J-M, Mach B (1990) Interferon gamma drastically modifies the regulation of interleukin 1 genes by endotoxin in U937 cells. J Clin Invest 85: 185–191

Van der Poll T, Bueller HR, ten Cate H, Wortel CH, Bauer KA, van Deventer SJH, Hack CE, Sauerwein HP, Rosenberg RD, ten Cate JW (1990) Activation of coagulation after administration of tumor necrosis factor to normal subjects. N Engl J Med 322: 1622–1627

Van der Poll T, van Deventer SJH, Hack CE, Wolbink GJ, Aarden LA, Büller HR, ten Cate JW (1991) Effects of leukocytes following injection of tumor necrosis factor into healthy humans. Blood 79: 693–698

Van der Poll T, van Deventer SJH, ten Cate H, Levi M, ten Cate JW (1994) Tumor necrosis factor is involved in the appearance of interleukin 1 receptor antagonist in endotoxemia. J Infect Dis 169: 665–667

Van Zee KJ, Kohno T, Fischer E, Rock CS, Moldawer LL, Lowry SF (1992) Tumor necrosis factor soluble receptors circulate during experimental and clinical inflammation and can protect against excessive tumor necrosis factor-α in vitro and in vivo. Proc Natl Acad Sci USA 89: 4845–4849

Vannier E, Dinarello CA (1993) Histamine enhances interleukin (IL)-1-induced IL-1 gene expression and protein synthesis via H2 receptors in peripheral blood mononuclear cells: comparison with IL-1 receptor antagonist. J Clin Invest 92: 281–287

Vannier E, Dinarello CA (1994) Histamine enhances interleukin (IL)-1-induced IL-6 gene expression and protein synthesis via H2 receptors in peripheral blood mononuclear cells. J Biol Chem 269: 9952–9956

Vannier E, Lefort J, Lellouch-Tubiana A, Terlain B, Vargaftig BB (1991a) Lipopolysaccharide from E. coli reduces antigen-induced bronchoconstriction in actively sensitized guinea pigs. J Clin Invest 87: 1936–1944

Vannier E, Miller LC, Dinarello CA (1991b) Histamine suppresses gene expression and synthesis of tumor necrosis factor α via histamine H2 receptors. J Exp Med 174: 281–284

Vannier E, Miller LC, Dinarello CA (1992) Coordinated anti-inflammatory effect of IL-4: IL-4 down regulates IL-1 synthesis but up regulates IL-1ra production. Proc Natl Acad Sci USA 89: 4076–4080

Vincent J-L, Bakker J, Marècaux G, Schandene L, Kahn RJ, Dupont E (1992) Administration of anti-TNF antibody improves left ventricular function in septic shock patients. Chest 101: 810–815

Waring PM, Waring LJ, Billington T, Metcalf D (1995) Leukemia inhibitory factor protects against experimental lethal *Escherichia coli* septic shock in mice. Proc Natl Acad Sci USA 92: 1337–1341

Watanabe N, Kobayashi Y (1994) Selective release of a processed form of interleukin-lα. Cytokine 6: 597–601

Young PR, McDonnell PC, Laydon JT, Gallagher TF, Green D, Blumenthal MJ, McNulty D, Strickler JE, Kumar S, Heys JR, Adams JL, Landvatter SW, Lee JC (1994) A novel MAP kinase regulates the production of IL-1 and TNF in LPS activated human monocytes. Cytokine 6: 564

Zheng H, Fletcher D, Kozak W, Jiang M, Hofmann K, Conn CC, Siszynski D, Grabiec C, Trumbauer MA, Shaw AR, Kostura MJ, Stevens K, Rosen H, North RJ, Chen HY, Tocci MJ, Kluger MJ, Van der Ploeg LHT (1995) Resistance to fever induction and impaired acute-phase response in interleukin-1β deficient mice. Immunity (in press)

Tumor Necrosis Factor and the Systemic Inflammatory Response Syndrome

P. BROUCKAERT and W. FIERS

1 Introduction 167
2 TNF: Structure and Mechanism of Action 168
3 TNF Is Involved in SIRS 171
4 Two Receptors for TNF 172
5 Seed and Soil: Other Factors Determine the Outcome of a Challenge with TNF 174
5.1 Sensitization 174
5.2 Tolerance 175
5.3 Desensitization 175
6 TNF and Other Cytokine Mediators of SIRS 176
6.1 IL-1 176
6.2 IL-6 177
6.3 IL-12 177
6.4 IFN-γ 178
7 TNF and Nitric Oxide 178
8 α_1-Acid Glycoprotein 179
9 Conclusions and Prospects 180
References 181

1 Introduction

Tumor necrosis factor (TNF) was discovered in the middle 1970s by the research group of OLD (CARSWELL et al. 1975; OLD 1988) as the result of a long search for the mechanisms underlying endotoxin-induced antitumor effects. However, it was only after the cloning of the genes for TNF in 1984–1985 (PENNICA et al. 1984; WANG et al. 1985; FRANSEN et al. 1985; MARMENOUT et al. 1985; SHIRAI et al. 1985) that sufficient amounts of pure material became available for detailed investigation of its effects. Studies in a syngeneic murine melanoma model confirmed the antitumor effectiveness of TNF, especially in combination with interferon (IFN)-γ, but also showed that when the homologous murine (m) TNF is used, the effective dose is close to the lethal dose in systemic administrations

Laboratory of Molecular Biology, Flanders Interuniversity Institute for Biotechnology, University of Ghent, K.L. Ledeganckstraat 35, 9000 Ghent, Belgium

(Brouckaert et al. 1986). Subsequent clinical trials have largely confirmed the major conclusions drawn from those preclinical experiments. When locoregional treatments such as the isolated limb perfusion technique are used for treatment of melanoma, a rise in the complete response rate from 50% to 80%–100% is observed when TNF and IFN-γ are added to the classical treatment with melphalan (Lejeune et al. 1994). Also in the case of sarcomas a 91% limb salvage rate has been obtained (Eggermont et al. 1993). Systemic treatments with these cytokines, however, are still impossible due to the severe systemic inflammatory response syndrome (SIRS) that occurs when even subtherapeutic doses of TNF are administered. Hypotension and hepatotoxicity are the most prominent dose-limiting toxicities observed in clinical trials (Fraker et al. 1995). This chapter discusses only the role of TNF as an inducer and mediator of SIRS. Current knowledge regarding the antitumor effects of TNF has recently been reviewed elsewhere (Brouckaert et al. 1993a, 1994; Fiers 1995).

The role of TNF as a mediator of endotoxic shock was first described by Beutler et al. (1985a,b) and Tracey et al. (1986), who cloned TNF as the result of their search for a cachexia-inducing factor. Subsequent studies in animals, patients, and human volunteers have yielded a large body of evidence for such a role of TNF. This has been reviewed extensively (e.g., Tracey and Cerami 1993; Beutler and Grau 1993). This chapter focuses on observations that question the early paradigm that TNF is both a necessary and a sufficient mediator of sepsis. In fact, since 1985 quite a number of studies have shown that the outcome of a challenge with TNF is highly dependent on the presence or absence of other factors. More recent results obtained in transgenic TNF receptor knockout mice even raise the question of whether TNF is indeed a necessary and important factor in endotoxic shock.

The data that form the basis for our knowledge of the relationship between TNF and SIRS originate mainly from two distinct research areas: the development of TNF as an anticancer agent and the development of TNF inhibitors as a treatment for sepsis. To understand the approaches taken in different studies it is important to note some major differences between the two fields. When administering TNF as an anticancer drug, it is (at least theoretically) possible to avoid synergistic factors. Moreover, drugs that are active only when administered before TNF remain very valuable and useful, and this is in contrast to the situation in sepsis. On the other hand, inhibitors that completely destroy the activity of TNF might be applicable in sepsis, while this is obviously not valid in antitumor treatment.

2 TNF: Structure and Mechanism of Action

The mRNA for TNF codes for a presequence of 76 amino acids (79 amino acids in the mouse) followed by a mature protein of 157 amino acids (156 amino acids in the mouse) and is transcribed from a single gene located in the middle of the MHC locus (SPIES et al. 1991). It is expressed as a 26-kDa integral membrane protein from which a 17-kDa mature TNF protein subunit is released by proteolytic cleavage. Although the protease responsible has not yet been identified, it has recently been shown that synthetic metalloproteinase inhibitors can specifically interfere with this step, both in vitro and in vivo (GEARING et al. 1994; MCGEEHAN et al. 1994; MOHLER et al. 1994). Both the membrane-bound form and the secreted form of TNF are biologically active (PEREZ et al. 1990). The native TNF molecule is a trimer consisting of three identical subunits held together by hydrophobic forces (WINGFIELD et al. 1987) which can be dissociated to inactive monomers by mild detergents (SMITH and BAGLIONI 1987) or suramin (ALZANI et al. 1993). The three-dimensional structure of TNF has been solved (JONES et al. 1989; ECK and SPRANG 1989), and the receptor-binding site has been located in the lower half of the cleft between two subunits of this triangular pyramid (YAMAGISHI et al. 1990; VAN OSTADE et al. 1991), implying the existence of three receptor-binding sites on each TNF molecule.

In addition to activated macrophages, which form the major source of TNF, a number of other cell types are also able to produce TNF: T lymphocytes, natural killer (NK) cells, endothelial cells, mast cells, glial cells, Kupffer cells, granulosa cells, and smooth muscle cells. The production of TNF is regulated at multiple levels: transcription, mRNA stability, and translation (BEUTLER et al. 1992; KRUYS et al. 1993). Pentoxifylline (BEUTLER et al. 1992) and thalidomide (SAMPAIO et al. 1991) inhibit synthesis at the transcriptional level, while glucocorticoids are active mainly at the posttranscriptional level, as glucocorticoid-treated macrophages accumulate the TNF message in abundance but fail to produce the TNF protein (BEUTLER et al. 1986). In addition to glucocorticoids, which are themselves induced by TNF, a number of other endogenous molecules suppress the induction of TNF: prostaglandin E_2 (BEUTLER et al. 1992), interleukin (IL)-4 (HART et al. 1989), transforming growth factor-β (FLYNN and PALLADINO 1992), and most predominantly IL-10 (FIORENTINO et al. 1991).

TNF exerts its activities by clustering two distinct types of receptor: TNF-R55 and TNF-R75 (VANDENABEELE et al. 1995a). Most interestingly, human (h) TNF does not interact with mTNF-R75 (LEWIS et al. 1991). Mutants of hTNF have been engineered that recognize selectively either hTNF-R55 or hTNF-R75 (VAN OSTADE et al. 1993, 1994; LOETSCHER et al. 1993). The biological significance of these receptors is discussed in a later section. In the liver a third type of TNF-R has been reported, but its significance remains unclear (SCHWALB et al. 1993).

Both TNF-Rs can give rise to circulating TNF-binding proteins following proteolytic cleavage at the base of their extracellular domains. These circulating proteins, called soluble (s) TNF-R, play a major role in the clearing of TNF. Depending on the situation, they may act as an inhibitor or as a reservoir. It should be noted that while the interaction between TNF and TNF-R55 is rather stable, that between TNF and TNF-R75 is very labile. This has important consequences for the use of sTNF-R immunoglobulin constructs as potential inhibitors of TNF-induced pathology in vivo.

Activation of the receptors, which is brought about by binding of the trimeric TNF and subsequent clustering, results in both cytoplasmic and nuclear effects. The way in which the signals are transduced is currently being unraveled (Vandenabeele et al. 1995a). Evidence has been presented that proteins associate with the intracellular domains of TNF-R55 and TNF-R75, and that both receptors can be phosphorylated by an associated kinase (Darnay et al. 1994a,b). Also, two new putative signal transducers, TNF-R associated factors 1 and 2 have been identified as being associated with the cytoplasmic domain of TNF-R75 (Rothe et al. 1994).

A number of intracellular events have been described to occur after triggering of TNF-Rs. It is unclear at present, however, which ones are particular for a given cell line, which are epiphenomena, and what the relationship between the various signals is. In addition to the activation of phospholipases, sphingomyelinases, and kinases, especially of the mitogen-activating protein kinase pathways, the perturbation of the electron transport chain in mitochondria may be of particular relevance for the situation in SIRS (Schulze-Osthoff et al. 1992, 1993). TNF also induces the production and/or activation of a number of transcription factors such as nuclear factor κB, AP-1, interferon-regulator factor 1, nuclear factor–IL-6, cAMP response element binding protein, and serum-responsive factor (Fiers 1995). This in turn leads to gene induction and synthesis of a number of proteins which might well be involved in SIRS: other cytokines (such as IL-1, IL-6, IL-8, granulocyte-macrophage colony-stimulating factor, granulocyte colony-stimulating factor), adhesion molecules (such as ICAM-1, E-selectin, CD11/18), proteins involved in the coagulation and fibrinolytic pathway (such as tissue factor, urokinase, plasminogen-activator inhibitor types 1 and 2), inducible nitric oxide (NO) synthase and proteases, to name only a few (Fiers 1995).

These intracellular events lead to the activation of the target cells. Although almost every cell type bears receptors for TNF, leukocytes and endothelial cells may be considered to be the most important target cell types for the SIRS-inducing effects of TNF. The activation of these two cell types leads to a number of pro-inflammatory and procoagulant changes. TNF induces the expression of E-selectin, ICAM-1, and VCAM-4 and promotes leukocyte-endothelial cell adhesion (Bevilacqua el al. 1987; Gamble et al. 1992). It also induces neutrophil activation. It contributes to the procoagulant climate by induction of tissue factor and tissue plasminogen activator inhibitor and by the inhibition of protein C and tissue plasminogen activators (Bevilacqua et al. 1986; Clauss et al. 1992). Also important is the induction of NO in endothelial cells and vascular smooth muscle cells (Geng et al. 1992), which is responsible for the observed hypotension.

In addition to its role is SIRS and as a potential antitumor therapeutic, TNF is also involved in the pathogenesis of several other diseases, such as cerebral malaria and rheumatoid arthritis (Tracey and Cerami 1994). Further details about the structure and function of TNF and its activities on various cell types can be found in a recent review (Fiers 1995).

3 TNF Is Involved in SIRS

There is ample evidence for the role of TNF as an inducer of SIRS and as a mediator of SIRS induced by, for example, bacteria or α-CD3 treatment. When injected into animals, TNF can induce a number of symptoms that closely mimic those observed in sepsis. TNF injected into rats leads to hypotension, tachycardia, tachypnea, lactic acidosis, hemoconcentration, hyperkaliemia and hyperglycemia, followed by hypoglycemia (TRACEY et al. 1986). Studies in other animals and in human volunteers receiving TNF have largely confirmed that TNF can induce metabolic and pathological effects very similar to those observed after endotoxin (TRACEY and CERAMI 1993; BEUTLER and GRAU 1993).

Although early reports suggested a relationship between TNF levels and the severity of sepsis (WAAGE et al. 1987; GIRARDIN et al. 1988), this has not been confirmed by further studies. A review of the available data shows that TNF levels do not discriminate between surviving and nonsurviving patients (LOWRY et al. 1994). Moreover, patients treated with TNF using the isolated limb perfusion technique had in fact much higher levels of TNF in their general circulation than septic shock patients, but the resulting toxicity does not mimic septic shock (LEJEUNE et al. 1994). In particular, neither adult repiratory distress syndrome nor multiple organ failure were observed, and all toxicities were readily reversible. This may be due to the fact that these patients received high bolus amounts of TNF in circulation while sepsis patients could be exposed to a more prolonged production of lower amounts of TNF. It has been suggested that it is indeed the more sustained presence of low doses of TNF that is correlated with a fatal outcome in sepsis (OFFNER et al. 1990). Another element contributing strongly to the observed differences is without doubt that in the case of sepsis a number of other synergistic or sensitizing factors such as endotoxin, IL-1, or IL-12 are also present. It is worth noting that also in patients suffering from leakage during isolated limb perfusion no correlation is observed between levels of TNF in the general circulation and severity of clinical symptoms (LEJEUNE et al. 1994)

The administration of monoclonal antibodies against TNF prevents lethality in a number of sepsis models (TRACEY and CERAMI 1993). sTNF-R immunoglobulin fusion proteins also provide protection, but important differences have recently been observed between the effects of such constructs with sTNF-R55 vs. sTNF-R75. Because of the higher dissociation rate of the sTNF-R75/TNF it did not prevent lethality in an animal model of gram-negative sepsis (EVANS et al. 1994a). Whether such approaches or the use of inhibitors of the synthesis or the maturation cleavage of TNF will be of value in clinical sepsis remains to be determined. It may also be recalled that in models of bacterial peritonitis TNF proves to be a protective rather than a deleterious factor (SHEPPARD et al. 1989; ECHTENACHER et al. 1990)

4 Two Receptors for TNF

Our knowledge regarding the function of each of the receptor types for TNF is derived from four types of experiments. Studies in mice and on murine cells exploit the fact that hTNF is a selective agonist of mTNF-R55 as it does not interact with mTNF-R75 (LEWIS et al. 1991). The knowledge that TNF-R become activated by clustering allowed studies with agonistic monoclonal antibodies. Finally, the engineering of TNF mutants that are selective agonists of either hTNF-R55 or hTNF-R75, and of mice carrying a targeted deletion in either the TNF-R55 or TNF-R75 gene are two important tools that have recently become available.

The earliest results were obtained in in vivo models in mice (reviewed in BROUCKAERT et al. 1993b). It was observed that hTNF is lethal only at 50-fold higher concentrations than mTNF (BROUCKAERT et al. 1992b). Hence it seemed that triggering of TNF-R55 is not sufficient to induce lethality in mice. This, however, holds only for healthy mice receiving TNF as a single agent. Coadministration of agents such as galactosamine (GalN; LEHMANN et al. 1987; BROUCKAERT et al. 1992b), the glucocorticoid receptor antagonist RU38486 (BROUCKAERT et al. 1992a), IL-1 (EVERAERDT et al. 1989), or lipopolysaccharide (LPS; ROTHSTEIN and SCHREIBER 1988), or the presence of some infections or some tumors (see Sect. 5.1) considerably reduces the lethal dose of hTNF and abolishes partially or completely the species specificity. When one analyzes more closely the changes in quantitative parameters induced after TNF administration, one observes that the initial peak changes are very similar after hTNF and mTNF, but that while hTNF induces only transient changes, mTNF induces sustained, irreversible effects (Fig. 1; BROUCKAERT et al. 1992a; TAKAHASHI et al. 1995a).

Studies in TNF-R55$^{0/0}$ mice (PFEFFER et al. 1993; ROTHE et al. 1993) led to the conclusion that such mice are resistant to an otherwise lethal challenge with TNF or LPS in GalN-sensitized mice or with the combination of mTNF and IL-1. This is in agreement with our previous findings since in these situations no differences between hTNF and mTNF were observed. Remarkably, TNF-R55$^{0/0}$ mice turned out to be equally sensitive to the lethal effects of endotoxin as control mice (ROTHE et al. 1993), which raises the question of whether TNF is really an essential factor for lethality in endotoxic shock. TNF-R75$^{0/0}$ mice were marginally resistant to endotoxin and were dramatically less sensitive, but not completely resistant, to the lethal effects of mTNF (ERICKSON et al. 1994). The data, however, do not report the degree of resistance in a quantitative way. TNF-R75$^{0/0}$ mice are also resistant to TNF-induced skin necrosis. This observation has been confirmed by the administration of inhibiting antibodies against TNF-R75 (SHEEHAN et al. 1995).

Although several reports show that, except in T lymphocytes, the triggering of TNF-R55 is sufficient to obtain all TNF-inducible in vitro effects in a qualitative way, more elaborate studies show that interaction with TNF-R75 substantially enhances TNF-R55 mediated effects. Regarding important proinflammatory ef-

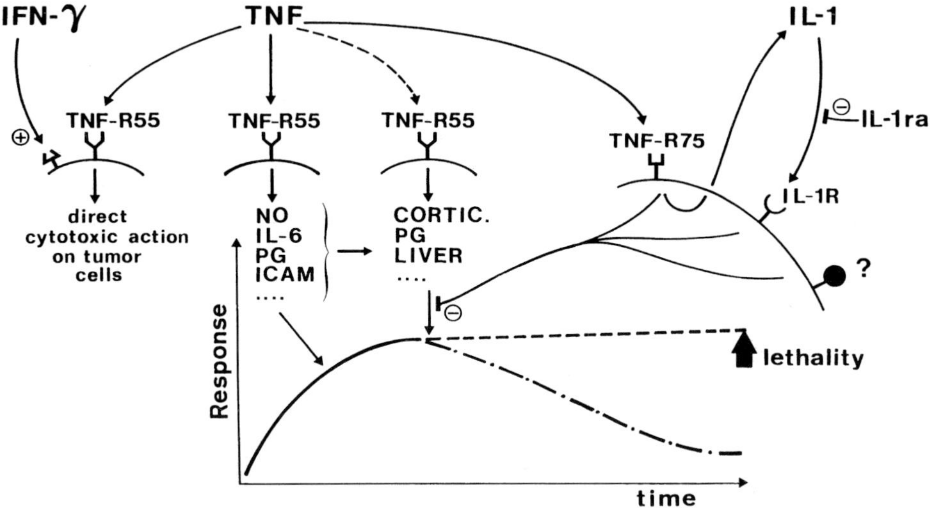

Fig. 1. Different kinetics of induction of mediators following administration of either hTNF or mTNF. Induced mediators, such as IL-6, NO, prostaglandins (*PG*) etc., reach peak levels after about 2–3 h following hTNF treatment, which acts only on mTNF-R55. Because of feedback mechanisms, at least partially based on induced glucocorticoids, the levels of these mediators subsequently start to decrease. After administration of mTNF, however, which also triggers TNF-R75, or when the IL-1 receptor is triggered, the feedback mechanism becomes inoperative, and the concentration of mediators, such as IL-6, plateaus or continues to increase until death ensues. Triggering of TNF-R55 is sufficient to cause cytotoxicity in many malignant cells, and this effect is strongly enhanced by IFN-γ cotreatment. (From Fiers 1995)

fects on endothelial cells and neutrophils, it was shown that although triggering of TNF-R75 alone had no effect, mutants of hTNF that interact only with TNF-R55 induce maximal effects only at almost 100-fold higher doses than wild-type TNF (Barbara et al. 1994). This cooperation may be the consequence either of ligand passing, where TNF-R75 traps TNF and transfers it to TNF-R55 (Tartaglia et al. 1993), or of true intracellular cooperation between both receptor-induced signaling pathways (Vandenabeele et al. 1995b). On T lymphocytes TNF-R75 seems to be the major signaling receptor (Vandenabeele et al. 1992). Whether TNF-R75 has a signal-transducing function on other cell types, especially as regards cytotoxicity, is subject to controversy.

Mutants of hTNF that are selective agonists of TNF-R55, are being developed as potential anticancer agents. The rationale is based on studies with mTNF in TNF-R75$^{0/0}$ mice and with hTNF in normal mice which indicate that TNF-R55 selective agonists cause lethality only at much higher doses than wild-type TNF. Furthermore, they retain their antitumor effect against human tumors in nude mice (Van Ostade et al. 1993). Studies in baboons, however, reveal that they display a different pharmacokinetic profile, and that their half-life in circulation is three times longer than that of wild-type TNF (Van Zee et al. 1994). This not only reveals an important function for sTNF-R75 in clearing TNF but also makes the interpretation of other data more difficult (Brouckaert et al. 1994).

The slower clearance may be an important advantage, as prolonged exposure of malignant cells to TNF is required for maximal cytotoxic efficacy.

5 Seed and Soil: Other Factors Determine the Outcome of a Challenge with TNF

Although TNF undoubtedly can induce SIRS, albeit not identical to that seen in patients with clinical septic shock, it is the condition of the host rather than the dose of TNF that determines the final outcome of a challenge with TNF. This is the case not only in cancer patients (Sect. 3) but also in animal models. Depending on the condition or on the presence or absence of other substances, the LD_{50} of hTNF for mice can be as low as 0.1 μg/mouse or exceed 500 μg/mouse. The strongest sensitizer described is GalN, a hepatotoxin depleting UTP in the liver (LEHMANN et al. 1987). It should be noted, however, that although presented as a model for endotoxic shock in many papers, the lethality induced in GalN-sensitized animals is mediated by a fulminant hepatic necrosis due to massive apoptosis of individual hepatocytes (LEIST et al. 1995a,b), very similar to what is observed after the administration of α-Fas (OGASAWARA et al. 1993), and as such is of little significance for clinical sepsis. Rather, it may be considered a model for certain forms of hepatitis.

The importance for the outcome of which TNF-R type is triggered is discussed in Sect. 4. Section 6 describes the role of four particular cytokines (IL-1, IL-6, IL-12, and IFN-γ). Here we discuss some of the more complex situations such as sensitization and tolerance.

5.1 Sensitization

Mice carrying certain types of tumor or suffering from bacterial infections are extremely sensitive to the lethal effects of TNF. This is a problem for the treatment of cancer patients with TNF, but especially the sensitization induced by infection is a situation which might be quite relevant for the septic syndrome. Indeed, in this condition TNF is present in a host which may have been infected since some time. Recent studies have provided important new information on the possible mechanisms by which infection induces such a sensitization. It has been observed that IFN-$\gamma R^{0/0}$ mice can no longer be sensitized to the lethal effects of either LPS (KAMIJO et al. 1993) or TNF (CAUWELS et al. 1995) by an infection with bacillus Calmette-Guérin (BCG). However, these mice remain as sensitive to the lethal effects of mTNF in uninfected animals as controls (CAUWELS et al. 1995). IL-12 was found to be a sufficient and necessary mediator (CAUWELS et al., 1996). It therefore seems that BCG and other infections bring about this sensitization by an IL-12 driven production of IFN-γ in NK cells

(CAUWELS et al. 1995, 1996). The target of INF-γ is presently unknown, but it seems that upregulation or activation of adhesion molecules is a central feature since α-Leukocyte-function associated antigen (LFA)-1 is able to prevent TNF-induced lethality in sensitized mice, although not in healthy mice challenged with mTNF (Cauwels et al., submitted). Most unexpectedly, this is not due to an inhibition of LFA-1/ICAM-1 interaction since ICAM-1$^{0/0}$ mice remain as sensitive as control mice (Cauwels et al., submitted). Neither E-selectin nor P-selectin seems to be involved in this process (Cauwels et al., submitted).

When sensitization is induced by the presence of some tumors, such as Lewis lung carcinoma, α-LFA-1 can still protect (Cauwels et al., submitted), but neither NK cells, IL-12, nor IFN-γ seem to be involved (CAUWELS et al. 1995, 1996).

5.2 Tolerance

The repetitive administration of rather low doses of TNF to mice results in a condition in which mice have become tolerant to the lethal effects of mTNF (TAKAHASHI et al. 1991). This tolerance is selective as some effects of TNF are still inducible in tolerant animals. Importantly, the antitumor effect of the combination of TNF and IFN-γ (TAKAHASHI et al. 1991, 1995b) and hypertriglyceridemia (GRUNFELD et al. 1989) are retained in tolerant mice.

When changes in quantitative parameters are determined, one observes in the tolerant mice a pattern that closely resembles the differences seen after the administration of hTNF or mTNF. In tolerant animals the initial peak changes in hypothermia, IL-6 levels, hematocrit increase, and NO production are indistinguishable from those in control animals. However, while these changes are reversible in tolerant animals, they are irreversible in control mice and eventually result in death.

Regarding the mechanism of tolerance, trivial explanations such as changes in the bioavailability of TNF, the involvement of soluble receptors, neutralizing antibodies, and changes in the number of receptors can be excluded (TAKAHASHI et al. 1994). We have recently proposed that the mechanism of tolerance must be found in the functional ablation of (part of) the signaling pathway induced by triggering of TNF-R75 (TAKAHASHI et al. 1995a). Tolerance itself is induced by TNF-R55, while triggering of TNF-R75 or the addition of IL-1 or RU38486 inhibit tolerance induction, although they do not break an acquired tolerance (TAKAHASHI et al. 1995a).

5.3 Desensitization

A single administration of TNF or IL-1, given several hours before the administration of an otherwise lethal dose of TNF, can protect against TNF-induced lethality (WALLACH et al. 1988; LIBERT et al. 1991b). In contrast to what is observed in the case of tolerance, this insensitivity is short-lasting. It is mediated by protein synthesis in the liver, but IL-6 is not involved (LIBERT et al. 1991b). Whether α_1-acid glycoprotein (AGP; see Sect. 8) is the protein responsible for desensitization remains to be established.

6 TNF and Other Cytokine Mediators of SIRS

TNF is part of a cytokine network. It induces other cytokines which may not only mediate part of its in vivo effects but also work in concert with it. Regarding the SIRS-inducing effect of TNF, four cytokines – IL-1, IL-6, IL-12, and IFN-γ – have been studied in detail. Other cytokines such as transforming growth factor-β, IL-4, IL-8, and IL-10 may also affect this syndrome, but as their in vivo relationship with the effects of TNF is less well characterized, we do not discuss them here.

6.1 IL-1

IL-1 is a cytokine which has overlapping effects with TNF but also distinct effects. It is suspected to play an important role in sepsis since its levels are correlated with the severity of the disease (WAAGE et al. 1989), and since in animal models IL-1ra, a natural IL-1 receptor antagonist, reduces lethality in various shock models (OHLSSON et al. 1990; ALEXANDER et al. 1991; FISCHER et al. 1992). Recently it has also been shown that mice deficient in IL-1βconverting enzyme, which are defective in the production of mature IL-1β and also produce much less IL-1α, have a decreased sensitivity to the lethal effects of endotoxin (LI et al. 1995). Phase III clinical trials with IL-1ra in sepsis patients, however, have not confirmed the promising protective results obtained in pilot trials.

The influence of IL-1 on the toxicity of a challenge with TNF depends strictly on the sequence and timing of treatment. Administration of IL-1 several hours after a challenge with TNF does not influence the outcome (EVERAERDT et al. 1989). Pretreatment with IL-1 protects against a subsequent challenge with TNF (LIBERT et al. 1991b). When given simultaneously, an important synergism in inducing lethality is observed (WAAGE and ESPEVIK 1988; EVERAERDT et al. 1989). Furthermore, in the presence of IL-1 the triggering of TNF-R55 is sufficient to cause lethality, as the species-specific toxicity of TNF is abolished (EVERAERDT et al. 1989). This was confirmed in TNF-R55$^{0/0}$ mice (ROTHE et al. 1993). Regarding the mechanism of this synergism, we have observed that the glucocorticoid receptor antagonist RU38486 mimics closely IL-1 or TNF-R75 triggering in its sensitizing activities to the IL-6 inducing and lethal effects of TNF (BROUCKAERT et al. 1992a). It is noteworthy that IL-1ra protects against a lethal challenge with *Escherichia coli* but does not alter the host response to sublethal endotoxemia (FISCHER et al. 1992). This argues for a common effect of IL-1 and TNF-R75 triggering, both of which constitute a "lethal switch" from a reversible, nonlethal syndrome to an irreversible, lethal syndrome (BROUCKAERT et al. 1993b). Regarding the possible involvement of TNF-induced IL-1 in the effects of TNF, it has been observed that recombinant IL-1ra protects somewhat against the lethal effects of an LD_{100} challenge with mTNF. In infection- or tumor-sensitized mice, however, IL-1ra was without effect (EVERAERDT et al. 1994).

6.2 IL-6

This cytokine, which has important activities on B cells, bone marrow cells, and hepatocytes, is induced by TNF as well as by LPS or IL-1 (BROUCKAERT et al. 1989a,b, 1992a; LIBERT et al. 1990, 1991a). Its production is highly correlated with lethality, both in patients with sepsis and in animal models (WAAGE et al. 1989; BROUCKAERT et al. 1989b; LIBERT et al. 1990, 1991a; LOWRY et al. 1994). Nevertheless, from a theoretical point of view, IL-6 has generally been considered an anti-inflammatory cytokine since it is a potent inducer of glucocorticoids and of acute-phase proteins with antiprotease activity. Several groups, however, have observed that monoclonal antibodies directed against IL-6 or IL-6R diminish rather than enhance the lethality induced by TNF or LPS (STARNES et al. 1990; LIBERT et al. 1992; HEREMANS et al. 1992). In contrast, mice carrying a targeted deletion of the IL-6 gene are equally sensitive to the lethal effects of TNF or LPS as the controls (LIBERT et al. 1994a). A possible explanation for this apparent discrepancy is that the antibodies act as a reservoir rather than a neutralizing agent (HEREMANS et al. 1992) and so provide a much longer exposure to IL-6. Since soluble IL-6R (α-chain) promotes rather than neutralizes the function of IL-6, the same could hold for the antibodies directed against IL-6R. In any case, these results should inspire caution in interpreting data obtained with neutralizing antibodies or using mice with a targeted gene deletion.

6.3 IL-12

IL-12 is a heterodimeric cytokine mainly produced by phagocytic cells (TRINCHIERI 1995). It acts as an activator of NK cells and CTL and shifts the immune response to a T_h1-controlled response. As such it is a major inducer of IFN-γ (GATELY et al. 1994). Also TNF is a mediator of the activities of IL-12, although only at higher doses or levels of IL-12 (ORANGE et al. 1995). Recently IL-12 has been implicated as the priming factor of the Shwartzman reaction (OZMEN et al. 1994). Whether the antitumor properties of IL-12 are mediated by TNF and IFN-γ remains to be established.

We have observed that IL-12 is a necessary and sufficient mediator in sensitization to the lethal properties of TNF induced by a BCG infection (CAUWELS et al., 1996). Pretreatment with IL-12 can render mice extremely sensitive to the lethal effects of TNF; in fact it reduces the LD_{50} of hTNF at least 100-fold. Also, this method of sensitization can be alleviated by either methylene blue or α-LFA-1 treatment. The requirements regarding dose and kinetics of the pretreatment indicate that this sensitizing effect is mediated by the induction of IFN-γ, probably in synergism with endogenous TNF. Mice with a targeted deletion of the IFN-γ receptor were fairly, although not absolutely resistant to this IL-12 induced sensitization to hTNF. The conclusion is that all conditions which bring about the production of IL-12 render the host extremely sensitive to TNF-induced lethality.

6.4 IFN-γ

IFN-γ (Farrar and Schreiber 1993) is a cytokine produced by T lymphocytes and NK cells. In addition to its antiviral effects, it functions mainly as an immune stimulator. It activates macrophages and induces intercellular adhesion molecules. Endothelial cells may also be an important target for the actions of this cytokine. IFN-γ is a mediator of the Shwartzman reaction (Heremans et al. 1990; Ozmen et al. 1994) and is an important contributor to the LPS-induced lethality in sensitized animals (Kamijo et al. 1993).

IFN-γ synergizes with TNF for quite a number of effects (Fiers 1995). IFN-$\gamma R^{0/0}$ mice are resistant to the lethality induced by hTNF in mice sensitized by an infection with BCG but not by a growing tumor. They are equally as sensitive as wild-type mice to the lethality induced by mTNF in healthy animals. It may be concluded that (a) IFN-γ can sensitize the host to the shock-inducing properties of TNF, (b) the role of IFN-γ as a sensitizer in LPS-induced shock cannot be reduced to an enhanced production of TNF, and (c) TNF-induced IFN-γ does not play a role in the lethality observed after administration of mTNF. The mechanism by which IFN- γ causes sensitization is at present unknown, but our current research points to effects on endothelial cells rather than to effects on macrophages.

7 TNF and Nitric Oxide

Rightly or wrongly, hypotension has always been considered as the central feature of sepsis-induced lethality and as the dose-limiting toxicity of TNF. The discovery of NO as the mediator of TNF-induced hypotension (Kilbourn et al. 1990) was considered a major breakthrough which would lead to a quick solution for the problems of sepsis and TNF-based anticancer treatment. As expected, inhibitors of the synthesis of NO can inhibit or even revert the hypotension caused by LPS or TNF, both in animal models and in septic patients (Kilbourn et al. 1990; Thiemermann and Vane 1990; Petros et al. 1991; Geroulanos et al. 1992).

The unpleasant surprise however, was, that this led to no improvement in survival. In contrast, in several animal models the addition of NO synthase inhibitors rather worsened the observed lethality and pathology (COBB et al. 1992; EVANS et al. 1994b; WRIGHT et al. 1992). Indeed, both hepatic damage (HARBRECHT et al. 1992) and glomerular thrombosis (SHULTZ and RAIJ 1992) are increased when NO synthesis is inhibited. Furthermore, such inhibition increases leukocyte adherence and extravasation (KUBES et al. 1991). Recently it was shown that NO downregulates the expression of IFN-γ and TNF (FLORQUIN et al. 1994). This has inspired several investigators to propose that NO is also a molecule with both deleterious and protective properties.

To separate these activities and to obtain a selective inhibition of the deleterious activities of NO most groups attempt specifically to inhibit the inducible NO synthase (reviewed in GRIFFITHS et al. 1994). We took another approach and tried to inhibit in a selective way part of the effects of NO, in particular those mediated by soluble guanylate cyclase, which includes NO-mediated hypotension. For this we studied the effects of methylene blue in TNF-related murine models and observed that methylene blue protects against the lethality caused by mTNF in healthy mice and by hTNF in mice sensitized by a BCG infection, by a tumor, or by a pretreatment with IL-12 (CAUWELS et al., submitted). When an inhibitor of NO synthase is added to the treatment with methylene blue, the protection is abated in the cases of mTNF-induced lethality and hTNF-induced lethality in tumor-sensitized mice. These results strongly suggest that NO, in addition to its deleterious effects, has protective properties that seem indispensable to survive a challenge with TNF. These protective effects operate by mechanisms different from those by which guanylate cyclase is activated. In mice sensitized by IL-12 or BCG the protection provided by methylene blue is not reverted by NO synthase inhibitors. However, in these models the inhibitors are less effective, and considerable amounts of NO are still produced; this in fact adds further evidence to a necessary protective role for NO. Methylene blue has been shown to abate the hypotension in sepsis patients (SCHNEIDER et al. 1992; PREISER et al. 1994). In mice, however, we observed no protection by methylene blue against the lethality induced by endotoxin (unpublished results). The reason for this difference between the effects of methylene blue against challenges with TNF or endotoxin are unclear. It illustrates again that the mechanisms underlying sepsis cannot be reduced to the effects of TNF.

8 α_1-Acid Glycoprotein

Unlike some of the inhibitors mentioned above, AGP, an acute-phase protein, protects mice from the lethal shock induced by either TNF or endotoxin in galactosamine-sensitized mice and the lethality induced by mTNF, but not endotoxin, in healthy mice (LIBERT et al. 1994b). It also prevents lethality in BCG- or tumor-sensitized mice (unpublished observations) and as such is one of the most potent and broadest inhibitors known. AGP is a highly glycosylated, polyanionic protein with an apparent molecular mass of about 40 kDa. It is induced in hepatocytes by TNF, IL-1, and other cytokines. Glucocorticoids are a permissive factor in its induction. The physiology of AGP is poorly understood (Baumann et al. 1993).

AGP not only protects against lethality but also provides a complete inhibition of all TNF-induced metabolic changes: fall in body temperature, release of liver enzymes, and enhanced clotting time (Libert et al. 1994b). Whether AGP is really the protective liver protein responsible for the desensitization phenomenon (Sect. 5.3) or the protein that forms the relevant target for the sensitizing activity of GalN is not yet clear. The mechanism by which AGP exerts its protective activity still needs to be elucidated.

9 Conclusions and Prospects

Shortly after the cloning of TNF, a number of important observations were made that seemed to provide a clear mechanism of sepsis. Since TNF is induced by endotoxin and mimicks the cardinal features of sepsis, and since specific antibodies capable of neutralizing TNF prevent lethality in various shock models, it was proposed that endotoxin induces TNF, and that this TNF in turn causes the pathology and lethality of sepsis. Subsequent research has shown that the condition of the host and other factors induced by endotoxin and/or bacteria to a large extent determine the outcome of a challenge with TNF. Indeed, both LPS and IL-1 are potent synergistic factors. IFN-γ and IL-12 sensitize the host substantially to the lethal effects of TNF. ICAM-$1^{0/0}$ mice are resistant to the lethal effects of endotoxin but not to those of TNF despite TNF levels induced by endotoxin that are comparable to those in control mice (Xu et al. 1994). On the other hand, some endogenous factors, such as AGP, inhibit the lethality induced by TNF, but not by endotoxin. Methylene blue also protects against lethality by TNF, but not by endotoxin. More recently another important observation has questioned even the role of TNF as a necessary mediator. Mice with a targeted deletion of TNF-R55, which are resistant to the lethal effects of TNF, remain as sensitive as the control animals to the lethal effects of endotoxin. Moreover, mice deficient in IL-1-β converting enzyme are resistant to endotoxic shock, despite high levels of LPS-induced TNF in circulation (Li et al. 1995).

It is quite probable that TNF is involved in the pathogenesis of septic shock. Sepsis should, however, not be reduced to the effects of this single contributor. Studies using various kinds of inhibitors suggest an important role for various cytokines, such as TNF, IL-1, and IFN-γ, in the events leading to lethality. It may be that at a minimal dose of the challenge sufficient to cause 100% lethality, death is the result of the synergistic activity of various cytokines which are all limiting, but that at higher doses the redundancy of the cytokine effects becomes more prominent, and therefore inhibition of a single cytokine no longer prevents lethality. Strategies aiming at interfering with common intracellular effects of these cytokines may possibly be more effective. Another problem that may hamper anticytokine strategies in sepsis is the fact that at least some of these cytokines, albeit at lower levels than observed in sepsis, may have indispensable protective properties.

Regarding SIRS induced as a side effect of antitumor treatment with TNF, it should be noted that it displays a number of features which are distinct from those observed in sepsis. Because fewer factors are involved in TNF-induced SIRS than in sepsis, and because preventive measures can be applied as the moment of the challenge is known, it may be more suitable for effective remedying. The use of receptor-selective mutants and the induction of tolerance can, at least in mice, effectively broaden the therapeutic margin, showing that the antitumor and SIRS-inducing effects of TNF are not necessarily linked. Recently we have identified two new potent inhibitors of TNF-induced SIRS: AGP and methylene blue. Their influence on the antitumor efficacy of TNF is under study, and it is hoped that they, or similar agents, will allow the systemic administration of therapeutic doses of receptor-specific TNF.

Acknowledgements. This research was supported by the IUAP, ASLK, "Centrum voor Studie en Behandeling van Gezwelziekten," and an EU Biotech grant. P.B. is a research associate with the NFWO.

References

Alexander HR, Doherty GM, Buresh CM, Venzon DJ, Norton JA (1991) A recombinant human receptor antagonist to interleukin 1 improves survival after lethal endotoxemia in mice. J Exp Med 173: 1029–1032

Alzani R, Corti A, Grazioli L, Cozzi E, Ghezzi P, Marcucci F (1993) Suramin induces deoligomerization of human tumor necrosis factor α. J Biol Chem 268: 12526–12529

Barbara JAJ, Smith WB, Gamble JR, Van Ostade X, Vandenabeele P, Tavernier J, Fiers W, Vadas MA, Lopez AF (1994) Dissociation of TNF-α cytotoxic and proinflammatory activities by p55 receptor- and p75 receptor-selective TNF-α mutants. EMBO J 13: 843–850

Baumann H, Prowse KR, Won K, Marinkovic-Pajovic S (1993) Regulation of α_1-acid glycoprotein genes and relationship to other type 1 acute phase plasma proteins. In: Mackiewicz A, Kushner I, Baumann H (eds) Acute phase proteins. Molecular biology, biochemistry, and clinical applications. CRC Press, Boca Raton, pp 409–423

Beutler B, Grau GE (1993) Tumor necrosis factor in the pathogenesis of infectious diseases. Crit Care Med 21: S423–S435

Beutler B, Greenwald D, Hulmes JD, Chang M, Pan YC, Mathison J, Ulevitch R, Cerami A (1985a) Identity of tumour necrosis factor and the macrophage-secreted factor cachectin. Nature 316: 552–554

Beutler B, Milsark IW, Cerami AC (1985b) Passive immunization against cachectin/tumor necrosis factor protects mice from lethal effect of endotoxin. Science 229: 869–980

Beutler B, Krochin N, Milsark IW, Luedke C, Cerami A (1986) Control of cachectin (tumor necrosis factor) synthesis: mechanisms of endotoxin resistance. Science 232: 977–980

Beutler B, Han J, Kruys V, Giroir BP (1992) Coordinate regulation of TNF biosynthesis at the levels of transcription and tanslation. Patterns of TNF expression in vivo. In: Beutler B (ed) Tumor necrosis factors. The molecules and their emerging role in medicine. Raven New York, pp 561–574

Bevilacqua MP, Pober JS, Majeau GR, Fiers W, Cotran RS, Gimbrone MA (1986) Recombinant tumor necrosis factor induces procoagulant activity in cultured human vascular endothelium: characterization and comparison with the actions of interleukin 1. Proc Natl Acad Sci USA 83: 4533–4537

Bevilacqua MP, Wheeler ME, Pober JS, Fiers W, Mendrick DL, Cotran RS, Gimbrone MA (1987) Endothelial-dependent mechanisms of leukocyte adhesion: regulation by interleukin-1 and tumor necrosis factor. In: Movat HZ (ed) Leukocyte emigration and its sequelae. Karger, Basel, pp 79–93

Brouckaert PGG, Leroux-Roels GG, Guisez Y, Tavernier J, Fiers W (1986) In vivo antitumour activity of recombinant human and murine TNF, alone and in combination with murine IFN-gamma, on a syngeneic murine melanoma. Int J Cancer 38: 763–769

Brouckaert P, Spriggs DR, Demetri G, Kufe DW, Fiers W (1989a) Circulating interleukin 6 during a continuous infusion of tumor necrosis factor and interferon γ. J Exp Med 169: 2257–2262

Brouckaert PG, Libert C, Everaerdt B, Takahashi N, Fiers W (1989b) A role for interleukin 1 in the in vivo actions of tumor necrosis factor. Lymphokine Res 8: 269–274

Brouckaert P, Everaerdt B, Fiers W (1992a) The glucocorticoid antagonist RU38486 mimics interleukin-1 in its sensitization to the lethal and interleukin-6-inducing properties of tumor necrosis factor. Eur J Immunol 22: 981–986

Brouckaert P, Libert C, Everaerdt B, Fiers W (1992b) Selective species specificity of tumor necrosis factor for toxicity in the mouse. Lymphokine Cytokine Res 11: 193–196

Brouckaert P, Everaerdt B, Libert C, Takahashi N, Cauwels A, Fiers W (1993a) Strategies to broaden the therapeutic margin of tumor necrosis factor. In: Fiers W, Buurman WA (eds) Tumor necrosis factor: molecular and cellular biology and clinical relevance. Karger, Basel, pp 226–232

Brouckaert P, Libert C, Everaerdt B, Takahashi N, Cauwels A, Fiers W (1993b) Tumor necrosis factor, its receptors and the connection with interleukin 1 and interleukin 6. Immunobiology 187: 317–329

Brouckaert P, Ameloot P, Cauwels A, Everaerdt B, Libert C, Takahashi N, Van Molle W, Fiers W (1994) Receptor-selective mutants of tumour necrosis factor in the therapy of cancer: preclinical studies. Circ Shock 43: 185–190

Carswell EA, Old LJ, Kassel RL, Green S, Fiore N, Williamson B (1975) An endotoxin-induced serum factor that causes necrosis of tumors. Proc Natl Acad Sci USA 72: 3666–3670

Cauwels A, Brouckaert P, Grooten J, Huang S, Aguet M, Fiers W (1995) Involvement of IFN-γ in bacillus Calmette-Guérin-induced but not in tumor-induced sensitization to TNF-induced lethality. J Immunol 154: 2753–2763

Cauwels A, Fiers W, Brouckaert P (1996) Murine IL-12 is involved in Bacillus Calmette-Guérin-induced sensitization and is by itself sufficient to sensitize mice to the lethal effects of human TNF. J Immunol 156 (in press)

Clauss M, Ryan J, Stern D (1992) Modulation of endothelial cell hemostatic properties by TNF: insights into the role of endothelium in the host response to inflammator stimuli. In: Beutler B (ed) Tumor necrosis factors. The molecules and their emerging role in medicine. Raven , New York, pp 49–63

Cobb JP, Natanson C, Hoffman WD, Lodato RF, Banks S, Koev CA, Solomon MA, Elin RJ, Hosseini JM, Danner RL (1992) N^{ω}-amino-L-arginine, an inhibitor of nitric oxide synthase, raises vascular resistance but increases mortality rates in awake canines challenged with endotoxin. J Exp Med 176: 1175–1182

Darnay BG, Reddy SAG, Aggarwal BB (1994a) Physical and functional association of a serine-threonine protein kinase to the cytoplasmic domain of the p80 form of the human tumor necrosis factor receptor in human histiocytic lymphoma U-937 cells. J Biol Chem 269: 19687–19690

Darnay BG, Reddy SAG, Aggarwal BB (1994b) Identification of a protein kinase associated with the cytoplasmic domain of the p60 tumor necrosis factor receptor. J Biol Chem 269: 20299–20304

Echtenacher B, Falk W, Männel DN, Krammer PH (1990) Requirement of endogenous tumor necrosis factor/cachectin for recovery from experimental peritonitis. J Immunol 145: 3762–3766

Eck MJ, Sprang SR (1989) The structure of tumor necrosis factor-α at 2.6 Å resolution Implications for receptor binding. J Biol Chem 264: 17595–17605

Eggermont AMM, Liénard D, Schraffordt Koops H, Rosenkaimer F, Lejeune FJ (1993) Treatment of irresectable soft tissue sarcomas of the limbs by isolation perfusion with high dose TNFα in

combination with interferon-γ and melphalan. In: Fiers W, Buurman WA (eds) Tumor necrosis factor: molecular and cellular biology and clinical relevance. Karger, Basel pp 239–243
Erickson SL, de Sauvage FJ, Kikly K, Carver-Moore K, Pitts-Meek S, Gillett N, Sheekan KCF, Schreiber RD, Goeddel DV, Moore MW (1994) Decreased sensitivity to tumour-necrosis factor but normal T-cell development in TNF receptor-2-deficient mice. Nature 372: 560–563
Evans TJ, Moyes D, Carpenter A, Martin R, Loetscher H, Lesslauer W, Cohen J (1994a) Protective effect of 55- but not 75-kD soluble tumor necrosis factor receptor-immunoglobulin G fusion proteins in an animal model of gram-negative sepsis. J Exp Med 180: 2173–2179
Evans TJ, Carpenter A, Silva A, Cohen J (1994b) Inhibition of nitirc oxide synthase in experimental gram-negative sepsis. J Infect Dis 169: 343–349
Everaerdt B, Brouckaert P, Shaw A, Fiers W (1989) Four different interleukin-1 species sensitize to the lethal action of tumour necrosis factor. Biochem Biophys Res Commun 163: 378–385
Everaerdt B, Brouckaert P, Fiers W (1994) Recombinant IL-1 receptor antagonist protects against TNF-induce lethality in mice. J Immunol 152: 5041–5049
Farrar M, Schreiber R (1993) The molecular cell biology of interferon-γ and its receptor. Annu Rev Immunol 11: 571–611
Fiers W (1995) Biologic therapy with TNF: preclinical studies. In: DeVita VT Jr, Hellman S, Rosenberg SA (eds) Biologic therapy of cancer, 2nd edn. Lippincott, Philadelphia, pp 295–327
Fiorentino DF, Zlotnik A, Mosmann TR, Howard M, O'Garra A (1991) IL-10 inhibits cytokine production by activated macrophages. J Immunol 147: 3815–3822
Fischer E, Marano MA, Van Zee KJ, Rock CS, Hawes AS, Thompson WA, DeForge L, Kenney JS, Remick DG, Bloedow DC, Thompson RC, Lowry SF, Moldawer LL (1992) Interleukin-1 receptor blockade improves survival and hemodynamic performance in Escherichia coli shock, but fails to alter host responses to sublethal endotoxemia. J Clin Invest 89: 1551–1557
Florquin S, Amraoui Z, Dubois C, Decuyper J, Goldman M (1994) The protective role of endogenously synthesized nitric oxide in staphylococcal enterotoxin B-induced shock in mice. J Exp Med 180: 1153–1158
Flynn RM, Palladino MA (1992) TNF and TGF-β: the opposite sides of the avenue? In: Beutler B (ed) Tumor necrosis factors. The molecules and their emerging role in medicine Raven, New York, pp 131–144
Fraker DL, Alexander HR, Pass HI (1995) Biologic therapy with TNF: systemic administration and isolation-perfusion. In: DeVita VT Jr, Hellman S, Rosenberg SA (eds) Biologic therapy of cancer, 2nd edn. Lippincott, Philadelphia, pp 329–345
Fransen L, Müller R, Marmenout A, Tavernier J, Van der Heyden J, Kawashima E, Chollet A, Tizard R, Van Heuversweyn H, Van Vliet A, Ruysschaert MR, Fiers W (1985) Molecular cloning of mouse tumour necrosis factor cDNA and its eukaryotic expression. Nucleic Acids Res 13: 4417–4429
Gamble JR, Smith WB, Vadas MA (1992) TNF modulation of endothelial and neutrophil adhesion. In: Beutler B (ed) Tumor necrosis factors. The molecules and their emerging role in medicine. Raven, New York, pp 65–86
Gately MK, Warrier RR, Honasonge S, Carvajal DM, Faherty DA, Connaughton SE, Anderson TD, Sarmiento U, Hubbard BR, Murphy M (1994) Administration of recombinant IL-12 to normal mice enhances cytolytic lymphocyte activity and induces production of IFN-γ in vivo Int Immunol 6: 157–167
Gearing AJH, Beckett P, Christodoulou M, Churchill M, Clements J, Davidson AH, Drummond AH, Galloway WA, Gilbert R, Gordon JL, Leber TM, Mangan M, Miller K, Nayee P, Owen K, Patel S, Thomas W, Wells G, Wood LM, Wolley K (1994) Processing of tumour necrosis factor-α precursor by metalloproteinases. Nature 370: 555–557
Geng Y, Hansson GK, Holme E (1992) Inferferon-γ and tumor necrosis factor synergize to induce nitric oxide production and inhibit mitochondrial respiration in vascular smooth muscle cells. Circ Res 71: 1268–1276
Geroulanos S, Schilling J, Cakmakci M, Jung HH, Largiader F (1992) Inhibition of NO synthesis in septic shock. Lancet 339: 435
Girardin E, Grau GE, Dayer JM, Roux-Lombard P, The J5 Study Group, Lambert PH (1988) Tumor necrosis factor and interleukin-1 in the serum of children with severe infectious purpura. N Engl J Med 319: 397–400
Griffiths MJD, Curzen NP, Sair M, Evans TW (1994) Nitric oxide synthase inhibitors in septic shock: theoretical considerations. Clin Intensive Care 5: 29–36

Grunfeld C, Wilking H, Neese R, Gavin LA, Moser AH, Gulli R, Serio MK, Feingold KR (1989) Persistence of the hypertriglyceridemic effect of tumor necrosis factor despite development of tachyphylaxis to its anorectic/cachectic effects in rats. Cancer Res 49: 2554–2560

Harbrecht BG, Billiar TR, Stadler J, Demetris AJ, Ochoa JB, Curran RD, Simmons RL (1992) Nitric oxide synthesis serves to reduce hepatic damage during acute murine endotoxemia. Crit Care Med 20: 1568–1574

Hart PH, Vitti GF, Burgess DR, Whitty GA, Piccoli DS, Hamilton JA (1989) Potential antiinflammatory effects of interleukin 4: suppression of human monocyte tumor necrosis factor α, interleukin 1, and prostaglandin E_2. Proc Natl Acad Sci USA 86: 3803–3807

Heremans H, Van Damme J, Dillen C, Dijkmans R, Billiau A (1990) Interferon γ, a mediator of lethal lipopolysaccharide-induced Shwartzman-like shock reactions in mice. J Exp Med 171: 1853–1869

Heremans H, Dillen C, Put W, Van Damme J, Billiau A (1992) Protective effect of anti-interleukin (IL)-6 antibody against endotoxin, associated with paradoxically increased IL-6 levels. Eur J Immunol 22: 2395–2401

Jones EY, Stuart DI, Walker NPC (1989) Structure of tumour necrosis factor. Nature 338: 225–228

Kamijo R, Le J, Shapiro D, Havell EA, Huang S, Aguet M, Bosland M, Vilček J (1993) Mice that lack the interferon-γ receptor have profoundly altered responses to infection with bacillus Calmette-Guérin and subsequent challenge with lipopolysaccharide. J Exp Med 178: 1435–1440

Kilbourn RG, Gross SS, Jubran A, Adams J, Griffith OW, Levi R, Lodato RF (1990) N^G-methyl-L-arginine inhibits tumor necrosis factor-induced hypotension: implications for the involvement of nitric oxide. Proc Natl Acad Sci USA 87: 3629–3632

Kruys V, Thompson P, Beutler B (1993) Tumor necrosis factor biosynthesis: from gene to protein. In: Fiers W, Buurman WA (eds) Tumor necrosis factor: molecular and cellular biology and clinical relevance. Karger, Basel, pp 10–18

Kubes P, Suzuki M, Granger DN (1991) Nitric oxide: an endogenous modulator of leukocyte adhesion. Proc Natl Acad Sci USA 88: 4651–4655

Lehmann V, Freudenberg MA, Galanos C (1987) Lethal toxicity of lipopolysaccharide and tumor necrosis factor in normal and D-galactosamine-treated mice. J Exp Med 165: 657–663

Leist M, Gantner F, Bohlinger I, Tiegs G, Germann PG, Wendel A (1995a) Tumor necrosis factor-induced hepatocyte apoptosis precedes liver failure in experimental murine shock models. Am J Pathol 146: 1220–1234

Leist M, Gantner F, Jilg S, Wendel A (1995b) Activation of the 55 kDa TNF receptor is necessary and sufficient for TNF-induced liver failure, hepatocyte apoptosis, and nitrite release. J Immunol 154: 1307–1316

Lejeune F, Liénard D, Eggermont A, Schraffordt Koops H, Kroon B, Gérain J, Rosenkaimer F, Schmitz P (1994) Clinical experience with high-dose tumor necrosis factor alpha in regional therapy of advanced melanoma. Circ Shock 43: 191–197

Lewis M, Tartaglia LA, Lee A, Bennett GL, Rice GC, Wong GHW, Chen EY, Goeddel DV (1991) Cloning and expression of cDNAs for two distinct murine tumor necrosis factor receptors demonstrate one receptor is species specific. Proc Natl Acad Sci USA 88: 2830–2834

Li P, Allen H, Banerjee S, Franklin S, Herzog L, Johnston C, McDowell J, Paskind M, Rodman L, Salfeld J, Towne E, Tracey D, Wardwell S, Wei F, Wong W, Kamen R, Seshadri T (1995) Mice deficient in IL-1β-converting enzyme are defective in production of mature IL-1β and resistant to endotoxic shock. Cell 80: 401–411

Libert C, Brouckaert P, Shaw A, Fiers W (1990) Induction of interleukin 6 by human and murine recombinant interleukin 1 in mice. Eur J Immunol 20: 691–694

Libert C, Van Bladel S, Brouckaert P, Fiers W (1991a) The influence of modulating substances on tumor necrosis factor and interleukin-6 levels after injection of murine tumor necrosis factor or lipopolysaccharide in mice. J Immunother 10: 227–235

Libert C, Van Bladel S, Brouckaert P, Shaw A, Fiers W (1991b) Involvement of the liver, but not of IL-6, in IL-1-induced desensitization to the lethal effects of tumor necrosis factor. J Immunol 146: 2625–2632

Libert C, Vink A, Coulie P, Brouckaert P, Everaerdt B, Van Snick J, Fiers W (1992) Limited involvement of interleukin-6 in the pathogenesis of lethal septic shock as revealed by the effect of monoclonal antibodies against interleukin-6 or its receptor in various murine models. Eur J Immunol 22: 2625–2630

Libert C, Takahashi N, Cauwels A, Brouckaert P, Bluethmann H, Fiers W (1994a) Response of interleukin-6-deficient mice to tumor necrosis factor-induced metabolic changes and lethality. Eur J Immunol 24: 2237–2242

Libert C, Brouckaert P, Fiers W (1994b) Protection by α_1-acid glycoprotein against tumor necrosis factor-induced lethality. J Exp Med 180: 1571–1575

Loetscher H, Stueber D, Banner D, Mackay F, Lesslauer W (1993) Human tumor necrosis factor α (TNFα) mutants with exclusive specificity for the 55-kDa or 75-kDa TNF receptors. J Biol Chem 268: 26350–26357

Lowry SF, Moldawer LL, Calvano SE (1994) Cytokine markers of the human response to sepsis. In: Vincent J-L (ed) Yearbook of intensive care and emergency medicine. Springer, Berlin Heidelberg New York, pp 14–23

Marmenout A, Fransen L, Tavernier J, Van der Heyden J, Tizard R, Kawashima E, Shaw A, Johnson MJ, Semon D, Müller R, Ruysschaert MR, Van Vliet A, Fiers W (1985) Molecular cloning and expression of human tumor necrosis factor and comparison with mouse tumor necrosis factor. Eur J Biochem 152: 515–522

McGeehan GM, Becherer JD, Bast RC Jr, Boyer CM, Champion B, Connolly KM, Conway JG, Furdon P, Karp S, Kidao S, McElroy AB, Nichols J, Pryzwansky KM, Schoenen F, Sekut L, Truesdale A, Verghese M, Warner J, Ways JP (1994) Regulation of tumour necrosis factor-α processing by a metalloproteinase inhibitor. Nature 370: 558–561

Mohler KM, Sleath PR, Fitzner JN, Cerretti DP, Alderson M, Kerwar SS, Torrance DS, Otten-Evans C, Greenstreet T, Weerawarna K, Kronheim SR, Petersen M, Gerhart M, Kozlosky CJ, March CJ, Black RA (1994) Protection against a lethal dose of endotoxin by an inhibitor of tumour necrosis factor processing. Nature 370: 218–220

Offner F, Philippé J, Vogelaers D, Colardyn F, Baele G, Baudrihaye M, Vermeulen A, Leroux-Roels G (1990) Serum tumor necrosis factor levels in patients with infectious disease and septic shock. J Lab Clin Med 116: 100–105

Ogasawara J, Watanabe-Fukunaga R, Adachi M, Matsuzawa A, Kasugal T, Kitamura Y, Itoh N, Suda T, Nagata S (1993) Lethal effect of the anti-Fas antibody in mice. Nature 364: 806–809

Ohlsson K, Björk P, Bergenfeldt M, Hageman R, Thompson RC (1990) Interleukin-1 receptor antagonist reduces mortality from endotoxin shock. Nature 348: 550–552

Old LJ (1988) Tumor necrosis factor, Sci Am 258: 59–75

Orange JS, Salazar-Mather TP, Opal SM, Spencer RL, Miller AH, McEwen BS, Biron CA (1995) Mechanism of interleukin 12-mediated toxicities during experimental viral infections: role of tumor necrosis factor and glucocorticoids. J Exp Med 181: 901–914

Ozmen L, Pericin M, Hakimi J, Chizzonite RA, Wysocka M, Trinchieri G, Gately M, Garotta G (1994) Interleukin 12, interferon γ, and tumor necrosis factor α are the key cytokines of the generalized Shwartzman reaction. J Exp Med 180: 907–915

Pennica D, Nedwin GE, Hayflick JS, Seeburg PH, Derynck R, Palladino MA, Kohr WJ, Aggarwal BB, Goeddel DV (1984) Human tumour necrosis factor: precursor structure, expression and homology to lymphotoxin. Nature 312: 724–729

Perez C, Albert I, DeFay K, Zachariades N, Gooding L, Kriegler M (1990) A nonsecretable cell surface mutant of tumour necrosis factor (TNF) kills by cell-to-cell contact. Cell 63: 251–258

Petros A, Bennett D, Vallance P (1991) Effect of nitric oxide synthase inhibitors on hypotension in patients with septic shock. Lancet 338: 1557–1558

Pfeffer K, Matsuyama T, Kündig TM, Wakeham A, Kishihara K, Shahinian A, Wiegmann K, Ohashi PS, Krönke M, Mak TW (1993) Mice deficient for the 55 kd tumor necrosis factor receptor are resistant to endotoxic shock, yet succumb to L. monocytogenes infection. Cell 73: 457–467

Preiser JC, Zhang H, Vincent JL (1994) Administration of methylene blue in septic shock: is it beneficial? In: Vincent J-L (ed) Yearbook of intensive care and emergency medicine. Springer, Berlin Heidelberg New York, pp 90–100

Rothe J, Lesslauer W, Lötscher H, Lang Y, Koebel P, Köntgen F, Althage A, Zinkernagel R, Steinmetz M, Bluethmann H (1993) Mice lacking the tumour necrosis factor receptor 1 are resistant to TNF-mediated toxicity but highly susceptible to infection by Listeria monocytogenes. Nature 364: 798–802

Rothe M, Wong SC, Hensel WJ, Goeddel DV (1994) A novel family of putative signal transducers associated with the cytoplasmic domain of the 75 kDa tumor necrosis factor receptor. Cell 78: 681–692

Rothstein JL, Schreiber H (1988) Synergy between tumor necrosis factor and bacterial products causes hemorrhagic necrosis and lethal shock in normal mice. Proc Natl Acad Sci USA 85: 607–611

Sampaio EP, Sarno EN, Galilly R, Cohn ZA, Kaplan G (1991) Thalidomide selectively inhibits tumor necrosis factor α production by stimulated human monocytes. J Exp Med 173: 699–703

Schneider F, Lutun P, Hasselmann M, Stoclet JC, Tempe JD (1992) Methylene blue increases systemic vascular resistance in human septic shock. Preliminary observations. Intensive Care Med 18: 309–311
Schulze-Osthoff K, Bakker AC, Vanhaesebroeck B, Beyaert R, Jacob WA, Fiers W (1992) Cytotoxic activity of tumor necrosis factor is mediated by early damage of mitochondrial functions. Evidence for the involvement of mitochondrial radical generation. J Biol Chem 267: 5317–5323
Schulze-Osthoff K, Beyaert R, Vandevoorde V, Haegeman G, Fiers W (1993) Depletion of the mitochondrial electron transport abrogates the cytotoxic and gene-inductive effects of TNF. EMBO J 12: 3095–3104
Schwalb DM, Han H-M, Marino M, Warren R, Porter A, Goh C, Fair WR, Donner DB (1993) Identification of a new receptor subtype for tumor necrosis factor-α. J Biol Chem 268: 9949–9954
Sheehan KCF, Pinckard JK, Arthur CD, Dehner LP, Goeddel DV, Schreiber RD (1995) Monoclonal antibodies specific for murine p55 and p75 tumor necrosis factor receptors: identification of a novel in vivo role for p75. J Exp Med 181: 607–617
Sheppard BC, Fraker DL, Norton JA (1989) Prevention and treatment of endotoxin and sepsis lethality with recombinant human tumor necrosis factor. Surgery 106: 156–162
Shirai T, Yamaguchi H, Ito H, Todd CW, Wallace RB (1985) Cloning and expression in Escherichia coli of the gene for human tumour necrosis factor. Nature 313: 803–806
Shultz PJ, Raij L (1992) Endogenously synthesized nitric oxide prevents exdotoxin-induced glomerular thrombosis. J Clin Invest 90: 1718–1725
Smith RA, Baglioni C (1987) The active form of tumor necrosis factor is a trimer. J Biol Chem 262: 6951–6954
Spies T, Blanck G, Bresnahan M, Sands J, Strominger JL (1991) A new cluster of genes within the human major histocompatibility complex. Science 243: 214–217
Starnes HF Jr, Pearce MK, Tewari A, Yim JH, Zou J-C, Abrams JS (1990) Anti-IL-6 monoclonal antibodies protect against lethal Escherichia coli infection and lethal tumor necrosis factor-α challenge in mice. J Immunol 145: 4185–4191
Takahashi N, Brouckaert P, Fiers W (1991) Induction of tolerance allows separation of lethal and antitumor activities of tumor necrosis factor in mice. Cancer Res 51: 2366–2372
Takahashi N, Brouckaert P, Bemelmans MHA, Buurman WA, Fiers W (1994) Mechanism of induction of tolerance to tumour necrosis factor (TNF): no involvement of modulators of TNF bioavailability or receptor binding. Cytokine 6: 235–242
Takahashi N, Brouckaert P, Fiers W (1995a) Mechanism of tolerance to tumor necrosis factor: receptor-specific pathway and selectivity. Am J Physiol 269: R398–R405
Takahashi N, Fiers W, Brouckaert P (1995b) Anti-tumor activity of tumor necrosis factor in combination with interferon-γ is not affected by prior tolerization. Int J Cancer 63: 846–854
Tartaglia LA, Weber RF, Figari IS, Reynolds C, Palladino MA Jr, Goeddel DV (1991) The two different receptors for tumor necrosis factor mediate distinct cellular responses. Proc Natl Acad Sci USA 88: 9292–9296
Tartaglia LA, Pennica D, Goeddel DV (1993) Ligand Passing: the 75-kDa tumor necrosis factor (TNF) receptor recruits TNF for signaling by the 55-kDa TNF receptor. J Biol Chem 268: 18542–18548
Thiemermann C, Vane J (1990) Inhibition of nitric oxide synthesis reduces the hypotension induced by bacterial lipopolysaccharides in the rat in vivo. Eur J Pharmacol 182: 591–595
Tracey KJ, Cerami A (1993) Tumor necrosis factor: an updated review of its biology. Crit Care Med 21: S415–S422
Tracey KJ, Cerami A (1994) Tumor necrosis factor: a pleiotropic cytokine and therapeutic target. Annu Rev Med 45: 491–503
Tracey KJ, Beutler B, Lowry SF, Merryweather J, Wolpe S, Milsark IW, Hariri RJ, Fahey TJIII, Zentella A, Albert JD, Shires GT, Cerami A (1986) Shock and tissue injury induced by recombinant human cachectin. Science 234: 470–474
Trinchieri G (1995) Interleukin-12: a proinflammatory cytokine with immunoregulatory functions that bridge innate resistance and antigen-specific adaptive immunity. Annu Rev Immunol 13: 251–276
Van Ostade X, Tavernier J, Prangé T, Fiers W (1991) Localization of the active site of human tumour necrosis factor (hTNF) by mutational analysis. EMBO J 10: 827–836
Van Ostade X, Vandenabeele P, Everaerdt B, Loetscher H, Gentz R, Brockhaus M, Lesslauer W, Tavernier J, Brouckaert P, Fiers W (1993) Human TNF mutants with selective activity on the p55 receptor. Nature 361: 266–269

Van Ostade X, Tavernier J, Fiers W (1994) Structure-activity studies of human tumour necrosis factors. Protein Eng 7: 5–22

Van Zee KJ, Stackpole SA, Montegut WJ, Rogy MA, Calvano SE, Hsu KC, Chao M, Meschter CL, Loetscher H, Stüber D, Ettlin U, Wipf B, Lesslauer W, Lowry SF, Moldawer LL (1994) A human tumor necrosis factor (TNF) α mutant that binds exclusively to the p55 TNF receptor produces toxicity in the baboon. J Exp Med 179: 1185–1191

Vandenabeele P, Declercq W, Vercammen D, Van de Craen M, Grooten J, Loetscher H, Brockhaus M, Lesslauer W, Fiers W (1992) Functional characterization of the human tumor necrosis factor receptor p75 in a transfected rat/mouse T cell hybridoma. J Exp Med 176: 1015–1024

Vandenabeele P, Declercq W, Beyaert R, Fiers W (1995a) Two tumour necrosis factor receptors: structure and function. Trends Cell Biol 5: 392–399

Vandenabeele P, Declercq W, Vanhaesebroeck B, Grooten J, Fiers W (1995b) Both TNF receptors are required for TNF-mediated induction of apoptosis in PC60 cells. J Immunol 154: 2904–2913

Waage A, Espevik T (1988) Interleukin 1 potentiates the lethal effect of tumor necrosis factor α/cachectin in mice. J Exp Med 167: 1987–1992

Waage A, Halstensen A, Espevik T (1987) Association between tumour necrosis factor in serum and fatal outcome in patients with meningococcal meningitis. Lancet 1: 355–357

Waage A, Brandtzaeg P, Halstensen A, Kierulf P, Espevik T (1989) The complex pattern of cytokines in serum from patients with meningococcal septic shock. Association between interleukin 6, interleukin 1, and fatal outcome. J Exp Med 169: 333–338

Wallach D, Holtmann H, Engelmann H, Nophar Y (1988) Sensitization and desensitization to lethal effects of tumor necrosis factor and IL-1. J Immunol 140: 2994–2999

Wang AM, Creasey AA, Ladner MB, Lin LS, Strickler J, van Arsdell JN, Yamamoto R, Mark DF (1985) Molecular cloning of the complementary DNA for human tumor necrosis factor. Science 228: 149–154

Wingfield P, Pain RH, Craig S (1987) Tumour necrosis factor is a compact trimer. FEBS Lett 211: 179–184

Wright CE, Rees DD, Moncada S (1992) Protective and pathological roles of nitric oxide in endotoxin shock. Cardiovasc Res 26: 48–57

Xu H, Gonzalo JA, St Pierre Y, Williams IR, Kupper TS, Cotran RS, Springer TA, Gutierrez-Ramos JC (1994) Leukocytosis and resistance to septic shock in intercellular adhesion molecule 1-deficient mice. J Exp Med 180: 95–109

Yamagishi J, Kawashima H, Matsuo N, Ohue M, Yamayoshi M, Fukui T, Kotani H, Furuta R, Nakano K, Yamada M (1990) Mutational analysis of structure-activity relationships in human tumor necrosis factor-alpha. Protein Eng 3: 713–719

Mediators: Nitric Oxide and Other Toxic Oxygen Species

T.J. EVANS and J. COHEN

1	Introduction	189
2	Historical Background	190
3	Reduced Oxygen Species	191
4	Nitric Oxide: Mechanisms of Production	191
4.1	Enzymes Producing Nitric Oxide	191
4.2	Inducers of Nitric Oxide Synthesis	192
4.2.1	Proinflammatory Cytokines	192
4.2.2	Bacterial Products	193
4.2.3	Down-Regulation of iNOS Activation	193
5	Nitric Oxide: Pathophysiological Effects	193
5.1	Hypotension	193
5.2	Cytotoxicity	195
5.3	Immunomodulation	197
5.4	Myocardial Effects	197
5.5	Cerebral Effects	199
6	Nitric Oxide: Clinical Data in Sepsis	199
6.1	Measurement of Nitrite/Nitrate in Sepsis	199
6.2	Clinical Studies with Inhibitors of Nitric Oxide	201
References		203

1 Introduction

Amongst the many mediators that have been implicated in the pathophysiological changes seen in sepsis and septic shock, nitric oxide (NO) is one of the most recent to be described. Its role in mammalian cell biology is enormous (reviewed in NATHAN 1992; BREDT and SNYDER 1994), but this chapter focuses on its role in sepsis. While the functions of NO are legion, we concentrate on a few central roles that NO plays in sepsis. In particular, we discuss its hypotensive, cytotoxic, immunomodulatory, cardiovascular, and cerebral effects. In common with other mediators in sepsis there are problems in extending the results obtained in animal models to humans. We therefore discuss the clinical data per-

Department of Infectious Diseases and Bacteriology, Royal Postgraduate Medical School, Hammersmith Hospital, Du Cane Road, London W12 0NN, UK

taining to the role of NO in sepsis separately. NO is one example of a potentially toxic oxygen compound, and many others have been identified as possible mediators in sepsis which we do not have space to review. However, it is clear that NO can react with a number of these compounds to generate very toxic molecules, which are described here.

2 Historical Background

A number of different lines of research have revealed the important roles that NO plays in mammalian cell biology. Since 1977 it has been known that NO activates soluble guanylate cyclase, producing a rise in intracellular cyclic GTP (Arnold et al. 1977). This in turn can produce vascular smooth muscle cell relaxation and accounts for the vasorelaxatory properties of a number of different compounds such as organic nitrates and sodium nitroprusside (Gruetter et al. 1979; Ignarro et al. 1981). The mechanism of action of other vasodilators which do not themselves generate NO remained obscure but was shown to be dependent on an intact endothelium, leading to the concept of an endothelium-derived relaxing factor (EDRF; Furchgott and Zawadzki 1980). Further work by two groups led to the discovery that the chemical nature of EDRF is in fact NO, or a closely related oxide of nitrogen (Palmer et al. 1987; Ignarro et al. 1987).

A completely separate area of research concerned the generation of inorganic oxides of nitrogen by mammalian cells, a process thought to be of potential importance in carcinogenesis. Both rodents and humans were found to have endogenous mechanisms responsible for the generation of nitrate which increased in intercurrent infection (Green et al. 1981a,b). This activity was found also to reside in macrophages removed from suitably treated animals (Stuehr and Marletta 1985); biochemical studies showed that the generation of such reactive nitrogen intermediates depends on the oxidation of the terminal guanidino nitrogen atom of L-arginine (Iyengar et al. 1987). In addition, such reactive nitrogen intermediates were also found to be responsible for the cytostatic and tumoristatic activities of activated rodent macrophages (Hibbs et al. 1987). Subsequent to the identification of EDRF as NO it was demonstrated that the reactive nitrogen intermediate released from activated murine macrophages is NO, which subsequently forms nitrate and nitrite ions in aqueous solution (Stuehr et al. 1989).

Thus NO was found to be of key importance in vasodilatation and macrophage killing of micro-organisms and tumor cells. Subsequent work has shown that NO is important in mediating a diverse number of biological effects, including acting as a neurotransmitter in non-adrenergic, non-cholinergic nerves (Bult et al. 1990), inhibiting platelet aggregation (Mellion et al. 1981), and acting as a negative inotrope and chronotrope (Finkel et al. 1992). The chemical nature of EDRF has also been analysed more thoroughly. More precise chemical analyses have established that the main oxide of nitrogen produced by mammalian cells is NO, although undoubtedly under certain conditions other reactive nitrogen oxide species may be produced (IGNARRO 1990). In addition, longer lasting 'storage' forms of NO exist, such as S-nitroso protein adducts (STAMLER et al. 1992).

3 Reduced Oxygen Species

A large number of other toxic oxygen metabolites have been suggested as potential mediators of tissue damage in sepsis. These include superoxide ($O_2^{-\cdot}$), hydroxyl radical ($OH^{\cdot}$), and hydrogen peroxide (H_2O_2). We do not have space here to review these areas in detail (see BRIGHAM 1991; FLOHÉ and GIERTZ 1987). However, it is clear that some of these compounds can react with NO to produce even more toxic molecules (see Sect. 5.2). Moreover, in the absence of arginine, brain NO synthase can actually form superoxide and hydrogen peroxide (POU et al. 1992). Thus there is considerable interaction between the systems generating reactive nitrogen and reactive oxygen species.

4 Nitric Oxide: Mechanisms of Production

4.1 Enzymes Producing Nitric Oxide

NO is produced by a family of enzymes known as NO synthases (SESSA 1994). There are three NO synthases, each encoded by a separate gene. Direct comparisons of protein sequence similarity between the different human isoforms shows that they have about 60% sequence identity and share a number of similar properties which reflect their structural similarities (FORSTERMANN et al. 1994). All of the enzymes catalyse the same reaction, which is the oxidation of L-arginine by molecular oxygen to L-citrulline and NO. In each case this involves the transfer of electrons from the co-factor NADPH. In addition, each of the enzymes contains tightly bound flavins, flavin mononucleotide (FMN) and flavin adenine dinucleotide (FAD), as well as tetrahydrobiopterin, heme and calmodulin, which are involved in the intramolecular transfer of electrons from NADPH and tetrahydrobiopterin. These similarities in enzymatic action are reflected in the presence of common structural motifs within the proteins, which possess consensus sequences for the binding of FAD, FMN, NADPH and calmodulin. The presence of binding sites for both FMN and FAD is relatively unusual; the only other example of a mammalian enzyme with such properties is cytochrome P450 reductase. Indeed, the C terminal portion of the NO synthases has about 30% identity with this cytochrome (BREDT et al. 1991; SESSA 1994).

The NO synthases differ most significantly in their dependence on calcium ion concentration and their cellular and tissue distributions. Two of these NO synthases, the brain and endothelial isoforms, are constitutive, that is, they are constantly expressed in tissues where they are present. These isoforms are also highly dependent on calcium ion concentration for their enzymatic activity (BREDT and SNYDER 1990; POLLOCK et al. 1991). In common with most calmodulin-binding enzymes, calmodulin is bound to these isoforms only when free calcium levels are elevated. Hence NO synthesis from the brain and endothelial isoforms occurs only when calcium levels within the cell are elevated, with EC_{50} values for Ca^{2+} of approximately 0.2–0.4 μM (NATHAN 1992). Thus in cells possessing these isoforms the concentration of calcium within the cell provides a virtually instantaneous control mechanism over NO synthesis.

This is not the case with the third isoform of NO synthase, the inducible form. This was originally identified within macrophages (STUEHR and MARLETTA 1985), but it is now known that a huge variety of different cells can make this isoform under suitable conditions (XIE and NATHAN 1994). Under resting conditions cells do not produce the inducible NO synthase, but after suitable stimulation with agents such as pro-inflammatory cytokines and lipopolysaccharide (LPS) they can make large amounts of this enzyme. Importantly, although this NO synthase isoform contains very tightly bound calmodulin, under most conditions it appears that this binding is essentially independent of calcium ion concentration (CHO et al. 1992; EVANS et al. 1992a). Thus, at least at the calcium ion concentrations pertaining within the cell, inducible NO synthase activity is not affected by changes in calcium ion concentration. Once synthesised, the inducible NO synthase can make large amounts of NO for extended periods of time; in cell culture the NO output is limited only by the depletion of arginine within the tissue culture medium (VODOVOTZ et al. 1994).

4.2 Inducers of Nitric Oxide Synthesis

The inducible isoform of NO synthase is responsible for most of the increased output of NO during sepsis (GREEN et al. 1981b; BUTTERY et al. 1994). Once synthesised, this isoform can make large amounts of NO over a long period (VODOVOTZ et al. 1994). Thus the induction of this form of the enzyme is a critical step in the production of NO in sepsis. An enormous number of different agents can act as inducers, but many of these depend on the synthesis of more proximal activators.

4.2.1 Proinflammatory Cytokines

Sepsis is characterised by the release of a large number of cytokines which initiate and augment the inflammatory response (GLAUSER et al. 1991). Principal amongst these are tumour necrosis factor (TNF)-α, interleukin (IL)-1, and interferon (IFN)-γ. Combinations of two or more cytokines are much more effective inducers of inducible NO synthase (iNOS) synthesis, with marked synergy often demonstrable within a variety of different cell types (NATHAN 1992). Human cells particularly require the presence of a number of different pro-inflammatory cytokines to produce significant iNOS activation (NUSSLER et al. 1992).

4.2.2 Bacterial Products

LPS is an extremely potent activator of iNOS production, both on its own and as an agent synergising with pro-inflammatory cytokines (STUEHR and MARLETTA 1985; NAKAYAMA et al. 1992). In vivo of course many of its effects are due to the intermediate synthesis of a variety of cytokines. Products of gram-positive organisms can also induce iNOS synthesis. These include cell wall components such as lipoteichoic acid (LONCHAMPT et al. 1992) as well as secreted extracellular toxins such as the staphylococcal protein toxic shock syndrome toxin-1 (ZEMBOWICZ and VANE 1992). The intracellular pathways involved in the induction of iNOS by these agents are not well characterised.

4.2.3 Down-Regulation of iNOS Activation

A number of counter-regulatory cytokines such as IL-4 (BOGDAN et al. 1994) and transforming growth factor-β (VODOVOTZ et al. 1993) can prevent iNOS induction if present at the same time as the inducing agents. Corticosteroids are also very effective at preventing iNOS induction (RADOMSKI et al. 1990). None of these agents appears to have a significant effect on iNOS activity once it has been induced.

5 Nitric Oxide: Pathophysiological Effects

5.1 Hypotension

As outlined above, NO has been identified as the chemical identity of EDRF, an extremely potent vasodilator. NO is able to relax vascular smooth muscle by activation of the enzyme guanylate cyclase by nitrosation of its heme group (ARNOLD et al. 1977). This in turn leads to increased levels of cGMP within the cell. cGMP has many effects within cells, but its critical action in vasorelaxation depends on its activation of cGMP-dependent kinase, which in turn increases the conductance of a potassium (K^+) channel within the smooth muscle cell membrane, the calcium-dependent K^+ channel (ARCHER et al. 1994). Increased K^+ efflux from the cell then leads to a hyperpolarisation of the cell membrane which inhibits muscle contraction. NO is a potent vasodilator, active at concentrations as low as 10^{-8} M (PALMER et al. 1987). In a whole animal therefore vasorelaxation leads to decreased peripheral vascular resistance and a fall in systemic blood pressure.

Hypotension is of course one of the cardinal features of septic shock and leads to ischaemic organ damage and multi-organ failure (GLAUSER et al. 1991). What evidence is there that the overproduction of NO in sepsis causes this vasodilatation and hypotension? Four lines of evidence in experimental animals support this assertion; the evidence for the role of NO in human sepsis is considered separately (Sect. 6).

1. Animals which have been treated with inflammatory agents such as LPS or infected with live bacteria show an increased production of nitrite and nitrate, stable end-products of NO in aqueous solution (GREEN et al. 1981b; EVANS et al. 1992b). This output is correlated with the development of hypotension and multi-organ failure in treated animals. Thus, increased NO production does occur in animals with septic shock.

2. The source of this increased NO production is from iNOS. In cell culture LPS and pro-inflammatory cytokines are potent stimuli for the induction of iNOS which can produce large amounts of NO (NATHAN 1992). This can occur not only in inflammatory cells such as macrophages but also directly in vascular smooth cells, where the local production of NO leads to rapid vasodilatation and hypotension (BUSSE and MÜLSCH 1990). In whole-animal models of sepsis the administration of LPS or live bacteria can also increase iNOS expression. This has been demonstrated by enzyme assays for iNOS which show marked induction of the enzyme in a wide variety of tissues, notably the liver, lung, heart and gut (KNOWLES et al. 1990). Immunochemical techniques have allowed detailed localisation of the sites of iNOS production (BUTTERY et al. 1994; SATO et al. 1995). Although induction of the enzyme within tissue macrophages and hepatocytes has been easily demonstrated, the production of iNOS within vascular smooth muscle cells following LPS or live bacterial treatment of animals has been more difficult to demonstrate. Indeed, a number of studies have failed to show the presence of iNOS in vascular smooth muscle following LPS treatment of whole animals. However, in cultured vascular smooth muscle cells (BUTTERY et al. 1994) or whole arterial preparations (THORIN TRESCASES et al. 1995) treated with LPS and cytokines such immunoreactive iNOS can be demonstrated. An additional source of NO of relevance in the development of hypotension in shock is the vascular endothelium, which in addition to the constitutive NOS can also produce the inducible isoform following stimulation with cytokines and/or LPS (GROSS et al. 1991).

3. Pharmacological blockade of NO production has demonstrated its importance in the hypotension which develops in animals following LPS or bacterial injection. A number of different inhibitors of the NO synthases have been developed (MARLETTA 1995). Most work has utilised drugs which are N-substituted L-arginine derivatives, such as N^G-monomethyl-L-arginine (L-NMMA) and N^G-nitro-L-ariginine-methyl ester (L-NAME). Although these do differ in their rank order of potencies of inhibition on the different NOS isoforms, they effectively do not discriminate between them. More recently a novel NOS inhibitor has been de-

veloped which does have some selectivity for the inducible isoform – this is S-methylisothiourea sulfate, which is 10 to 30-fold more potent an inhibitor of iNOS than the endothelial form (SOUTHAN et al. 1995).

A large number of experiments using different inhibitors, animals and LPS doses have been performed. The results are often confusing, with significant differences between species. However, a number of common themes have evolved. Firstly, these inhibitors can attenuate the hypotension consequent upon LPS or bacterial injection into an animal (THIEMERMANN and VANE 1990). However, the effects on overall survival are much more variable and depend critically upon dose of the inhibitor used and the animal species (WRIGHT et al. 1992; NAVA et al. 1992; EVANS et al. 1994). In part this is due to the relative non-selectivity of most of the inhibitors used, so that not only the iNOS isoform is inhibited but also the endothelial and brain isoforms. Since a basal level of NO production from the constitutive endothelial isoform is essential for maintaining basal vasorelaxation and hence organ perfusion, it is not surprising that inhibition of all NO production can lead to the vascular collapse and rapid death of treated animals that has been seen in some models (WRIGHT et al. 1992; NAVA et al. 1992). In addition, inhibition of brain NOS may account for the CNS toxicity of some of these drugs (COBB et al. 1992). The selective agent S-methylisothiourea sulfate not only reverses the LPS-induced hypotension in rats but also improves their long-term survival, suggesting that such targetted NO inhibition may be beneficial clinically (SZABO et al. 1994).

4. The final and most direct evidence that NO production from the iNOS isoform is essential to the development of hypotension following LPS challenge comes from experiments using transgenic mice in which the gene for iNOS has been disrupted (MACMICKING et al. 1995; WEI et al. 1995). Such animals make none of the inducible enzyme on LPS challenge and do not show any increase in nitrite/nitrate production, compared to the large increases seen in the wild-type parent strain. Blood pressure in the knock-out mice does not fall significantly following LPS administration, in marked difference to the wild-type animals. In addition, with large doses of LPS the knock-out mice are protected from its lethal effects compared to the wild type. Interestingly, when these mice are sensitised to the effects of LPS with *Propionibacterium acnes*, they show no protection from the massive hepatic necrosis following LPS challenge or any improvement in survival compared to the wild-type strain. These experiments provide the most direct evidence that NO production from the iNOS is essential to the production of hypotension following LPS treatment. They also demonstrate, however, that while NO is an important mediator of the toxicity of LPS, under certain circumstances it clearly is not pivotal in determining survival.

5.2 Cytotoxicity

The potentially toxic effects of NO were recognised in the early phases of its discovery. As a product of activated macrophages it was found to be cytotoxic to certain tumour cells (HIBBS et al. 1987) and also to participate in the killing of a number of different micro-organisms (NATHAN and HIBBS 1991; OSWALD et al. 1994). However, the toxic effects of NO may not just be beneficial, but in addition may be damaging when overproduced during sepsis. The evidence for the production of NO in sepsis is reviewed above. What evidence is there for NO-mediated toxicity being important in the pathology of sepsis?

NO cytotoxicity occurs through a number of different mechanisms. At high enough concentrations NO can inhibit a range of critical enzymes which possess iron-sulfur centres, by forming iron-sulfur-nitrosyl derivatives. Thus its inhibition of the mitochondrial enzymes aconitase and NADPH:ubiquinone oxidoreductase leads to cessation of oxidative metabolism in the affected cell and prevents protein synthesis (STADLER et al. 1991). A more potent mechanism of cytotoxicity, however, can occur through the reaction of NO with superoxide anions. Superoxide ($O_2^{-\cdot}$) is produced by activated macrophages and a variety of epithelial and endothelial cells under very similar conditions which induce the production of NO (BRIGHAM 1991). Superoxide itself can be directly toxic, but in fact it has limited bioreactivity (BECKMAN et al. 1990). However, a number of studies have shown that intravenous treatment with superoxide dismutase ameliorates endothelial injury seen in a number of models of sepsis (FREEMAN et al. 1985). This indicates that superoxide plays a role in this process. Combination of NO with superoxide leads to the production of the potent oxidising agent, peroxynitrite ($ONOO^-$). Peroxynitrite may be a crucial intermediate in the induction of oxidant damage in sepsis (BECKMAN et al. 1990, 1994).

Endothelial injury is critical in many areas of sepsis, particularly in the lung, where the breakdown of the endothelial barrier that this injury produces leads to the increased influx of fluid and inflammatory cells that is typical of adult respiratory distress syndrome (ARDS). Peroxynitrite has a number of different toxic effects. It can initiate lipid peroxidation in biological membranes (RADI et al. 1991a) and produce nitration of aromatic amino residues (ISCHIROPOULOS et al. 1992) as well as oxidation of protein sulfhydryl groups (RADI et al. 1991b). It is more potent than superoxide in producing alveolar type II cell injury and damage to surfactant apoproteins. As well as being directly toxic, peroxynitrite can decompose under acid conditions to yield the very reactive hydroxyl radical, $OH^{\cdot}$ (BECKMAN et al. 1994). All these effects combine to make peroxynitrite a very attractive candidate as a mediator of the oxidative damage to lung in sepsis.

One method of analysing for the production of peroxynitrite is to assay for the presence of nitrotyrosine, since only peroxynitrite and not NO or superoxide can nitrate the phenolic ring of this aminoacid (ISCHIROPOULOS et al. 1992). Specific polyclonal antibodies to nitrotyrosine have been used to immunostain the lungs of patients with ARDS and rats that have sustained hyperoxic damage (HAUSCHILDT et al. 1992). In both cases a clear increase in nitrotyrosine staining was seen over control sections. Additional studies have shown increased levels of nitrotyrosine in the lungs of rats after LPS treatment (WIZEMANN et al. 1994). These data demonstrate that peroxynitrite is generated in experimental animal models of lung damage and in human cases of ARDS and may play a significant role in its pathology.

5.3 Immunomodulation

NO production is stimulated by pro-inflammatory cytokines produced during sepsis. NO itself may also influence the production of these cytokines. Production of different cytokines by T cells has allowed two functional subsets to be defined, at least in mice: those that secrete cytokines such as IL-2 and IFN-γ, known as T_h1 cells, and those that produce IL-4 and IL-5, known as T_h2 cells (MOSMANN et al. 1986). In work on the immune response to infection with the pathogen *Leishmania* it is apparent that T_h1 cells produce NO which exerts a negative feedback effect on the same cells to inhibit the production of IL-2 and IFN-γ (LIEW et al. 1991; TAYLOR-ROBINSON et al. 1994). Such a negative regulatory pathway may also operate in sepsis, such that the production of NO down-regulates the production of T_h1 cytokines, which include the important mediators IFN-γ and TNF-α.

Most evidence for the operation of this feedback pathway has come from the use of NOS inhibitors. In a model of gram-negative sepsis in mice, NOS inhibition did not alter the levels of TNF-α following infection (EVANS et al. 1994) although other studies in models of gram-negative infection have observed increases in TNF-α levels. The effect of NO in these systems may depend on the degree to which the cytokine cells are stimulated; clearly, it there is maximal stimulation, NO inhibition does not produce any additional effect. Important mediators of gram-positive sepsis include a number of bacterial toxins which can act as potent stimulators of a large percentage of the T cell pool, so-called 'superantigens'. In mice given the staphylococcal superantigen staphylococcal enterotoxin B, NOS blockade markedly increased production of IFN-γ and TNF-α both in vitro and in vivo (FLORQUIN et al. 1994). The increased cytokine production in these mice was associated with an increased mortality.

Further work supporting these data has come from the use of transgenic mice in which the iNOS gene has been disrupted (WEI et al. 1995). Splenocytes from such animals produce more IFN-γ following antigenic challenge with *Leishmania major*. Concentration of NO may be important, however, since at low concentrations NO can enhance T cell proliferation. The pathophysiological effects of such feedback roles of NO on cytokine production are far from clear, especially in human studies; further work is necessary to establish the importance of this pathway and whether it limits the usefulness of NOS inhibitors in the treatment of sepsis.

5.4 Myocardial Effects

Although the majority of patients with septic shock have a normal or increased cardiac index, there is a sub-population who have reduced cardiac output (NISHIJIMA et al. 1973). Moreover, in patients with normal or increased cardiac output there is often profound reduction in ejection fraction with concomitant compensatory dilatation of the ventricles (PARKER et al. 1984). These abnormalities return to normal within 7–10 days after resolution of sepsis. The clinical significance of this myocardial depression is uncertain. In patients who have profound vasodilatation, depressed cardiac performance contributes to the difficulties of increasing blood pressure and tissue perfusion. In fact, however, no difference has been found between survivors and non-survivors of septic shock with respect to the degree of myocardial depression (PARKER et al. 1984). Although further work is required to clarify the exact role that myocardial depression plays in the outcome from septic shock, it is certainly a significant phenomenon which contributes to the altered cardiovascular responses seen in sepsis.

Initial studies using sera from patients with septic shock identified a circulating factor which reduced the extent of contraction of spontaneously beating neonatal rat cardiac myocytes (PARRILLO et al. 1985). The discovery of pro-inflammatory cytokines which are produced in large quantities in sepsis made these agents likely candidates for this myocardial depressant factor. The first report of the direct effect of pro-inflammatory cytokines on myocardial function found that cytokines such as TNF-α were able to produce a rapid (2–3 min) reduction in contractility of isolated hamster papillary muscle (FINKEL et al. 1992). This effect was blocked by inhibitors of NOS, suggesting that the negative inotropic effect was mediated by NO. A separate study, however, found that the immediate effects of TNF-α in adult cat myocytes do not depend on increased NO output (YOKOYAMA et al. 1993).

Such rapid effects of cytokines suggested that the induction of iNOS is unlikely to be involved. Certainly it is known that the resting cardiac myocyte contains constitutive NOS activity (FORSTERMANN et al. 1994), and it is possible that these effects result from activation of this isoform. More recent studies have analysed the properties of myocytes isolated from endotoxin-treated animals or treated for more prolonged times with cytokines. In endotoxin-treated rats, isolated cardiac myocytes show a reduction in contractility that is NO dependent (BRADY et al. 1992). NO is also involved in mediating the negative chronotropic effect of IL-1β on neonatal rat cardiac myocytes (ROBERTS et al. 1992). In both these cases the iNOS isoform was involved. A more detailed study of NO output of individual adult rat cardiac myocytes found that the sustained rise in NO output following treatment with IL-1β or IFN-γ is due to induction of the iNOS (BALLIGAND et al. 1994). Production of NO by this isoform is responsible for reducing the contractility of the muscle cells. Taking these results together, it would seem that the major effect of cytokines on cardiac myocytes is to induce iNOS. Production of NO by this isoform then mediates negative inotropic and chronotropic effects. Clearly under some conditions, however, cytokines can produce rapid effects which are likely mediated by the constitutive NOS(s).

The role that cardiac NO plays in the whole animal is also somewhat confusing. Studies using NOS inhibitors in LPS-treated animals have demonstrated in a number of cases clear reductions in cardiac output following inhibitor treatment (COBB et al. 1992). Obviously, if NO is a mediator of a negative inotropic effect in sepsis, the predicted effect of NOS inhibitors would be to increase cardiac output. Negative inotropic effects of NOS inhibitors have also been found in humans with septic shock (see Sect. 6). It is difficult to explain this apparent contradiction. Either compensatory cardiovascular reflexes produce cardiac depression on blocking NO synthesis, or NO production actually protects the heart by suppressing its metabolic activity, such that on removing NO production damaging factors produced in sepsis can further reduce cardiac output.

5.5 Cerebral Effects

NOS is widely distributed within the central nervous system, where it plays a number of roles (Bredt and Snyder 1994). One of these is in the synaptic transmission of the neurotransmitter glutamate, via *N*-methyl-D-aspartate subtype receptors. Excess levels of NO are neurotoxic, and it is possible therefore that the excess production of NO in sepsis has deleterious effects on brain function (Izumi et al. 1992; Iadecola et al. 1995).

There are not large amounts of data which address this question directly. Following exposure to LPS and/or inflammatory cytokines both astrocytes and neurones can produce iNOS (Wallace and Bisland 1994; Galea et al. 1994; Minc Golomb et al. 1994). Thus there is potentially a source a high-output NO production in the brain in sepsis, although immunochemical studies have not shown appreciable CNS staining for iNOS in a number of animal models of sepsis (Buttery et al. 1994; Sato et al. 1995). NOS inhibitors can block the neural damage seen after cerebral artery occlusion in a number of animals. No specific study has been made of pathophysiological brain changes in sepsis following NOS inhibition. However, the NOS inhibitor *N*-aminoarginine produces seizure activity and increased mortality in a model of sepsis in dogs (Cobb et al. 1992). Clearly, CNS effects of NOS inhibition may limit the usefulness of this approach in the therapy of sepsis; human data are at present lacking.

6 Nitric Oxide: Clinical Data in Sepsis

The clinical evidence that NO is involved in the pathology of sepsis comes from two sources; first, direct measurement of NO (or, more usually, surrogate markers of its presence) in septic patients, and, second, data from clinical trials in which various NO blocking drugs have been used.

6.1 Measurement of Nitrite/Nitrate in Sepsis

Three studies have made direct measurements of nitrite/nitrate (the stable end-products of NO biosynthesis) in septic patients. Ochoa et al. (1991) obtained sequential samples from a total of 39 critically ill patients: 17 general surgery patients who were clinically septic or had a severe inflammatory process, 14 trauma patients who had an uneventful recovery, and 8 trauma patients who became septic. The average NO_2^-/NO_3^- level in the septic patients was indeed higher than that the healthy controls (63.1 ± 6.5 μmol/l versus 28.9 μmol/l, $p<0.02$). Curiously, however, levels in the 14 trauma patients without complications (12.8 ± 1.5 μmol/l, $p<0.001$) were significantly *lower* than the controls, and even more surprisingly, NO_2^-/NO_3^- levels did not rise (15.6 ± 1.9 μmol/l) in trauma patients who developed sepsis. To determine whether increases in NO_2^-/NO_3^- were correlated with measurable changes in vascular tone, the levels in the septic patients were compared with the systemic vascular resistance index (SVRI) in the (unspecified) subset of patients in whom invasive hemodynamic monitoring had been carried out. In patients with NO_2^-/NO_3^- levels greater than 42 μmol/l there was a significant reduction in SVRI, although no effect on cardiac index. Interestingly, serum endotoxin levels were also significantly elevated in this subgroup of patients.

The second study (Evans et al. 1993) was performed in 12 septic patients, 7 of whom were receiving inotropes at the time that blood was drawn. The mean arterial pressure (MAP) of the 12 patients ranged between 47 and 73 mmHg, and the mean score on the Acute Physiological and Chronic Health Evaluation (APACHE) II was 24 (range 6–32). The mean NO_2^-/NO_3^- level in the septic patients was 124 μ*M* (range 20–187) compared to 36.4 μ*M* in healthy controls ($p<0.001$). Of interest, a third group of 7 postoperative patients with no evidence of sepsis also had elevated NO_2^-/NO_3^- levels (87.3 μ*M*), intermediate between the other two groups.

The third of this group of studies was similar in design and investigated two groups of septic patients (with or without shock), with a group of non-septic patients as controls. The findings were consistent with those from the other two reports: septic patients with shock had significantly higher NO_2^-/NO_3^- levels (mean 52.9 ± 9.15 nmol/ml) than either the non-shock or non-septic groups. Furthermore, within the subgroup of 13 septic shocked patients, there was significant correlation between NO_2^-/NO_3^- levels and plasma endotoxin, and a negative correlation with systolic blood pressure. In the 8 patients in whom hemodynamic measurements were made NO_2^-/NO_3^- levels were also correlated with cardiac output and SVRI.

The only other study which attempted to correlate NO_2^-/NO_3^- levels with the clinical response was carried out not in sepsis but in patients receiving IL-2 as tumour immunotherapy (Ochoa et al. 1992). Toxicity to IL-2 is manifest principally by the vascular leak syndrome, hypotension, and a hyperdynamic response with low systemic vascular resistance. In 12 patients plasma nitrate levels were significantly elevated, and the authors concluded that NO was a likely cause of the toxicity.

In summary, these data indicate that plasma NO_2^-/NO_3^- levels are elevated in patients with septic shock, and that the levels are correlated directly with physiological variables such as blood pressure and systemic vascular resistance.

However convincing this may seem, the data do not prove causation; nitrate levels are also elevated in healthy controls as well as non-shocked patients, and they are not elevated in shocked trauma patients. Patients with sepsis are extremely complex to evaluate, and it is notoriously difficult to identify cause and effect. This problem was well illustrated by the initial report that plasma nitrate levels showed a highly significant correlation with APACHE III scores, suggesting to the authors that NO_2^-/NO_3^- levels could be used as a predictor of morbidity in postoperative patients (VAN DISSEL et al. 1994). However, MACKENZIE et al. (1994) later pointed out that this finding is probably just an artefact of the correlation of nitrate with renal function and the fact that renal function and age make a very significant contribution to the calculation of the APACHE score.

The best way to address the issue of cause is to study the effects of specific inhibitors of NO in patients with sepsis.

6.2 Clinical Studies with Inhibitors of Nitric Oxide

PETROS et al. (1991) were the first to describe the use of arginine analogues in two patients with severe septic shock. Both L-NMMA and L-NAME produced a rise in MAP lasting between 10 min and several hours, although in the second patient this was accompanied by a fall in cardiac index (CI) from 4.5 to 2.7, $l\ min^{-1}\ m^{-2}$ and the patient died with multi-organ failure. Independently, SCHILLING et al. (1993) also tried L-NMMA in a patient with refractory septic shock, and they too noted a rise in MAP accompanied by a 30% fall in CI, although their patient survived.

Since these early anecdotal reports two more substantial studies of arginine analogues have been completed. PETROS and colleagues (1994) carried out a randomised, double-blind, placebo-controlled investigation of L-NMMA in 12 patients with septic shock. Since the numbers were small, comparison of the two groups is rather unhelpful; it is not surprising that there were major differences between them. The 6 patients receiving active drug were given two bolus injections of 0.3 mg/kg, then 1 mg/kg, followed by a continuous infusion of 1 mg/kg per/hour for 6 h. The findings confirmed the earlier observations, in that there was a sustained rise in MAP and systemic vascular resistance, accompanied by a fall in heart rate and cardiac output.

In the second study LORENTE et al. (1993) carried out a more detailed investigation using a slightly different inhibitor, N^{ω}-nitro-L-arginine. The protocol they used was interesting: the first group of 8 patients were given a bolus injection of N^{ω}-nitro-L-arginine (20 mg/kg) followed by a bolus injection of L-arginine. The second group (7 patients) received L-arginine alone. NO is derived from the terminal guanidyl molecule of L-arginine, and to give L-arginine to a septic patient might be thought unwise. However, it is not as surprising as it seems at first sight. First, the early work of Petros and others had suggested that inhibiting NOS is potentially a "two-edged sword"; the rise in MAP is beneficial, but the fall in cardiac output is a significant disadvantage. Second, there was experimental evidence that L-arginine prevents death in animal models of sepsis (MADDEN et al. 1988; GIANOTTI et al. 1993). Hence the protocol adopted in this study was designed to address directly the causative role of NO in the hemodynamic changes that occur in sepsis, and indeed the results are very clear. The administration of N^{ω}-nitro-L-arginine caused a sustained rise in MAP from 89 ± 8 mmHg to a maximum of 140 ± 12 mmHg, accompanied by fall in CI from 3.51 ± 0.39 to 2.65 ± 0.21 l $min^{-1}m^{-2}$. These changes were reversed by the administration of L-arginine. In the second group of patients, who received L-arginine alone, there was transient hypotension and an increase in CI from 3.57 ± 0.15 to 4.74 ± 0.54 l $min^{-1}m^{-2}$.

Arginine analogues such as L-NMMA are competitive inhibitors of NOS; an alternative approach to limiting the effect of excess NO is to block soluble guanylate cyclase, the cellular target of NO. The dye methylene blue acts on soluble guanylate cyclase and has been in use for many years for the treatment of nitrate poisoning; therefore several investigators reasoned that it might also have a role in septic shock. SCHNEIDER et al. (1992) gave methylene blue to two patients in advanced septic shock. In each there was a modest rise in MAP (59–87 mmHg) and a fall in CI (7.8–6.6 l $min^{-1}m^{-2}$). Both patients died a few days later. In the second study (PREISER et al. 1995) 14 patients in shock were given an infusion of 2 mg/kg of methylene blue, and 6 of them were given a second dose 90 min later. The drug produced short-lived increases in MAP (from 61.1 ± 7.6 to 71.7 ± 12.0 mmHg) without any significant effect on heart rate or cardiac output. No adverse effects were seen, but 11 of the 14 patients subsequently died as a consequence of multi-organ failure.

To what extent then do these clinical studies with NO inhibitors of various types provide firm evidence implicating NO in the pathology of septic shock? Ideally, to be able to answer that question we need to know: (a) that the drugs specifically block the production (or the effect) of NO, and that they have no other pharmacological effect at the doses employed; (b) that they can be shown to significantly reduce the plasma levels of nitrite/nitrate; (c) that any haemodynamic changes can be confidently attributed to the intervention, and that no other changes in treatment occur simultaneously; and (d) that there is a sustained effect on the clinical course (ideally, although not necessarily associated with improved survival).

How do the studies described above match up to these goals? The various arginine analogues are specific inhibitors of NO synthase, although they are not specific for the inducible isoform which many believe is principally responsible for the excess NO production in sepsis. Methylene blue is clearly non-specific. The lack of nitrite/nitrate measurements is perhaps the most surprising omission from these clinical studies. None of the papers reviewed here included data showing that treated patients had lower NO_2^-/NO_3^- levels than controls or healthy normals; given the fact that such measurements are feasible (see above), this is an oversight which should be remedied. Indeed, the only *direct* evidence that the drugs were used at a clinically relevant concentration was in the two patients given methylene blue, in whom plasma cGMP levels were reduced (SCHNEIDER et al. 1992). Probably the best evidence that haemodynamic changes are the direct result of the intervention comes from the study by LORENTE et al. (1993), although in all the others the authors were careful to emphasise that no other changes to the treatment were made during the period of experimental observation.

Finally, what about efficacy? It is not our purpose here to discuss in detail the rationale for NO as a therapeutic target or to compare specific forms of treatment. However, in assessing whether there is good evidence that NO plays an important role in the disease or is present merely as a bystander, we must look at the effect these interventions have had. There is a consistent theme in the reports discussed above which suggests that modification of NO activity is reflected in changes in the haemodynamic state in septic patients. The fact that these changes are often merely transitory and may be accompanied by unwanted effects represent real challenges for the pharmacologist and clinician, but nevertheless do suggest that there is a true cause and effect relationship.

In summary, the clinical data presently available do suggest that excess NO is produced during septic shock, and that it is responsible for some of the clinical features of the disease. The data are not perfect; we have alluded to some of the caveats above. There is a lack of histochemical data in patients showing NOS activity in patients with shock, and the absence of specific, effective therapy limits the conclusions that can be drawn. Given the active state of investigation in this field, most of these issues are likely to be resolved in the near future.

References

Archer SL, Huang JMC, Hampl V, Nelson DP, Shultz PJ, Weir EK (1994) Nitric oxide and cGMP cause vasorelaxation by activation of a charybdotoxin-sensitive K channel by cGMP-dependent protein kinase. Proc Natl Acad Sci USA 91: 7583–7587

Arnold WP, Mittal CK, Katsuki S, Murad F (1977) Nitric oxide activates guanylate cyclase and increases guanosine 3′: 5′-cyclic monophosphate levels in various tissue preparations. Proc Natl Acad Sci USA 74: 3203–3207

Balligand JL, Ungureanu Longrois D, Simmons WW, Pimental D, Malinski TA, Kapturczak M, Taha Z, Lowenstein CJ, Davidoff AJ, Kelly RA et al. (1994) Cytokine-inducible nitric oxide synthase (iNOS) expression in cardiac myocytes. Characterization and regulation of iNOS expression and detection of iNOS activity in single cardiac myocytes in vitro. J Biol Chem 269: 27580–27588

Beckman JS, Beckman TW, Chen J, Marshall P, Freeman BA (1990) Apparent hydroxyl radical production by peroxynitrite: implications for endothelial injury from nitric oxide and superoxide. Proc Natl Acad Sci USA 87: 1620–1624

Beckman JS, Chen J, Ischiropoulos H, Crow JP (1994) Oxidative chemistry of peroxynitrite. Methods Enzymol 233: 229–240

Bogdan C, Vodovotz Y, Paik J, Xie QW, Nathan C (1994) Mechanism of suppression of nitric oxide synthase expression by interleukin-4 in primary mouse macrophages. J Leukoc Biol 55: 227–233

Brady AJB, Poole-Wilson PA, Harding SE, Warren JB (1992) Nitric oxide production within cardiac myocytes reduces their contractility in endotoxemia. Am J Physiol 263: H1963–H1966

Bredt DS, Snyder SH (1990) Isolation of nitric oxide synthase, a calmodulin-requiring enzyme. Proc Natl Acad Sci USA 87: 682–685

Bredt DS, Snyder DS (1994) Nitric oxide: a physiologic messenger molecule. Annu Rev Biochem 63: 175–195

Bredt DS, Hwang PM, Glatt CE, Lowenstein C, Reed R, Snyder SH (1991) Cloned and expressed nitric oxide synthase structurally resembles cytochrome P-450 reductase. Nature 351: 714–718

Brigham KL (1991) Oxygen radicals – an important mediator of sepsis and septic shock. Klin Wochenschr 69: 1004–1008

Bult H, Boeckxstaens GE, Pelckmans PA, Jordaens FH, Van Maercke YM, Herman AG (1990) Nitric oxide as an inhibitory non-adrenergic non-cholinergic neurotransmitter [see comments]. Nature 345: 346–347

Busse R, Mülsch A (1990) Induction of nitric oxide synthase by cytokines in vascular smooth muscle cells. FEBS Lett 275: 87–90

Buttery LD, Evans TJ, Springall DR, Carpenter A, Cohen J, Polak JM (1994) Immunochemical localization of inducible nitric oxide synthase in endotoxin-treated rats. Lab Invest 71: 755–764

Cho HJ, Xie QW, Calaycay J, Mumford RA, Swiderek KM, Lee TD, Nathan C (1992) Calmodulin is a subunit of nitric oxide synthase from macrophages. J Exp Med 176: 599–604

Cobb JP, Natanson C, Hoffman WD, Lodato RF, Banks S, Koev CA, Solomon MA, Elin RJ, Hosseini JM, Danner RL (1992) N-Omega-amino-L-argine, an inhibitor of nitric oxide synthase, raises vascular resistance but increases mortality rates in awake canines challenged with endotoxin. J Exp Med 176: 1175–1182

Evans T, Carpenter A, Cohen J (1992a) Purification of a distinctive form of endotoxin-induced nitric oxide synthase from rat liver. Proc Natl Acad Sci USA 89: 5361–5365

Evans T, Carpenter A, Silva A, Cohen J (1992b) Differential effects of monoclonal antibodies to tumor necrosis factor alpha and gamma interferon on induction of hepatic nitric oxide synthase in experimental gram-negative sepsis. Infect Immun 60: 4133–4139

Evans TJ, Carpenter A, Kinderman H, Cohen J (1993) Evidence of increased nitric oxide production in patients with the sepsis syndrome. Circ Shock 41: 77–81

Evans T, Carpenter A, Silva A, Cohen J (1994) Inhibition of nitric oxide synthase in experimental gram-negative sepsis. J Infect Dis 169: 343–349

Finkel MS, Oddis CV, Jacob TD, Watkins SC, Hattler BG, Simmons RL (1992) Negative inotropic effects of cytokines on the heart mediated by nitric oxide. Science 257: 387–389

Flohé L, Giertz H (1987) Endotoxins, arachidonic acid, and superoxide formation. Rev Infect Dis 9 Suppl 5: S553–S561

Florquin S, Amraoui Z, Dubois C, Decuyper J, Goldman M (1994) The protective role of endogenously synthesized nitric oxide in staphylococcal enterotoxin B-induced shock in mice. J Exp Med 180: 1153–1158

Forstermann U, Closs EI, Pollock JS, Nakane M, Schwarz P, Gath I, Kleinert H (1994) Nitric oxide synthase isozymes. Characterization, purification, molecular cloning, and functions. Hypertension 23: 1121–1131

Freeman BA, Turrens JF, Mirza Z, Crapo JD, Young SL (1985) Modulation of oxidant lung injury by using liposome-entrapped superoxide dismutase and catalase. Fed Proc 44: 2591–2595

Furchgott RF, Zawadzki JV (1980) The obligatory role of endothelial cells in the relaxation of arterial smooth muscle by acetylcholine. Nature 288: 373–376

Galea E, Reis DJ, Feinstein DL (1994) Cloning and expression of inducible nitric oxide synthase from rat astrocytes. J Neurosci Res 37: 406–414

Gianotti L, Alexander JW, Pyles T, Fukushima R (1993) Arginine-supplemented diets improve survival in gut-derived sepsis and peritonitis by modulating bacterial clearance. The role of nitric oxide. Ann Surg 217: 644–654

Glauser MP, Zanetti G, Baumgartner J-D, Cohen J (1991) Septic shock: pathogenesis. Lancet 338: 732–736

Green LC, De Luzuriaga KR, Wagner DA, Rand W, Istfan N, Young VR, Tannenbaum SR (1981a) Nitrate biosynthesis in man. Proc Natl Acad Sci USA 78: 7764–7768

Green LC, Tannenbaum SR, Goldman P (1981b) Nitrate synthesis in the germfree and conventional rat. Science 212: 56–58

Gross SS, Jaffe EA, Levi R, Kilbourn RG (1991) Cytokine-activated endothelial cells express an isotype of nitric oxide synthase which is tetrahydrobiopterin-dependent, calmodulin-independent and inhibited by arginine analogs with a rank-order of potency characteristic of activated macrophages. Biochem Biophys Res Comm 178: 823–829

Gruetter CA, Barry BK, McNamara DB, Gruetter DY, Kadowitz PJ, Ignarro L (1979) Relaxation of bovine coronary artery and activation of coronary arterial guanylate cyclase by nitric oxide, nitroprusside and a carcinogenic nitrosoamine. J Cyclic Nucleotide Res 5: 211–224
Hauschildt S, Scheipers P, Bessler WG, Mulsch A (1992) Induction of nitric oxide synthase in L929 cells by tumour-necrosis factor alpha is prevented by inhibitors of poly(ADP-ribose) polymerase. Biochem J 288: 255–600
Hibbs JB, Vavrin Z, Taintor RR (1987) L-arginine is required for expression of the activated macrophage effector mechanism causing selective metabolic inhibition in target cells. J Immunol 138: 550–565
Iadecola C, Zhang F, Xu X (1995) Inhibition of inducible nitric oxide synthase ameliorates cerebral ischemic damage. Am J Physiol 268: R286–R292
Ignarro LJ (1990) Biosynthesis and metabolism of endothelium-derived nitric oxide. Annu Rev Pharmacol Toxicol 30: 535–560
Ignarro LJ, Lippton H, Edwards JC, Baricos WH, Hyman AL, Kadowitz PJ, Gruetter CA (1981) Mechanism of vascular smooth muscle relaxation by organic nitrates, nitrites, nitroprusside and nitric oxide: evidence for the involvement of S-nitrosothiols as active intermediates. J Pharmacol Exp Ther 218: 739–749
Ignarro LJ, Buga GM, Wood KS, Byrns RE, Chaudhuri G (1987) Endothelium-derived relaxing factor produced and released from artery and vein is nitric oxide. Proc Natl Acad Sci USA 84: 9265–9269
Ischiropoulos H, Zhu L, Chen J, Tsai M, Martin JC, Smith CD, Beckman JS (1992) Peroxynitrite-mediated tyrosine nitration catalyzed by superoxide dismutase. Arch Biochem Biophys 298: 431–437
Iyengar R, Stuehr DJ, Marletta MA (1987) Macrophage synthesis of nitrite, nitrate, and N-nitrosamines: precursors and role of the respiratory burst. Proc Natl Acad Sci USA 84: 6369–6373
Izumi Y, Benz AM, Clifford DB, Zorumski CF (1992) Nitric oxide inhibitors attenuate N-methyl-D-aspartate excitotoxicity in rat hippocampal slices. Neurosci Lett 135: 227–230 [published erratum appeared in Neurosci Lett (1992) 140(1): 135]
Knowles RG, Merrett M, Salter M, Moncada S (1990) Differential induction of brain, lung and liver nitric oxide synthase by endotoxin in the rat. Biochem J 270: 833–836
Liew FY, Li Y, Severn A, Millott S, Schmidt J, Salter M, Moncada S (1991) A possible novel pathway of regulation by murine T helper type-2 (Th2) cells of a Th1 cell activity via the modulation of the induction of nitric oxide synthase on macrophages. Eur J Immunol 21: 2489–2494
Lonchampt MO, Auguet M, Delaflotte S, Goulin Schulz J, Chabrier PE, Braquet P (1992) Lipoteichoic acid: a new inducer of nitric oxide synthase. J Cardiovasc Pharmacol 20 Suppl 12: S145–S147
Lorente JA, Landín L, De Pablo R, Renes E, Liste D (1993) L-Arginine pathway in the sepsis syndrome. Crit Care Med 21: 1287–1295
Mackenzie IM, Young JD, Garrard CS (1994) Serum nitrates as markers of postoperative morbidity. Lancet 344: 410
MacMicking JD, Nathan C, Hom G, Chartrain N, Fletcher DS, Trumbauer M, Stevens K, Xie Q-W, Sokol K, Hutchinson N, Chen H, Mudgett JS (1995) Altered responses to bacterial infection and endotoxic shock in mice laking inducible nitric oxide synthase. Cell 81: 641–650
Madden HP, Breslin RJ, Wasserkrug HL, Efron G, Barbul A (1988) Stimulation of T cell immunity by arginine enhances survival in peritonitis. J Surg Res 44: 658–663
Marletta MA (1995) Approaches toward selective inhibition of nitric oxide synthase. J Med Chem 37: 1899–1907
Mellion BT, Ignarro LJ, Ohlstein EH, Pontecorvo EG, Hyman AL, Kadowitz PJ (1981) Evidence for the inhibitory role of guanosine 3', 5'-monophosphate in ADP-induced human platelet aggregation in the presence of nitric oxide and related vasodilators. Blood 57: 946–955
Minc Golomb D, Tsarfaty I, Schwartz JP (1994) Expression of inducible nitric oxide synthase by neurones following exposure to endotoxin and cytokine. Br J Pharmacol 112: 720–722
Mosmann TR, Cherwinski H, Bond MW, Giedlin MA, Coffman RL (1986) Two types of murine helper T cell clone. I. Definition according to profiles of lymphokine activities and secreted proteins. J Immunol 136: 2348–2357
Nakayama DK, Geller DA, Lowenstein CJ, Chern HD, Davies P, Pitt BR, Simmons RL, Billiar TR (1992) Cytokines and lipopolysaccharide induce nitric oxide synthase in cultured rat pulmonary artery smooth muscle. Am J Respir Cell Mol Biol 7: 471–476 [published erratum appeared in Am J Respir Cell Mol Biol (1993) 9(2): 229]
Nathan C (1992) Nitric oxide as a secretory product of mammalian cells. FASEB J 6: 3051–3064
Nathan C, Hibbs JB Jr (1991) Role of nitric oxide synthesis in macrophage antimicrobial activity. Curr Opin Immunol 3: 65–70

Nava E, Palmer RMJ, Moncada S (1992) The role of nitric oxide in endotoxic shock: effects of NG-monomethyl-L-arginine. J Cardiovasc Pharmacol 20(S12): S132–134
Nishijima H, Weil MH, Shubin H, Cavanilles J (1973) Hemodynamic and metabolic studies on shock associated with gram negative bacteremia. Medicine (Baltimore) 52: 287–294
Nussler A, Di Silvio M, Billiar T, Hoffman RA, Geller DA, Selby R, Madariaga J, Simmons RL (1992) Stimulation of the nitric oxide synthase pathway in human hepatocytes by cytokines and endotoxin. J Exp Med 176: 261–264
Ochoa JB, Udekwu AO, Billiar TR, Curran RD, Cerra FB, Simmons RL, Peitzman AB (1991) Nitrogen oxide levels in patients after trauma and during sepsis. Ann Surg 214: 621–626
Ochoa JB, Curti B, Peitzman AB, Simmons RL, Billiar TR, Hoffman R, Rault R, Longo DL, Urba WJ, Ochoa AC (1992) Increased circulating nitrogen oxides after human tumor immunotherapy: correlation with toxic hemodynamic changes. J Natl Cancer Inst 84: 864–867
Oswald IP, Wynn TA, Sher A, James SL (1994) NO as an effector molecule of parasite killing: modulation of its synthesis by cytokines. Comp Biochem Physiol Pharmacol Toxicol Endocrinol 108: 11–18
Palmer RMJ, Ferrige AG, Moncada S (1987) Nitric oxide release accounts for the biological activity of endothelium-derived relaxing factor. Nature 327: 524–526
Parker MM, Shelhamer JH, Bacharach SL, Green MV, Natanson C, Frederick TM, Damske BA, Parrillo JE (1984) Profound but reversible myocardial depression in patients with septic shock. Ann Intern Med 100: 483–490
Parrillo JE, Burch C, Shelhamer JH, Parker MM, Natanson C, Schuette W (1985) A circulating myocardial depressant substance in humans with septic shock. J Clin Invest 76: 1539–1553
Petros A, Bennett D, Vallance P (1991) Effect of nitric oxide synthase inhibitors on hypotension in patients with septic shock. Lancet 338: 1557–1558
Petros A, Lamb G, Leone A, Moncada S, Bennett D, Vallance P (1994) Effects of a nitric oxide synthase inhibitor in humans with septic shock, Cardiovasc Res 28: 34–39
Pollock JS, Förstermann U, Mitchell JA, Warner TD, Schmidt HHHW, Nakane M, Murad F (1991) Purification and characterization of particulate endothelium-derived relaxing factor synthase from cultured and native bovine aortic endothelial cells. Proc Natl Acad Sci USA 88: 10480–10484
Pou S, Pou WS, Bredt DS, Snyder SH, Rosen GM (1992) Generation of superoxide by purified brain nitric oxide synthase. J Biol Chem 267: 24173–24176
Preiser J-C, Lejeune P, Roman A, Carlier E, De Backer D, Leeman M, Kahn RJ, Vincent J-L (1995) Methylene blue administration in septic shock: a clinical trial. Crit Care Med 23: 259–264
Radi R, Beckman JS, Bush KM, Freeman BA (1991a) Peroxynitrite-induced membrane lipid peroxidation: the cytotoxic potential of superoxide and nitric oxide. Arch Biochem Biophys 288: 481–487
Radi R, Beckman JS, Bush KM, Freeman BA (1991b) Peroxynitrite oxidation of sulfhydryls. The cytotoxic potential of superoxide and nitric oxide. J Biol Chem 266: 4244–4250
Radomski MW, Palmer RM, Moncada S (1990) Glucocorticoids inhibit the expression of an inducible, but not the constitutive, nitric oxide synthase in vascular endothelial cells. Proc Natl Acad Sci USA 87: 10043–10047
Roberts AB, Vodovotz Y, Roche NS, Sporn MB, Nathan CF (1992) Role of nitric oxide in antagonistic effects of transforming growth factor-beta and interleukin-1 beta on the beating rate of cultured cardiac myocytes. Mol Endocrinol 6: 1921–1930
Sato K, Miyakawa K, Takeya M, Hattori R, Yui Y, Sunamoto M, Ichimori Y, Ushio Y, Takahashi K (1995) Immunohistochemical expression of inducible nitric oxide synthase (iNOS) in reversible endotoxic shock studied by a novel monoclonal antibody against rat iNOS. J Leukoc Biol 57: 36–44
Schilling J, Cakmakci M, Bättig U, Geroulanos S (1993) A new approach in the treatment of hypotension in human septic shock by N^G-monomethyl-L-arginine, an inhibitor of the nitric oxide synthetase. Intensive Care Med 19: 227–231
Schneider F, Lutun P, Hasselman M, Stoclet JC, Tempé JD (1992) Methylene blue increases systemic vascular resistance in human septic shock. Preliminary observations. Intensive Care Med 18: 309–311
Sessa WC (1994) The nitric oxide synthase family of proteins. J Vasc Res 31: 131–143
Southan GJ, Szabo C, Thiemermann C (1995) Isothioureas: potent inhibitors of nitric oxide synthases with variable isoform selectivity. Br J Pharmacol 114: 510–516
Stadler J, Billiar TR, Curran RD, Stuehr DJ, Ochoa JB, Simmons RL (1991) Effect of exogenous and endogenous nitric oxide on mitochondrial respiration of rat hepatocytes. Am J Physiol 260: C910–C916

Stamler JS, Jaraki O, Osborne J, Simon DI, Keaney J, Vita J, Singel D, Valeri CR, Loscalzo J (1992) Nitric oxide circulates in mammalian plasma primarily as an S-nitroso adduct of serum albumin. Proc Natl Acad Sci USA 89: 7674–7677

Stuehr DJ, Marletta MA (1985) Mammalian nitrate biosynthesis: mouse macrophages produce nitrite and nitrate in response to Escherichia coli lipopolysaccharide. Proc Natl Acad Sci USA 82: 7738–7742

Stuehr DJ, Gross SS, Sakuma I, Levi R, Nathan CF (1989) Activated murine macrophages secrete a metabolite of arginine with the bioactivity of endothelium-derived relaxing factor and the chemical reactivity of nitric oxide. J Exp Med 169: 1011–1020

Szabo C, Southan GJ, Thiemermann C (1994) Beneficial effects and improved survival in rodent models of septic shock with S-methylisothiourea sulfate, a potent and selective inhibitor of inducible nitric oxide synthase. Proc Natl Acad Sci USA 91: 12472–12476

Taylor-Robinson A, Liew FY, Severn A, Xu D, McSorley SJ, Garside P, Padron J, Phillips RS (1994) Regulation of the immune response by nitric oxide differentially produced by T helper type 1 and T helper type 2 cells. Eur J Immunol 24: 980–984

Thiemermann C, Vane J (1990) Inhibition of nitric oxide synthesis reduces the hypotension induced by bacterial lipopolysaccharides in the rat in vivo. Eur J Pharmacol 182: 591–595

Thorin Trescases N, Hamilton CA, Reid JL, McPherson KL, Jardine E, Berg G, Bohr D, Dominiczak AF (1995) Inducible L-arginine/nitric oxide pathway in human internal mammary artery and saphenous vein. Am J Physiol 268: H1122–H1132

Van Dissel JT, Groeneveld PH, Maes B, van Furth R, Frolich M, Feuth HD (1994) Nitric oxide: a predictor of morbidity in postoperative patients? Lancet 343: 1579–1580

Vodovotz Y, Bogdan C, Paik J, Xie QW, Nathan C (1993) Mechanisms of suppression of macrophage nitric oxide release by transforming growth factor beta. J Exp Med 178: 605–613

Vodovotz Y, Kwon NS, Pospischil M, Manning J, Paik J, Nathan C (1994) Inactivation of nitric oxide synthase after prolonged incubation of mouse macrophages with IFN-gamma and bacterial lipopolysaccharide. J Immunol 152: 4110–4118

Wallace MN, Bisland SK (1994) NADPH-diaphorase activity in activated astrocytes represents inducible nitric oxide synthase. Neuroscience 59: 905–919

Wei X-Q, Charles I, Smith A, Ure J, Feng G-J, Huang F-P, Xu D, Muller W, Moncada S, Liew FY (1995) Altered immune responses in mice lacking inducible nitric oxide synthase. Nature 375: 408–411

Wizemann TM, Gardner CR, Laskin JD, Quinones S, Durham SK, Goller NL, Ohnishi ST, Laskin DL (1994) Production of nitric oxide and peroxynitrite in the lung during acute endotoxemia. J Leukoc Biol 56: 759–768

Wright CE, Rees DD, Moncada S (1992) Protective and pathological roles of nitric oxide in endotoxin shock [see comments]. Cardiovasc Res 26: 48–57

Xie Q-W, Nathan C (1994) The high output nitric oxide pathway: role and regulation. J Leukoc Biol 56: 576–582

Yokoyama T, Vaca L, Rossen RD, Durante W, Hazarika P, Mann DL (1993) Cellular basis for the negative inotropic effects of tumor necrosis factor-alpha in the adult mammalian heart. J Clin Invest 92: 2303–2312

Zembowicz A, Vane JR (1992) Induction of nitric oxide synthase activity by toxic shock syndrome toxin 1 in a macrophage-monophage-monocyte cell line. Proc Natl Acad Sci USA 89: 2051–2055

Multiple Organ Failure in Septic Shock

L.G. THIJS, A.B.J. GROENEVELD, and C.E. HACK

1 Introduction 209
2 Definitions 211
3 Risk Factors and Mortality 212
4 Incidence and Prevalence 213
5 Is There a Sequence of Organ Failure? 215
6 Clinical Manifestations 216
6.1 Adult Respiratory Distress Syndrome 217
6.2 Renal Dysfunction 217
6.3 Central Nervous System Failure 218
6.4 Hematological Failure 218
6.5 Liver Dysfunction 219
7 Pathophysiology–Pathogenesis 219
7.1 Mediators 220
7.1.1 The Cytokine Network 221
7.1.2 Complement System 223
7.1.3 Contact System, Coagulation, Fibrinolysis 224
7.1.4 Neutrophils 225
7.2 Ischemia and Reperfusion 226
References 229

1 Introduction

In the late 1960s several reports appeared describing remote organ failure (e.g., pulmonary or liver failure) as a complication of severe sepsis (CLOWES et al. 1968; SKILLMAN et al. 1969). Stimulated by the observation of TILNEY et al. (1973) that sequential failure of initially uninvolved organ systems may follow operations for ruptured aortic aneurysm, the concept emerged that severe injury can result in damage to distant organ systems. In a classical editorial by BAUE (1975) entitled "Multiple, Progressive or Sequential Systems Failure: A Syndrome of the 1970s," this concept was formulated as the basis of a "new" clinical syndrome. Several terms were coined thereafter, such as multiple organ failure (EISEMAN

Medical Intensive Care Unit, Free University Hospital, 1066 AD Amsterdam, The Netherlands et al. 1977), multiple system(s) organ failure (BORDER et al. 1976), and multiple organ system failure (BELL et al. 1983), to describe this evolving clinical syndrome of otherwise unexplained progressive physiological failure of several interdependent organ systems. More recently the term multiple organ dysfunction syndrome has been proposed as a more appropriate description (BONE et al. 1992).

Major advances in the field of surgical and medical treatment and resuscitation techniques, widespread introduction of monitoring and life-support technologies, and the evolution of intensive care units (ICUs) have allowed the survival of critically ill patients who previously would have died. However, it also became increasingly evident that in a number of patients, in particular septic patients, progressive deterioration of multiple organ function occurs, notwithstanding optimal therapeutic efforts and interventions.

The increasing incidence of morbidity and mortality caused by multiple organ failure (MOF) has paralleled improvements in life-support technologies and their application to an increasingly high-risk patient population. The emergence of the MOF syndrome, usually defined as failure of at least two organ systems, has therefore been a direct consequence of the evolution of ICU capabilities.

MOF is presently the leading cause of death in the ICU and is responsible for 50%–80% of all ICU deaths, with a major impact on ICU and hospital resources (BARTON and CERRA 1989). Critically ill patients today rarely die as a consequence of the disease process progressing to the point that admission to the ICU is required but rather from MOF, which is increasingly recognized as the final common pathway leading to death in these patients (MARSHALL 1994). Initial studies stressed the relationship between uncontrolled sepsis and remote organ dysfunction (POLK and SHIELDS 1977; FRY et al. 1980). It became increasingly evident, however, that remote organ system dysfunction can evolve in the absence of indentifiable infection (MEAKINS et al. 1980; FAIST et al. 1983; GORIS et al. 1985). MOF has now been documented as a complication in a variety of noninfectious conditions such as acute pancreatitis, major surgery, severe trauma, burns, and hypovolemic shock. No obvious differences in clinical signs and symptoms of MOF can be identified between patients with or without a bacterial infection (GORIS et al. 1985). Moreover, a similar syndrome can be reproduced in the animal model by infusion of various mediators of inflammation (TRACEY et al. 1986). These observations have led to the concept of a generalized inflammatory host response to variable insults as the pathophysiological basis of MOF. The fact that organ failure occurs simultaneously or in a certain sequence also suggests a common or at least similar pathogenesis. In addition, a complex interrelationship among individual organ systems may promote a domino effect in that failure of one organ may establish an amplification process that hastens injury to another (EISEMAN et al. 1977; MATUSCHAK and RINALDO 1988; CALLERY et al. 1991).

2 Definitions

Although the term MOF has been widely used since its introduction, the criteria used to characterize it emerged in an arbitrary and retrospective fashion and differed widely among the various studies. Many authors regard organ *failure* as a dichotomous event, either present or absent, and define the severity of MOF by the number of failing organ systems (FRY et al. 1980; BELL et al. 1983; KNAUS et al. 1985; HENAO et al. 1991). This concept of organ failure reflects insufficient physiologic function to sustain the normal needs of the host without exogenous intervention (BAUE 1975).

Others, however, support the concept that organ *dysfunction* is a continuum with incremental degrees of physiologic derangements in individual organ systems, a process rather than an event (GORIS et al. 1985; MARSHALL et al. 1988). Alteration in organ function can vary widely from some mild degree of organ dysfunction such as a moderate elevation in creatinine or bilirubin levels, which have no clinical consequences, to frank organ failure such as severe respiratory failure or coma. The degree of organ dysfunction therefore has a major clinical impact.

Recently the term *multiple organ dysfunction syndrome* (MODS) has been proposed, defined as a clinical syndrome characterized by the development of progressive but potentially reversible physiological dysfunction in two or more organs or organ systems induced by a variety of acute insults, including sepsis (BONE et al. 1992).

The term dysfunction identifies this process as a phenomenon in which organ function is not able to maintain homeostasis and recognizes it as a dynamic continuum of deranged physiological function rather than an event which is either present or absent. In the continuous process of the host inflammatory response MODS may be understood to represent the more severe end of the spectrum of the severity of illness (BONE et al. 1992). If MODS represents a continuum of severity, ICU mortality would be related not only to the number of failing organs but also to the degree of dysfunction of any particular organ system (MARSHALL 1994). Support for this concept has been provided by the observation of a strong correlation between increasing dysfunction within organ systems and the probability of death (MARSHALL 1994, 1995). There is, however, no clear consensus on what specific dysfunctioning (or failing) organ systems should constitute the syndrome. If looked for, abnormalities in physiological function can be found in virtually any organ system in critically ill (septic) patients. Moreover, there is no consensus on the best descriptors of organ dysfunction or the most appropriate assessment of the degree of dysfunction. Also, if the term MODS is generally accepted, it remains to be decided whether the term organ failure should still be used for the most severe degrees of dysfunction. Authors have used a wide variety of indicators of organ dysfunction, including (biochemical) markers reflecting functional organ abnormalities with various cutoff points, the necessity of mechanical support (e.g., mechanical ventilation, hemodialysis) or pharmacological support (e.g., inotropes), and even disease categories such as cholecystitis and stress ulcer. Therapeutic measures as a descriptor of organ dysfunction are subject to different medical policies and local practices. Also, diagnostic categories may be based on subjective criteria, and one may question whether they really reflect remote organ dysfunction. Objective, indisputable, and quantifiable variables, widely used in clinical practice, therefore appear to be the best candidates to assess organ dysfunction.

In a review of 30 papers related to the subject MARSHALL (1995) showed that seven systems (respiratory, renal, gastrointestinal, hepatic, central nervous system, hematological, and cardiovascular) were included in the definition of MOF in more than half of the reports. It appeared that, although criteria for organ dysfunction varied markedly, definitions were based primarily on physiological abnormalities and to a lesser extent on the basis of therapeutic interventions. In this analysis valid single variables reflecting physiological dysfunction appeared to be available for five organ systems: PaO_2/FiO_2 ratio for the respiratory system, serum creatinine level for the renal system, serum bilirubin level for the hepatic system, platelet count for the hematological system, and the Glasgow Coma Score for the central nervous system. Descriptors of dysfunction for the cardiovascular system and the gastrointestinal system seem at present suboptimal (MARSHALL 1995). A single variable has the advantage of simplicity and provides a measure that can be very sensitive to progressive degrees of organ dysfunction, but it may lack specificity, is prone to measurement error, and can be influenced by therapeutic intervention (MARSHALL 1995).

The importance of this detailed analysis of the literature is that it provides a basis for more uniformly accepted definitions and criteria for MOF/MODS. Given the lack of uniform definitions, the term organ failure for specific organ systems is used below on the basis of the definitions used by the authors, which implies obvious limitations for conclusions.

3 Risk Factors and Mortality

Several studies have demonstrated that advanced age, preexisting chronic disease (including malnutrition and alcohol abuse), and sepsis are major (independent) risk factors for the development of MOF (PINE et al. 1983; KNAUS et al. 1985; TRAN et al. 1990). Underlying chronic disease and advanced age are usually associated with a depression in functional reserve of individual organs, reducing the tolerance to injury (GEE et al. 1990). Advanced age and preexisting chronic disease are in turn risk factors for sepsis (TRAN et al. 1990). In a large multicenter study it appeared that patients with septic shock (and those with cardiac arrest) had a much higher chance to develop MOF than patients with other diagnoses (KNAUS et al. 1985). A case-control study in a surgery and trauma patient population showed that hypovolemic shock, sepsis, and the time before adequate resuscitation were independent risk factors for MOF (HENAO et al. 1991). Therefore, sepsis and shock seem to be the most common conditions leading to MOF.

Mortality of patients with MOF is extremely high, ranging from 30% to 100%. Mortality is related significantly to the number of failing organ systems (FRY et al. 1980; PINE et al. 1983; KNAUS et al. 1985; GORIS et al. 1985; TRAN et al. 1990; HENAO et al. 1991; HEBERT et al. 1993) and to the duration of organ failure (KNAUS et al. 1985). Mortality almost uniformly approaches 90%–100% when three or more organ systems fail. In a large-scale multicenter study 99 patients had three or more failing organ systems after 72 h of intensive therapy, of whom only two survived (KNAUS et al. 1985). Most of these patients (80%) were non-operative admissions with diagnoses such as septic shock or cardiac arrest. In addition to the number of failing organs, advanced age and preexistent chronic disease are major determinants of overall mortality (KNAUS et al. 1985; TRAN et al. 1990).

4 Incidence and Prevalence

Depending on the type of ICU and its case mix, MOF can be observed to develop in about 10%–15% of the patients (FAIST et al. 1983; KNAUS et al. 1985). However, much higher incidences have been reported, even as high as about 40% (TRAN et al. 1990). There is a remarkable paucity of published epidemiological data and of studies focusing specifically on the incidence of MOF in patients with septic shock. Most studies have addressed MOF as an entity more or less regardless its cause. However, the prevalence of sepsis in most series is high, ranging from 50% to 90% (EISEMAN et al. 1977; FRY et al. 1980; HENAO et al. 1991).

In a study in patients with positive blood cultures who exhibited clinical signs of the sepsis syndrome the course of disease was complicated by the adult respiratory distress syndrome (ARDS) in 15%, by disseminated intravascular coagulation (DIC) in 5.4%, and by other manifestations of MOF (acute renal and hepatic failure) in 36% (KIEFT et al. 1993). A recent prospective evaluation of the natural history of the systemic inflammatory response syndrome (BONE et al. 1992) demonstrated that the incidence of end organ dysfunction increased with the more severe forms of sepsis (RANGEL-FRAUSTO et al. 1995). In patients with septic shock the attack rate of ARDS was 18%, of DIC 38%, and of acute renal failure 51% (Table 1). A higher incidence of organ failure was observed in another study in 52 septic patients: ARDS 79%, acute renal failure 67%, liver failure 65%, and 38 patients died of at least three organ failures (ESEN et al. 1992). It is not clear in these studies, however, how many patients developed dysfunction of two or more organ systems. In a survey of 393 patients with bacteremia over a 7-year period MOF was present in 22.6% of the patients

Table 1. Clinical trials in sepsis

Study	Type of patients	*n*	Organ failure (%)				
			Respiratory	Renal	Hematological	Hepatic	Neurological
Ziegler et al. (1991) (HA-IA study 1)	Gram-negative, bacteremia with, sepsis syndrome	200	10.5**	40.5	19.5*	22.5	–
Greenman et al. (1991) (E5-study 1)	Gram-negative sepsis	316	21.5**	22.5	27 *	–	–
	No shock	137	14.0	15.0	16.0		
	Shock	179	27.0	28.0	35.0		
Fisher et al. (1994) (IL-1ra study)	Sepsis syndrome	893	25.0**	30.0	14.0*	26.0	–
Dhainaut et al. (1994) (anti-PAF study)	Severe gram-negative sepsis	120	67.0	30.0	29.0	28.0	32.0
Brun-Buisson et al. (1994) (HA-1A study 2)	Sepsis syndrome	600	83.0	44.5	14.0	–	6.0
Abraham et al. (1995) (Norasept anti-TNF study)	Sepsis syndrome	971	19.5**	9.0	32 *	6.0	7.5
Bone et al. (1995) (E5 study 2)	Gram-negative sepsis (no shock)	530	9.5**	13.5	7.5*	–	–
Cohen and Carlet (submitted) (Intersept anti-TNF study)	Septic shock	420	19.5**	18.0	49 *	4.5	6.0
Rangel-Frausto et al. (1995)	Sepsis	649	6 **	19	16 *	–	–
	Severe sepsis	469	8 **	23	18 *	–	–
	Septic shock	110	18 **	51	38 *	–	–

*ARDS, **DIC

(Uzun et al. 1992). In a report from a medical ICU including 196 patients with sepsis, 27 (14%) did not have any organ failure at any time during their ICU stay, 48 had single-organ failure, and 121 (62%) developed MOF, but cardiovascular failure was included in the definition of organ failure (Tran et al. 1990). A study of 48 patients with severe septic shock found the mean number of failing organ systems 48 h after admission to be 3.3 for survivors and 4.0 for nonsurvivors (Ruokonen et al. 1991).

The recently completed clinical trials in sepsis provide a large data base, including the prevalence of organ dysfunction at the time of entry into the study (Table 1). There is, however, no published information available on the prevalance of MOF, i.e., simultaneous failure of at least two organ systems and some degree of organ dysfunction was usually included in the inclusion criteria. In the French HA-1A study 33% of patients had three or more organ failures at study entry (Brun-Buisson et al. 1994). Moreover, limited details are available on the development of organ dysfunction in the placebo group in these studies. In the second E5 study 9% of the placebo group patients developed ARDS, 10%

acute renal failure, 7% DIC, 18% central nervous dysfunction, and 6% hepatobiliary dysfunction after enrollment in the study (BONE et al. 1995). In the Intersept study 33.6% of the patients in septic shock had dysfunction of two organ systems, 10.3% of three and 2.3% of four are more organ system at the time of inclusion (data kindly provided by Bayer-Miles Inc.). Therefore, taken together, 46.2% of the patients had multiple organ failure early in the course of septic shock (COHEN and CARLET, submitted). During a 28-day follow-up an additional 31.2% of the placebo group developed acute renal failure, 14.3% ARDS, 23.4% DIC, 14.1% liver dysfunction, and 6.3% neurological dysfunction (Bayer-Miles). Although the definition of organ failure varied, the conclusion is warranted that the incidence of organ failure in sepsis and particularly septic shock is high, and this complication seems to occur mainly early in the course of disease. Because only the more severe forms of organ dysfunction are commonly included in the definition, the true incidence of organ dysfunction is most likely underestimated.

5 Is There a Sequence of Organ Failure?

It has been suggested that the evolution of MOF generally follows a predictable course, beginning with the lungs and followed by hepatic, intestinal, and renal failure, in that order (DEITCH 1992). This may be particularly clear in polytrauma patients but is often less so in other patient categories. The exact temporal pattern of organ dysfunction in MOF differs somewhat between studies (EISEMAN et al. 1977; FRY et al. 1980; GORIS et al. 1985). The temporal sequence of organ failure can be significantly modified by the presence of preexistent disease and by the nature of the precipitating event. For example, in patients with chronic renal disease the kidney tends to be the first organ to fail. This illustrates that the physiological reserve of a specific organ system is an important determinant of its failure following an acute insult. Also, acute renal failure may be an early event in patients who sustain a prolonged period of circulatory shock. Nevertheless, the respiratory system in general is almost invariably the first organ system to fail.

The pattern of failing organ systems may furthermore change over time. Nowadays, bleeding due to stress ulcers, in various reports a criterion for gastrointestinal failure, has become a rare event in the ICU, most likely due to improved resuscitation policies (COOK et al. 1994). There is only very limited information in the literature on the time sequence of failing organ systems in patients with septic shock. In this condition severe cardiovascular failure is by definition present from the very start. In a study in 48 patients with severe septic shock all had respiratory failure, defined as the clinical need for mechanical ventilation at the onset of shock (RUOKONEN et al. 1991). After 48 h the prevalence of acute renal failure was 50%, gastrointestinal failure 64%, hepatic failure 14.3%, and central nervous system failure 50%. This study, again, illustrates that organ failure may occur early and simultaneously in various organ system, especially in severe septic shock. In general, respiratory failure and ARDS are early complications of septic shock, followed by acute renal failure, whereas clinical jaundice is usually a late complication. Also, (severe) thombocytopenia can occur early in the course of septic shock, in some cases associated with DIC. Central nervous system dysfunction also may occur very early in the course of septic shock. (BOLTON et al. 1993) but is often difficult to evaluate as many patients are mechanically ventilated.

6 Clinical Manifestations

Circulatory failure, i.e., septic shock, complicates invasive bacterial infection in about 40%–50% of cases and may be the most impressive initial clinical expression of organ failure in many of these patients. Circulatory failure by itself can contribute significantly to the development of organ failure (e.g., acute renal failure) and may therefore also be considered a cause of MOF. There is, however, a wide variety in the time course and severity of complicating circulatory failure in septic patients, ranging from mild circulatory instability somewhere in the course of a septic period to severe progressive shock as its earliest and often fatal manifestation. MOF and circulatory failure however, are, often dissociated, and multiple organ dysfunction may complicate severe sepsis without clinical signs of shock. Also, before circulatory shock develops, sepsis-related dysfunction such as thrombocytopenia can already be observed, and acute respiratory distress may be its earliest manifestation (VITO et al. 1974). These observations emphasize that sepsis-induced multiple organ dysfunction is the result of a systemic disorder with varying clinical expression and severity, among other things related to preexisting chronic disease and/or limited organ physiological reserve.

Patients with septic shock usually respond initially to therapeutic management. However, some may die within 72 h of progressive nonresponsive hypotension (PARKER et al. 1987; RUOKONEN et al. 1991). Most patients survive the acute hypotensive episode with gradual improvement in clinical state (JORDAN et al. 1987). In the days following resuscitation the clinical condition can deteriorate due to increasing distant organ dysfunction. Several clinical stages in the development of MOF have been described in postoperative patients, starting with a phase with relative hemodynamic stability in which only very mild abnormalities can be detected, followed by stages of progressive clinical and laboratory deterioration (CARRICO et al. 1986; BARTON and CERRA 1989). A more fulminant course may be followed in septic shock, and MOF is often an early complication (RUOKONEN et al. 1991). Nevertheless, in patients surviving the early acute episode progressive multi-organ involvement becomes the primary clinical problem and cause of mortality (PARKER et al. 1987; RUOKONEN et al. 1991). It appears therefore that in general the septic patient can exhibit both an acute phase with severe physiological derangements in a few organ systems, which resolve when the patient responds to treatment, and a more chronic phase with functional deterioration of multiple organ systems which may resolve upon adequate therapy but often becomes progressive and resistant to treatment (JORDAN et al. 1987). These patients are usually hypermetabolic and have a hyperdynamic circulation with an elevated cardiac output and a low systemic vascular resistance (DEITCH 1992; BARTON and CERRA 1989). This hypermetabolism–organ failure syndrome (BARTON and CERRA 1989) is characterized by persistent (low-grade) fever, leukocytosis, tachycardia, tachypnea, hyperglycemia, hyperlactatemia, increased urea production, and increased total body oxygen uptake.

6.1 Adult Respiratory Distress Syndrome

Septic shock is the major risk factor for ARDS (PEPE et al. 1982), and this complication occurs in 20%–60% of cases. The basic pathogenetic mechanism of ARDS is pulmonary microvascular injury with increased permeability and resultant flooding of alveoli with protein-rich fluid which in addition damages surfactant. This has major pathophysiological consequences: alveolar collapse, increased intrapulmonary shunting, progressive hypoxemia, decreased compliance, decreased functional residual capacity, and increased work of breathing.

Tachypnea and hyperventilation with respiratory alkalosis occur almost uniformly in early septic shock and do not necessarily indicate significant pulmonary damage. However, increased pulmonary microvascular leakage of protein-risk fluid may occur even at this stage. Clinical signs of evolving ARDS are progressive dyspnea, use of auxilliary respiratory muscles, flaring of the nostrils, and central cyanosis. ARDS may develop within hours after the onset of septic shock. The chest roentgenogram shows diffuse alveolar infiltrates compatible with pulmonary edema, but significant changes in pulmonary gas exchange usually occur before pulmonary edema becomes radiographically visible. When not supported by mechanical ventilation, respiratory muscle fatigue develops, resulting in progressive (hypercapnic) ventilatory failure and death. The increased work of breathing, diminished oxygen supply to respiratory muscles, and metabolic factors are its pathogenetic basis.

6.2 Renal Dysfunction

Oliguria is a very common initial sign in septic shock and is often included in its definition. It is a functional compensatory response to hypotension with sodium retention and a rise in serum urea in excess of serum creatinine. Proteinuria is not uncommon and is often of tubular origin (RICHMOND et al. 1982). In some patients, inappropriate polyuria is observed, which may be related to a little understood tubular dysfunction (LUCAS 1976). When shock is not readily and adequately reversed, acute tubular necrosis with progressive uremia ensues. Most cases of acute renal failure in septic shock are oliguric, but increasing incidence of nonoliguric cases is observed (KLEINKNECHT 1990). Renal hypoperfusion is the most important pathogenetic mechanism, also in septic acute renal failure, with decrease in outer cortical blood flow and other intrarenal hemodynamic alterations (GROENEVELD 1994). However, nonhemodynamic factors including sepsis-induced humoral mediators (e.g., thromboxane A_2, endothelin) and activated cellular elements (e.g., granulocytes, coagulation products) may contribute significantly to the development of renal dysfunction (GROENEVELD 1994).

6.3 Central Nervous System Failure

Changes in mental state may be an early and sometimes the sole initial sign of severe sepsis. Elderly patients seem to be more prone to central nervous system dysfunction, which ranges from mild disorientation, impaired orientation and concentration, agitation, confusion, or lethargy to frank obtundation (GLECKMAN and HIBERT 1982; HOLLOWAY and REINHARDT 1984). These clinical symptoms are often included in the definition of septic shock as a sign of organ hypoperfusion. Focal signs and seizures may occur but are rare. In a prospective study YOUNG et al. (1992) observed encephalopathy in 70% of a group of 69 septic patients, being severe in most. Mild clinical abnormalities were observed even within hours of a positive finding on blood culture. Severe encephalopathy manifesting as delirium and then coma was evident within 2 weeks.

The liquor cerebrospinalis usually does not show abnormalities except for a mildly elevated protein level. The EEG is a sensitive indicator of encephalopathy and may show any one of four main patterns of increasing severity (BOLTON et al. 1993). The pathogenesis of septic encephalopathy has not been clarified and is presumably multifactorial. Cerebral hypoperfusion, hypoxemia, dysfunction of the blood-brain barrier, direct effects of cytokines on brain function, focal thrombosis (or bleeding), vasogenic cerebral edema, and disturbances of amino acid metabolism have been implicated (BOLTON et al. 1993). Multiple microabcesses in the brain have been demonstrated in a significant number of patients (JACKSON et al. 1985), but it is unclear whether these can account for the observed mental state abnormalities.

6.4 Hematological Failure

Leukocytosis with a left shift, a common finding in septic shock, reflects the release of less mature leukocytes from bone marrow storage pools. Leukopenia is sometimes observed in the early phase of severe septic shock. Microvascular aggregation of granulocytes apparently then predominates, a phenomenon that may herald the development of ARDS (THOMMASEN et al. 1984). Varying degrees of leukocytosis or leukopenia have been included in the definition of hematological failure. However, most often DIC or (severe) thrombocytopenia is used as a marker of hematological failure. Isolated thrombocytopenia due to increased platelet turnover is a frequent companion of septic conditions and can be an early clue to sepsis. Thrombocytopenia can be extremely pronounced and is clinically apparent by petechiae and (mucosal) bleeding. In cases of renal failure these clinical signs can develop at higher platelet counts due to uremic thrombocytopathy. The main cause of isolated thrombocytopenia is probably endothelial cell injury resulting in a procoagulant surface with platelet aggregation. Sepsis is the most common cause of acute DIC, a syndrome characterized by intravascular clotting and microvascular fibrin deposition and consumption of clotting factors. Prolonged clotting times, hypofibrinogenemia, thrombocytopenia and the presence of fibrin degradation products are laboratory signs of the syndrome. Both hemorrhagic complications and thrombotic small vessel occlusions are its clinical manifestations.

6.5 Liver Dysfunction

Liver dysfunction is a virtually universal occurrence in patients with severe sepsis and may range from minor elevation of liver enzymes to pronounced jaundice in the late stages. A rising bilirubin level disproportional to levels of hepatocellular enzymes or alkaline phosphatase appears to be associated with a poor prognosis (Banks et al. 1982; Barton and Cerra 1989). It has been suggested in fact that the development of clinical hepatic failure defines MOF (Barton and Cerra1989). Mild elevations in serum bilirubin may be noticed even early in septic shock (Franson et al. 1985). Characteristically bilirubin is almost entirely of the conjugated type, but nonconjugated bilirubin can appear in associated hemolysis (Zimmerman et al. 1979). In some cases marked elevations of alkaline phosphatase are observed (Fang et al. 1980). Pathological examination of the liver may reveal Kupffer's cell hyperplasia, focal hepatic necrosis, and cholestatic changes (Zimmerman et al. 1979). The pathogenesis of liver dysfunction in sepsis is clearly multifactorial and includes such mechanism as liver ischemia, direct injury from endotoxin, and sepsis-induced mediators, drug toxicity, parenteral nutrition, and bacterial overgrowth.

7 Pathophysiology–Pathogenesis

The similarity in clinical, pathophysiological, and biochemical abnormalities in patients with sepsis-induced MOF and those with MOF associated with severe noninfectious conditions has led to the concept that MOF is the result of a systemic widespread inflammatory host response. A major conceptual change has been that the host is not an innocent bystander whose tissues are ravaged by invading bacteria but instead is an active participant in this destructive process (DEITCH 1992). The phrase "generalized autodestructive inflammation" has been coined to describe this process (GORIS et al. 1985). This shift in view has important therapeutic implications: if bacteria or their products directly cause tissue injury, the goal of immunomodulation would be to increase the host's immunoresponse, but if an uncontrolled inflammatory reaction damages tissues, inhibition of this response would limit tissue injury.

Sepsis is initiated by a nidus of infection which causes a local inflammatory response with local activation of macrophages and neutrophils, which is intended to contain and eradicate infecting micro-organisms and to clear damaged tissues of cell debris. Although generally beneficial to the host, inflammatory processes are intrinsically destructive to surrounding tissues (DEITCH 1992). When this infection cannot be controlled locally, bacteria or bacterial products may induce a generalized and uncontrolled inflammatory response with the vascular endothelium as essential target (BONE 1991). The clinical correlates are sepsis, severe sepsis, or septic shock, indicating clinical syndromes of progressive severity which can all be complicated by MOF (BONE et al. 1992; RANGEL-FRAUSTO et al. 1995).

In septic shock excessive generation of *mediators* and microcirculatory *ischemia* are considered the main mechanisms in the generation of tissue injury. These two processes are closely interrelated and amplify each other. Septic shock itself is the clinical expression of an excessive inflammatory host response which sets the scene for further propagation and amplification of processes leading to multiple organ dysfunction or failure. It is therefore not surprising that MOF often complicates septic shock. Widespread endothelial inflammation and dysfunction, abnormalities in vascular tone, myocardial depression, metabolic derangements, coagulation abnormalities, and resulting organ tissue damage are hallmarks of severe septic shock.

7.1 Mediators

A large number of humoral mediator cascades and products released by various cells are involved in the exaggerated host response which causes most if not all of the clinical symptoms of sepsis and its more severe form septic shock. Invasive (bacterial) infection is a trigger that sets in motion an orchestra of mediators (HACK and THIJS 1991), which activates cell-cell interactions and induces metabolic cellular alterations, processes that ultimately lead to tissue damage and MOF. The major (initial) players in this orchestra are proinflammatory cytokines (tumor necrosis factor, TNF; interleukins, IL), the complement system, the contact system and the extrinsic pathway of coagulation, the fibrinolytic system, and cellular elements such as mononuclear cells, neutrophils, endothelial cells, and platelets. Activation of these cellular and humoral systems results in the release of bioactive products such as prostaglandins, leukotrienes, platelet activating factor (PAF), nitric oxide (NO), endothelin, oxygen free radicals, and proteinases. The neuroendocrine and the autonomous nervous system are activated in addition, as part of the general stress response resulting in changes in hormonal levels, and release of mediators such as endorphins. Significant metabolic alterations induced, among others, by cytokines and hormonal changes are part of the syndrome. However, the pathogenesis of the sepsis syndrome and consequently of MOF is far from elucidated. Biological effects of these mediator systems are diverse and complex, many can prompt release of others (and some also of themselves), or have synergistic effects, many may have different effects in combination than they have alone, and many interactions occur that may either augment or limit the inflammatory response (SHALABY et al. 1985; BEUTLER et al. 1986; DINARELLO et al. 1988; BALK et al. 1988; BUTLER et al. 1989; PETRAK et al. 1989; ELIAS et al. 1990). Moreover, actions of a mediator may differ from setting to setting (ELIAS et al. 1990), and the preexistent clinical condition of the patient can modify the effects of mediators (BONE 1991). Nevertheless, several mediator systems including cytokines may, when excessively and protactedly produced, directly or indirectly cause tissue injury and establish an self-amplification process which can progress to MOF (TRACEY and LOWRY 1990).

7.1.1 The Cytokine Network

Bacterial products once shed into the circulation can activate macrophages and monocytes with subsequent release of proinflammatory cytokines (TNF, IL-1, IL-6, IL-8) together with other products. In particular *TNF* and *IL-1* are considered the most important proximal mediators of the generalized inflammatory response. These have considerable overlapping actions although they bind to different receptors and act synergistically in many respects. They have strong biological activities. The interactions of TNF and IL-1 with endothelial cells, leukocytes, and macrophages are in particular important to understand why these cytokines can be instrumental in the induction of tissue injury. These cytokines exert direct toxic effects on *endothelial cells* and enhance permeability (MARTIN et al. 1988; SCHUGER et al. 1989; ROBAYE et al. 1991).

TNF and IL-1 can induce expression of tissue factor and downregulation of thrombomodulin, which reduces the anticoagulant properties of the endothelial surface and enhances its procoagulant activity (BEVILACQUA et al. 1986; NAWROTH and STERN 1986; LENTZ et al. 1991) as well as synthesis of plasminogen activator inhibitor (PAI-1; NACHMAN et al. 1986; VAN HINSBERGH et al. 1988). In vivo TNF can induce coagulation and fibrinolysis followed by inhibition of fibrinolysis resulting in a procoagulant state (VAN DER POLL et al. 1990, 1991; VAN HINSBERGH et al. 1990). Both TNF and IL-1 can promote expression of adhesion molecules in endothelial cells such as ICAM-1, ELAM-1, and VCAM-1, which mediate the adherence of leukocytes to the endothelium (GAMBLE et al. 1985; MANTOVANI and DEJANA 1989; BRISCOE et al. 1992).

Interaction of these cytokines with *endothelial cells* results in the production of hematopoietic growth factors, IL-1, IL-6, IL-8, monocyte chemoattractant protein 1 (MCP-1), PAF, and the prostaglandins E_2 and I_2 (prostacyclin) as well as the generation of vasodilating substances via induction of NO synthase and of the vasoconstrictor peptide endothelin-1 (KAWAKAMI et al. 1986; NAWROTH et al. 1986; KAUSHANSKY et al. 1988; SICA et al. 1990a,b; BEASLEY et al. 1991; MARSDEN and BRENNER 1992). These activities set the scene for amplification processes of mediators that have profound microcirculatory effects. Interaction of TNF with *neutrophils* triggers an increased expression of surface adherence glycoproteins, augmentation of superoxide production, phagocytosis, enzyme release, and aggregation of these cells (KLEBANOFF et al. 1986; LE and VILÇEK 1987; LARRICK et al. 1987; DE FORGE et al. 1990). TNF is able to activate and degranulate neutrophils in vivo (VAN DER POLL et al. 1992). The effects of IL-1 on neutrophil function are less clear. Interaction of TNF with *macrophages* increases their cytotoxic capacity and the production of oxygen radicals and induces the release of other cytokines such as IL-1, IL-6, IL-8, and PAF and prostaglandin E_2 (DE FORGE et al. 1990; ROCK and LOWRY 1991). Recently it has been demonstrated that TNF has myocardial depressant activities (PAGANI et al. 1992; FINKEL et al. 1992; YOKOYAMA et al. 1993) which may be mediated by NO (FINKEL et al. 1992). Also IL-1 is able to induce nitric oxide in myocardial cells (TSUJINO et al. 1994). Infusion of TNF and IL-1 in the animal model results in hemodynamic, metabolic, and histopathological abnormalities as observed in severe sepsis (TRACEY et al. 1986; BUTLER et al. 1989). Moreover, anti-TNF antibodies prevented the development of shock and vital organ failure in a sepsis model (TRACEY et al. 1987), indicating the pivotal role of this cytokine in the systemic inflammatory response. In vivo TNF can induce the release of IL-1, macrophage, and granulocyte-macrophage colony-stimulating factors (KAUSHANSKY et al. 1988), IL-6 (VAN DER POLL et al. 1992), IL-8 (VAN DEVENTER et al. 1993), and IL-10 (VAN DER POLL et al. 1994). TNF is also able to activate the complement and contact systems (NURNBERGER et al. 1993), although the underlying mechanisms are far from clear.

Increased plasma levels of TNF and IL-1, albeit inconsistently and variably, can be observed in septic patients (see review by THIJS and HACK 1995), but a correlation with the development of MOF has not been clearly established. Plasma concentration may not precisely reflect activation in tissues. However, persistent elevation of TNF levels have in some studies been associated with MOF (PINSKY et al. 1993; MARTIN et al. 1994), and high levels of TNF and IL-1, higher than in plasma, have been found in the bronchoalveolar lavage (BAL) fluid of patients with ARDS (SUTER et al. 1992). IL-6 can be released by many cell types and has a number of biological activities, of which perhaps the most important is the induction of the synthesis of acute-phase proteins by the liver (GEIGER et al. 1988). It is most consistently elevated in the plasma of patients with sepsis and related to outcome (HACK et al. 1989a), but its role in the pathophysiology of sepsis has not been defined. IL-8 is produced by a variety of cells including monocytes, macrophages, neutrophils and endothelial cells and shows a remarkable specifity for neutrophils. IL-8 has chemoattractant activity, is able to induce degranulation, and to elicit a respiratory burst (Collins et al. 1991, Schröder 1989). In addition, IL-8 may promote adherence of neutrophils to endothelium (Carveth et al. 1989; Huber et al. 1991) whereas, in contrast, under certain conditions this cytokine can also inhibit this process (Gimbrone et al. 1989). Elevated levels of this cytokine which are related to outcome have been found in the plasma of septic patients (Marty et al. 1994; Hack et al. 1992) and in the BAL fluid in patients with ARDS (Miller et al. 1992), indicating that local production of IL-8 may contribute to pulmonary damage. High plasma levels of IL-8 have recently been demonstrated in patients with sepsis-associated MOF, which was much more pronounced than in MOF not related to infection (Marty et al. 1994). The role of IL-8 in the pathogenesis of sepsis and its sequelae is, however, not clear.

7.1.2 Complement System

Bacteria and their products can activate the complement system and result in the generation of anaphylatoxins (C3a, C4a, C5a) which have strong biological activities. C5a is the most powerful anaphylatoxin, and binding of C5a to neutrophils stimulates an oxidative burst, degranulation, and enhanced adhesion of these cells to each other and to endothelium (Jacob et al. 1980). Complement-stimulated neutrophils damage endothelium by the release of reactive oxygen species and proteases (e.g., elastase, collagenase, cathepsins), favoring the lung as a target. Complement activation stimulates the release of other substances such as histamine, serotonin, PAF, prostaglandins, leukotrienes, and cytokines (TNF, IL-1, IL-6) by the interaction with neutrophils, platelets, mast cells, and monocytes (Cavaillon et al. 1990; Scholz et al. 1990). The pathophysiological consequences are among others vasodilation, cell aggregation, and increased microvascular permeability. Animal studies have demonstrated that complement activation may produce severe inflammatory responses with microvascular injury in the lungs and other organs (Goris et al. 1986; Nuytinck et al. 1986a). In neutropenic animals such changes could not be produced or were dramatically attenuated (Hosea et al. 1980; Hohn et al. 1980), indicating the important role of activated leukocytes in the generation of tissue damage.

A polyclonal antibody to C5a can markedly attenuate markers of ARDS in septic primates (Stevens et al. 1986). Some studies have suggested that complement activation per se is unable to fully simulate the effects of endotoxin, and that other factors are necessary to express its effects (Henson et al. 1982). Although activated complement factors are elevated in the plasma of most septic patients (Hack et al. 1989b), the relationship to the development of ARDS is inconsistent. Some investigators (Slotman et al. 1986; Tennenberg et al. 1987; Parsons et al. 1989) have found such a relationship, but others (Weinberg et al. 1984; Hack et al. 1989b; Ketai and Grum 1986) have not. It has been suggested that persistence of an elevated C4a level is associated with MOF (Heideman and Hugli 1984), but no correlation between the development of ARDS and this complement activation product was observed in another study (Langlois and

Gawryl 1988). Anaphylatoxin levels may, however, underestimate the extent of complement activation since they are rapidly cleared from the circulation. Instead, the more stable fluid-phase terminal complement complex could be a more useful marker of this process. Higher levels have been found in (septic) patients with ARDS than in those without ARDS (Langlois and Gawryl 1988; Heideman et al. 1988) but not in all series (Parsons and Giclas 1990). Activated complement products can be found in the BAL fluid of patients with ARDS (Modig et al. 1986; Zheutlin et al. 1987) which may be more sensitive markers of the effects of anaphylatoxins at the tissue level than blood levels.

7.1.3 Contact System, Coagulation, Fibrinolysis

The contact system of coagulation consists of the proteins factor XII, prekallikrein, high molecular weight kininogen, and factor XI. The activation products of the contact system have important biological effects: enhancement of vascular permeability, decrease of vascular tone, and aggregation and degranulation of neutrophils. Bradykinin, liberated from kininogen by plasma kallikrein is one of the most potent vasodilators (Colman 1989). The contact system is activated in sepsis and ARDS, indicated by lowered levels of factor XII and prekallikrein (Nuijens et al. 1988; Carvalho et al. 1988), which are correlated with blood pressure (Hack et al. 1990). Monoclonal antibodies to factor XII can reverse irreversible hypotension in a lethal sepsis model in baboons, indicating that bradykinin significantly contributes to the hypotensive response in sepsis (Pixley et al. 1993). Coagulation disorders and abnormalities of fibrinolysis are common in septic patients. Sepsis is the most common cause of acute DIC, and this syndrome is a frequent complication of sepsis. Widespread microvascular thrombosis in various organ systems is a feature of lethal septic shock (Coalson 1986) and can contribute to organ failure (Tanaka et al. 1990). DIC in septic patients is associated with MOF and is a predictor of both the occurrence of this syndrome and death (Fourrier et al. 1992).

Coagulation is initiated in vivo by tissue factor expressed on endothelial cells and monocytes (Furie and Furie 1992). This is followed by a sequence of proteolytic activation resulting in thrombin formation and fibrin generation. Thrombin-antitrombin complexes are sensitive markers of in vivo thrombin generation and high thrombin-antithrombin complex levels are common in septic patients (Hesselvik et al. 1991; Thijs et al. 1993; Lorente et al. 1993). Also, fibrinolysis is activated, resulting in the generation of plasmin and plasmin-antiplasmin complexes, indicators of this process, are found in plasma of septic patients (Thijs et al. 1993). Fibrinolysis is, however, inhibited by the release of PAI-1 (Pralong et al. 1989) probably resulting in impaired fibrinolysis and an imbalance between coagulation and fibrinolysis (Thijs et al. 1993). High levels of PAI-1 may in fact be associated with the development of MOF (Hesselvik et al. 1989). Also in the BAL fluid of patients with ARDS such a dysbalance was demonstrated (Idell et al. 1989). The role of coagulation in ARDS has recently been reviewed (Hasegawa et al. 1994). Importantly, modulation of coagulation in experimental sepsis protects the animals from organ damage and lethal effects (Taylor et al. 1987, 1988).

7.1.4 Neutrophils

A central mechanism in the pathogenesis of sepsis-induced tissue damage is the interaction between neutrophils and endothelial cells. Inflammation is associated with effusion and recruitment of neutrophils and monocytes and in sepsis these processes are generalized and take place in various organ systems, especially in the lung. The direct and indirect involvement of neutrophils in ARDS and MOF has been established by many studies (Flick et al. 1981; Nuytinck et al. 1986a). However, ARDS can occur in severe neutropenic patients (Ognibene et al. 1986) and in some animal studies neutrophil depletion had no effect on the degree of endotoxin-induced tissue injury (Winn et al. 1987). Other pathways may apparently also lead to tissue injury. Cytokines (TNF, IL-1) and endotoxin can induce expression of endothelial leukocyte adhesion molecules (ELAM-1, ICAM-1) which serve to bind neutrophils. ELAM-1 expression is one of the possible preconditions that result in leukocyte-related tissue damage. Endothelial expression of ELAM-1 has been demonstrated to occur in vivo in several organ systems after endotoxin administration in the experimental model (Engelberts et al. 1992a). Also, in patients with peritonitis skin biopsies revealed substantial endothelial ELAM-1 expression (Engelberts et al. 1992b). Neutrophils can be activated by a variety of mediators operative in sepsis, including cytokines (TNF, IL-1, IL-8, granulocyte-macrophage colony-stimulating factor), complement activation products (C5a), contact activation products (kallikrein, factor XIIa), and bacterial products (endotoxin). Neutrophil activation triggers upregulation of neutrophil adherence molecules, especially of the CD11b/CD18 complex that serves as a ligand for ICAM-1 on endothelial cells, a process which promotes adhesion of neutrophils to endothelium (Gamble et al. 1985).

In ARDS patients neutrophils circulating in the pulmonary artery show increased CD11b/CD18 expression and increased oxygen radical production (Simms and D'Amico 1991). Recently it was demonstrated that also circulating neutrophils in patients with sepsis show increased expression of CD11b/CD18 (Astiz et al. 1995). In the experimental model organ injury can be ameliorated by inhibition of CD18 dependent neutrophil adherence (Mileski et al. 1990). Also lymphocytes and monocytes bind to the endothelium (Arnaout et al. 1988). Cells of monocyte lineage can be activated by the potent MCP-1 which is released by various cells. Elevated concentrations of MCP-1 have recently been found in plasma of septic patients (Bossink et al. 1995). Activated neutrophils (and monocytes) attached to activated endothelium degranulate and release digesting proteases such as elastase, collagenase, and cathepsin G and toxic oxygen radicals which in concert with other products (e.g., leukotrienes, prostaglandin E_2, PAF) act to increase vascular permeability by damaging endothelial cells (Smedley et al. 1986; Varani et al. 1989). Prostaglandin E_2 strongly increases vascular permeability in vitro induced by C5a and neutrophils (Wedmore and

WILLIAMS 1981). In vivo TNF mediates leukocyte-endothelial adherence associated with microvascular protein leakage (EDWARDS et al. 1993).

This toxic damage of endothelial cells seems to be a self-perpetuing process since a microenvironment is created where toxic products act locally, and plasma inhibitors of elastase (α_1-antitrypsin, α_2-macroglobulin) gaining accesss to this area are inactivated by reactive oxygen radicals and other toxic products (BEATTY et al. 1980; REDDY et al. 1989). In this way elastase can exert its local activity for a longer period despite the presence of large amounts of inhibitors in plasma. There is no doubt that elastase is essential for neutrophil-mediated endothelial cell injury (SMEDLEY et al. 1986; HENSON and JOHNSTON 1987). Administration of elastase in the animal model results in pulmonary leukostasis, DIC, and increased pulmonary vascular resistance (STOKKE et al. 1986). Leukostasis in several organs is a prominant histopathological sign of the endothelial-leukocyte interaction as demonstrated in experimental hyperdynamic sepsis (REDL et al. 1991). Activation of neutrophils in vivo can be assessed by plasma levels of elastase–α_1-antitrypsin complexes and lactoferrin. In septic patients these levels are elevated and related to outcome (DUSWALD et al. 1985; NUIJENS et al. 1992). In trauma and sepsis patients, levels of elastase have been found to be correlated with the development and the severity of MOF (NUYTINCK et al. 1986b; TANAKA et al. 1991). In fact, all available studies show a positive relation between elastase and severity of sepsis, ARDS, and MOF (GORIS 1990). In patients with sepsis ARDS may be the first and often most important explicit clinical sign of endothelial damage. There is little doubt that increased pulmonary permeability is mediated by the toxic interaction between neutrophils and pulmonary endothelium presumably with a concerted action of other mediators such as thromboxane A_2, leukotriene B_4, PAF, activated complement products, and cytokines. It is likely that similar mechanisms are operative in other organ systems, although presumably the exact processes may vary. This, however, remains at present speculation as it has not been demonstrated in patients with sepsis.

7.2 Ischemia and Reperfusion

Several lines of evidence indicate that ischemia and reperfusion are strong mechanisms that can cause tissue injury. The impressive work by SHOEMAKER et al. (1992) has demonstrated that adequate tissue oxygenation is of paramount importance in maintaining homeostasis and preserving organ function. In one study the calculated oxygen deficit in high-risk surgical patients during and immediately after surgery appeared to be related to the development of MOF. Septic shock is characterized by a low systemic vascular resistance and usually an elevated cardiac output to meet the tissue's metabolic needs (GROENEVELD et al. 1986). However, a defect in effective oxygen extraction as manifested by a low arteriovenous oxygen difference may limit adequate supply of oxygen to the cells. In animal models of sepsis a diminished oxygen extraction capability has been demonstrated, not only for the whole body but also in various organ systems, with a decline in critical oxygen extraction and an increase in critical oxygen delivery (NELSON et al. 1988; ZHANG et al. 1995).

The concept of diminished oxygen supply to cells in severe sepsis has been challenged as increased levels of skeletal muscle pO_2 have in fact been found, suggesting that oxygen utilization is impaired rather than oxygen supply (BOEKSTEGERS et al. 1994). However, most investigators consider tissue ischemia an important pathogenetic mechanism, particularly in septic shock. Inadequate oxygen availability leads to cellular dysfunction, production of lactic acid, injury, and ultimately cell death. In clinical practice, blood lactate levels are used to globally assess the severity and duration of tissue ischemia. Increases of oxygen delivery may increase oxygen uptake even far above the normal critical oxygen delivery. This pathological supply dependency of oxygen uptake in the presence, but not in the absence of lactic acidemia during septic shock, suggests than an elevated lactate concentration is associated with an oxygen deficit and anaerobic metabolism in tissues (GILBERT et al. 1986; VINCENT et al. 1990; DE BACKER et al. 1994). This phenomenon is the main argument for human septic shock being accompanied by an oxygen deficit in the tissues. Admittedly, however, this supply dependency phenomenon is not observed in other studies and is certainly not beyond debate (see review GATTINONI et al. 1995). Other mechanisms, in addition to tissue ischemia such as a depressed activity of pyruvate dehydrogenase may, however, be involved in the rise in blood lactate concentration (VARY et al. 1986; HOTCHKISS and KARL 1992). Nevertheless, serial blood lactate levels are important in predicting the development of MOF in septic shock (BAKKER et al. 1996).

The mechanisms of impaired oxygen extraction in sepsis are not completely unraveled. Most likely the maldistribution of microcirculatory blood flow resulting in regional over- and underperfusion relative to demand and perfusional shunting can account for the paradox of a high total body oxygen supply and a low oxygen uptake for tissue needs in the presence of an inappropriately low oxygen extraction and pathological supply dependency of oxygen uptake (CAIN 1984; HOTCHKISS and KARL 1992). At least three mediator-induced mechanisms seem to be operative and to act in concert at the microcirculatory level: inappropriate vasodilation and vasoconstriction, microembolisation, and endothelial cell injury (CAIN 1984). Excessive, inappropriate vasodilation and loss of vascular reactivity may cause maldistribution of flow not adapted to local tissue needs (GROENEVELD et al. 1987). Mediator-induced microaggregation of neutrophils, platelets, and fibrin results in loss of autoregulation, limitation of recruitable capillary surface, increased inhomogeneity of flow, and diminished reactive hyperemia. Experimental embolization with microspheres abolishes reactive hyperemia, reduces oxygen extraction, and induces supply dependency of oxygen uptake (ELLSWORTH et al. 1981; CAIN et al. 1988). A reduced reactive hyperemia has been observed in septic animals (NELSON et al. 1988) as well as in patients (HARTL et al. 1988; ASTIZ et al. 1995). Endothelial cell injury may cause endothelial swelling and increases microvascular permeability with subsequent tissue edema, which further impairs oxygen diffusion. Therefore, these concerted microcirculatory abnormalities are most likely mechanisms by which oxygen supply to tissue cells is impaired.

In the animal model marked interorgan redistribution of blood flow induced by sepsis is observed, and this probably also occurs in humans (see review by THIJS 1994). Such a redistribution may not subserve adequate adaptation of oxygen supply to tissue needs in specific organ systems. The splanchnic tissues seem to be an important "target" in septic shock. Several studies have demonstrated an increase in splanchnic blood flow in hyperdynamic sepsis and shock (DAHN et al. 1987; FINK 1991; RUOKONEN et al. 1993). However, increases in splanchnic oxygen demand, related to sepsis-induced metabolic processes such as synthesis of acute-phase proteins and gluconeogenesis, may exceed the increase in oxygen supply (DAHN et al. 1987; RUOKONEN et al. 1993), which makes the splanchnic tissues very sensitive to alterations in oxygen supply. The ultimate balance between oxygen supply and demand determines the occurrence and severity of ischemia (ARVIDSSON et al. 1991; RASMUSSEN and HAGLUND 1992; BREUILLE et al. 1994). The villous countercurrent system present along the entire gut renders the gut mucosa extremely sensitive to ischemia (FALK et al. 1985; FINK 1991). Gut ischemia may damage its barrier function, thereby promoting bacterial or endotoxin translocation (discussed elsewhere in this volume). In septic patients a reduced gastric intramucosal pH has been found, even when global hemodynamics seemed adequate, suggesting splanchnic ischemia in spite of relatively high cardiac output (SILVERMAN and TUMA 1992; MARIK 1993; GUTIERREZ et al. 1994). This index of local ischemia may be a better predictor of MOF and death during sepsis than commonly used oxygen-derived variables (MARIK 1993).

Although reperfusion is necessary for restoration of cellular metabolic activity, it can by itself trigger or aggravate the extent of ischemic tissue injury (DEITCH 1992). Reperfusion injury is an important additional mechanism of tissue injury by the generation of toxic reactive oxygen species (GRANGER 1988; REILLY et al. 1991). During ischemia there occurs the breakdown of high-energy phosphates to hypoxanthine, which in the presence of oxygen delivered with reperfusion is converted by xanthine oxidase into oxygen radicals. Endothelium is rich in the precursor enzyme xanthine dehydrogenase. High-energy phosphate degradation products have been found in the circulation of patients with septic shock (GRUM et al. 1985). Oxygen free radicals react with all biological substances, but most susceptible are polyunsaturated fatty acids in cell membranes, and this reaction leads to lipid peroxidation and ultimtely to membrane disintegration. Elevated plasma levels of lipid peroxides and depressed concentrations of antioxidants have been found in patients with septic shock and with ARDS (OGILVIE et al. 1991; RICHARD et al. 1990). The respiratory condensate from mechanically ventilated patients with ARDS contain elevated concentrations of hydrogen peroxide (BALDWIN et al. 1986; SZNAJDER et al. 1989), indicating that oxygen radicals may be involved in the pathogenesis of the syndrome.

Hypoxia can also activate macrophages to produce cytokines such as TNF and IL-1 (COLLETTI et al. 1990; GHEZZI et al. 1991), IL-8 (METINKO et al. 1992), induce complement activation (Bengtsson et al. 1987), and production of prostaglandins (Paterson et al. 1989). These mediators may initiate or amplify injury also in organs remote from the site initially involved (Colletti et al. 1990). Hence, reperfusion may exert a direct injury effect on the specific area that was reperfused, as well as an indirect injury on organs remote from the ischemic tissue due to excessive mediator generation.

References

Abraham E, Wunderink R, Silverman H et al. (1995) Efficacy and safety of monoclonal antibody to human tumor necrosis factor α in patients with sepsis syndrome. A randomized, controlled, double-blind, multicenter clinical trial. JAMA 273: 934–941

Arnaout MA, Lanier LL, Faller DV (1988) Relative contribution of the leukocyte molecules Mol, LFA-1 and P150, 95 (Leu M5) in adhesion of granulocytes and monocytes to vascular endothelium is tissue-and stimulus-specific. J Cell Physiol 137: 305–309

Arvidsson D, Rasmussen I, Almqvist P, Niklasson F, Haglund U (1991) Splanchnic oxygen consumption in septic and hemorrhagic shock. Surgery 109: 190–197

Astiz ME, DeGent GE, Lin RY, Rackow EC (1995) Microvascular function and rheologic changes in hyperdynamic sepsis. Crit Care Med 23: 265–271

Bakker J, Gris P, Coffernils M, Kahn J, Vincent J-L (1996) Serial blood lactate levels can predict the development of multiple organ failure following septic shock. Arch Surg (in press)

Baldwin SR, Grum CM, Boxer LA et al. (1986) Oxidant activity in expired breath of patients with adult respiratory distress syndrome. Lancet 1: 11–14

Balk RA, Jacobs RF, Tryka F, Townsend JW, Walls RC, Bone RC (1988) Effects of ibuprofen on neutrophil function and acute lung injury in canine endotoxin shock. Crit Care Med 16: 1121–1127

Banks JG, Fowlis AK, Ledingham IM et al. (1982) Liver function in septic shock. J Clin Pathol 35: 1249–1262

Barton R, Cerra FB (1989) The hypermetabolism multiple organ failure syndrome. Chest 96: 1153–1160

Baue AE (1975) Multiple, progressive, or sequential systems failure. Arch Surg 110: 779–781

Beasley D, Schwartz JH, Brenner BM (1991) Interleukin 1 induces prolonged 1-arginine-dependent cyclic guanosine monophosphate and nitrite production in rat vascular smooth muscle cells. J Clin Invest 87: 602–608

Beatty K, Bieth B, Travis J (1980) Kinetics of association of serine proteases with native and oxidized α1-protease inhibitor and α1-antichymotrypsin. J Biol Chem 255: 3931–3934

Bell RC, Coalson JJ, Smith JD, Johanson WG (1983) Multiple organ system failure and infection in adult respiratory distress syndrome. Ann Intern Med 99: 293–298

Bengtsson A, Lannsjo W, Heideman M (1987) Complement and anaphylatoxin responses to cross clampling of the aorta studies during general anaesthesia with or without extradural blockade. Br J Anaesth 59: 1093–1097

Beutler B, Kronchin N, Milsark IW, Luedke C, Cerami A (1986) Control of cachetin (tumor necrosis factor) synthesis: mechanisms of endotoxin resistance. Science 232: 977–980

Bevilacqua MP, Pober JS, Majeau GR, Fiers W, Cotran RS, Gimbrone MA Jr (1986) Recombinant tumor necrosis factor induces procoagulant activity in cultured human vascular endothelium: characterization and comparison with the actions of interleukin 1. Proc Natl Acad Sci USA 83: 4533–4537

Boekstegers P, Weidenhöfer S, Kapsner T, Werdan K (1994) Skeletal muscle partial pressure of oxygen in patiens with sepsis. Crit Care Med 22: 640–650

Bolton CF, Young GB, Zochodne DW (1993) The neurological complications of sepsis. Ann Neurol 33: 94–100

Bone RC (1991) The pathogenesis of sepsis. Ann Intem Med 115: 457–469

Bone RC, Balk RA, Cerra FB et al. (1992) Definitions for sepsis and organ failure and guidelines for the use of innovative therapies in sepsis. Chest 101: 1644–1655
Bone RC, Balk RA, Fein AM et al. (1995) A second large controlled clinical study of E5, a monoclonal antibody to endotoxin: results of a prospective, multicenter, randomized, controlled trial. Crit Care Med 23: 994–1006
Border JR, Chenier R, McMenamy RH et al. (1976) Multiple systems organ failure: muscle fuel deficit with visceral protein malnutrition. Surg Clin North Am 56: 1147–1167
Bossink AWJ, Paemen L, Jansen PM, Hack CE, Thijs LG, van Damme J (1995) Plasma levels of the chemokines MCP-1 and MCP-2 (monocyte chemotactic protein) are elevated in human sepsis. Blood 86: 3841–3847
Breuille D, Rose F, Arnal M, Melin C, Obled C (1994) Sepsis modifies the contribution of different organs to whole-body protein synthesis in rats. Clin Sci 86: 663–669
Briscoe DM, Cotran RS, Pober JS (1992) Effects of tumor necrosis factor, lipopolysaccharide, and IL-4 on the expression of vascular cell adhesion molecule-1 in vivo. Correlation with CD3+ T cell infiltration. J Immunol 149: 2954–2960
Brun-Buisson C, Lepoutre A, Blot F (1994) The French national registry of HA-IA (Centoxin) in septic shock. Arch Intern Med 154: 2484–2491
Butler LD, Layman NK, Cain RL et al. (1989) Interleukin 1-induced pathophysiology: induction of cytokines, development of histiopathologic changes, and immunopharmacologic intervention. Clin Immunol' Immunopathol 53: 400–421
Cain SM (1984) Supply dependency of oxygen uptake in ARDS: myth or reality? Am J Med Sci 288: 119–124
Cain SM, King CE, Chapler CK (1988) Effects of time and microembolization on O_2 extraction by dog hindlimb in hypoxia. J Crit Care 3: 89–95
Callery MP, Kamei T, Mangino MJ, Flye MW (1991) Organ interactions in sepsis. Arch Surg 126: 28–32
Carrico CJ, Meakins JL, Marshall JC, Fry D, Maier RV (1986) Multiple-organ-failure syndrome. Arch Surg 121: 196–208
Carvalho AC, De Marinis S, Scott CF, Silver LD, Schmaier AH, Colman RW (1988) Activation of the contact system of plasma proteolysis in the adult respiratory distress syndrome. J Lab Clin Med 112: 270–277
Carveth HJ, Bohnsack JF, McIntyre TM, Baggiolini M, Prescott SM, Zimmerman GA (1989) Neutrophil activating factor (NAF) induces polymorphonuclear leukocyte adherence to endothelial cells and to subendothelial matrix proteins. Biochem Biophys Res Commun 162: 387–393
Cavaillon JM, Fitting C, Hauffner-Cavaillon N (1990) Recombinant C5a enhances interleukin-1 and tumor necrosis factor release by lipopolysaccharide-stimulated monocytes and macrophages. Eur J Immunol 20: 253–257
Clowes GHAJ, Zuschneid W, Turner M et al. (1968) Observations on the pathogenesis of the pneumonitis associated with severe infection in other parts of the body. Ann Surg 167: 630–650
Coalson JJ (1986) Pathology of sepsis, septic shock and multiple organ failure. In: Sibbald WJ, Sprung CL (eds) Perspectives on sepsis and septic shock. New Horizons, SCCM, Fullerton, pp 27–59
Colletti LM, Burtch GD, Remick DG, Kunkel SL, Stricter RM, Campbell DA (1990) Role of tumor necrosis factor alpha in the pathophysiologic alterations after hepatic ischaemia reperfusion injury in the rat. Transplantation 49: 268–272
Collins PD, Jose PJ, Williams TJ (1991) The sequential generation of neutrophil chemoattractant proteins in acute inflammation in the rabbit in vivo: relationship between C5a and proteins with the characteristics of IL-8/neutrophil-activating protein 1. J Immunol 146: 677–684
Colman RW (1989) The role of plasma proteases in septic shock. N Engl J Med 320: 1207–1209
Cook DJ, Fuller HD, Guyatt GH et al. (1994) Risk factors for gastrointestinal bleeding in critically ill patients. N Engl J Med 330: 377–381
Dahn MS, Lange P, Lobdell K et al. (1987) Splanchnic and total body oxygen consumption differences in septic and injured patients. Surgery 101: 69–80
De Backer D, Moraine J-J, Berre J, Kahn RJ, Vincent J-L (1994) Effects of dobutamine on oxygen consumption in septic patients. Direct versus indirect measurements. Am J Respir Crit Care Med 150: 95–100
De Forge LE, Nguyen DT, Kunkel SL, Remick DG (1990) Regulation of the pathophysiology of tumor necrosis factor. J Lab Clin Med 116: 429–438
Deitch EA (1992) Multiple organ failure. Pathophysiology and potential future therapy. Ann Surg 216: 117–134

Dhainaut J-FA, Tenaillon A, Le Tulzo Y et al. (1994) Platelet-activating factor receptor antagonist BN 52021 in the treatment of severe sepsis: a randomized, double-blind, placebo-controlled, multicenter clinical trial. Crit Care Med 22: 1720–1728

Dinarello CA, Cannon JG, Wolff SM (1988) New concepts on the pathogenesis of fever. Rev Infect Dis 10: 168–189

Duswald KH, Jochum M, Schram W, Fritz H (1985) Released granulocyte elastase: an indicator of pathobiochemical alterations in septicemia after abdominal surgery. Surgery 98: 892–899

Edwards MJ, Heniford BT, Miller FN (1993) Tumor necrosis factor mediates disseminated intravascular inflammation (DII) in the genesis of multiple organ edema. J Surg Res 54: 140–144

Eiseman B, Beart R, Norton L (1977) Multiple organ failure. Surg Gynecol Obstet 144: 323–326

Elias JA, Freundlich B, Kern JA, Rosenbloom J (1990) Cytokine networks in the regulation of inflammation and fibrosis in the lung. Chest 97: 1439–1445

Ellsworth ML, Goldfarb RD, Alexander RS, Bell DR, Powers SR (1981) Microembolization induced oxygen utilization impairment in the canine gracilis muscle. Adv Shock Res 5: 89–99

Engelberts I, Samyo SK, Leeuwenberg JFT, van der Linden CJ, Buurman WA (1992a) A possible role for ELAM-1 in the pathogenesis of MOF during septic shock. J Surg Res 53: 136–144

Engelberts I, van Hoof SCJ, Samyo SK, Buurman WA, van der Linden CJ (1992b) Generalized inflammation during peritonitis evidenced by intracutaneous ELAM-1 depression. Clin Immunol Immunopathol 65: 330–334

Esen F, Telci L, Akpir K, Kesecioglu J, Denkel T, Pembeci K (1992) Oxygen uptake/supply dependency in human sepsis: does it increase the risk of multisystem organ failure? Adv Exp Med Biol 317: 855–861

Faist E, Baue AE, Dittmer H, Heberer G (1983) Multiple organ failure in polytrauma patients. J Trauma 23: 775–787

Falk A, Redfors S, Myrvold H, Haglund U (1985) Small intestinal mucosal lesions in feline septic shock: a study on the pathogenesis. Circ Shock 17: 327–337

Fang MH, Ginsberg AL, Dobbins WO (1980) Marked elevation in serum alkaline phosphatase activity as a manifestation of systemic infection. Gastroenterology 78: 592–597

Fink MP (1991) Gastrointestinal mucosal injury in experimental models of shock, trauma, and sepsis. Crit Care Med 19: 627–641

Finkel MS, Oddis CV, Jacob TD, Watkins SC, Hattler BG, Simmons RL (1992) Negative inotropic effects of cytokines on the heart mediated by nitric oxide. Science 257: 387–389

Fisher CJ Jr, Dhainaut J-FA, Opal SM et al. (1994) Recombinant human interleukin 1 receptor antagonist in the treatment of patients with sepsis syndrome. Results from a randomized, double-blind, placebo-controlled trial. JAMA 271: 1836–1843

Flick MR, Perel A, Staub NC (1981) Leukocytes are required for increased lung microvascular permeability after microembolisation in sheep. Circ Res 48: 344–351

Fourrier F, Chopin C, Goudemand J et al. (1992) Septic shock, multiple organ failure, and disseminated intravascular coagulation. Chest 101: 816–823

Franson TR, Hierholzer WJ, LaBrecque DR (1985) Frequency and characteristics of hyperbilirubinemia associated with bacteremia. Rev Infect Dis 7: 1–9

Fry DE, Pearlstein L, Fulton RL, Polk HC Jr (1980) Multiple system organ failure. Arch Surg 115: 136–140

Furie B, Furie BC (1992) Molecular and cellular biology of blood coagulation. N Engl J Med 326: 800-806

Gamble JR, Harlan JM, Klebanoff SJ (1985) Stimulation of the adherence of neutrophils to umbilical vein endothelium by human recombinant tumor necrosis factor. Proc Natl Acad Sci USA 82: 8667–8671

Gattinoni L, Brazzi L, Pelosi P (1995) Increasing oxygen delivery in sepsis. In: Sibbald W, Vincent J-L (eds) Clinical trials for the treatment of sepsis. Springer, Berlin Heidelberg New York, pp 299–314 (Update of intensive care and emergency medicine, vol 19)

Gee MH, Gottlieb JE, Albetrine KH et al. (1990) Physiology of aging related to outcome in ARDS. J Appl Physiol 69: 822–829

Geiger T, Andus T, Klapproth J, Hirano T, Kishimoto T, Heinrich PC (1988) Induction of rat acute-phase proteins by interleukin-6 in vivo. Eur J Immunol 18: 717–721

Ghezzi P, Dinarello CA, Binachi M et al. (1991) Hypoxia increases production of interleukin-1 and tumor necrosis factor by human mononuclear cells. Cytokine 3: 189–194

Gilbert EM, Haupt MT, Mandanas RY, Huaringa AJ, Carlson RW (1986) The effect of fluid loading, blood transfusion, and catecholamine infusion on oxygen delivery and consumption in patients with sepsis. Am Rev Respir Dis 134: 873–878

Gimbrone MA Jr, Obin MS, Brock AF et al. (1989) Endothelial interleukin-8: a novel inhibitor of leukocyte-endothelial interactions. Science 246: 1601–1603

Gleckman R, Hibert D (1982) Afebrile bacteremia: a phenomenon in geriatric patients. JAMA 248: 1478–1481

Goris RJA (1990) Mediators of multiple organ failure. Intensive Care Med 16: S192–S196

Goris RJA, te Boekhorst TPA, Nuytinck JKS, Gimbrére JSF (1985) Multiple-organ failure. Generalized autodestructive inflammation? Arch Surg 120: 1109–1115

Goris RJA, Boekholtz WKF, van Bebber IPT, Nuytinck JKS, Schellings PHM (1986) Multiple organ failure and sepsis without bacteria. An experimental model. Arch Surg 121: 897–901

Granger DN (1988) Role of xanthine oxidase and granulocytes in ischemia-reperfusion injury. Am J Physiol 24: H1269–H1275

Greenman RL, Schein RMH, Martin MA et al. (1991) A controlled clinical trial of E5 Murine monoclonal IgM antibody to endotoxin in the treatment of gramnegative sepsis. JAMA 266: 1097–1102

Groeneveld ABJ (1994) Pathogenesis of acute renal failure during sepsis. Nephrol Dial Transplant 9 Suppl 4: 47–51

Groeneveld ABJ, Bronsveld W, Thijs LG (1986) Hemodynamic determinants of mortality in human septic shock. Surgery 91: 140–152

Groeneveld ABJ, Kester ADM, Nauta JJP, Thijs LG (1987) Relation of blood lactate to oxygen delivery and hemodynamic variables in human shock states. Circ Shock 22: 35–53

Grum CM, Simon RH, Dantzker DR, Fox IH (1985) Evidence for adenosine triphosphate degradation in critically-ill patients. Chest 88: 763–767

Gutierrez G, Clark C, Brown SD et al. (1994) Effect of dobutamine on oxygen consumption and gastric mucosal pH in septic patients. Am J Respir Crit Care Med 150: 324–329

Hack CE, Thijs LG (1991) The orchestra of mediators in the pathogenesis of septic shock: a review. In: Vincent J-L (ed) Update in intensive care and emergency medicine, vol 14. Springer, Berlin Heidelberg New York, pp 232–246

Hack CE, de Groot ER, Felt-Bersma RJF et al. (1989a) Increased plasma levels of interleukin-6 in sepsis. Blood 74: 1704–1710

Hack CE, Nuijens JH, Felt-Bersma RJF et al. (1989b) Elevated plasma levels of anaphylatoxin C3a and C4a are associated with a fatal outcome in sepsis. Am J Med 86: 20–26

Hack CE, Nuijens JH, Strack van Schijndel RJM, Abbink JJ, Eerenberg AJM, Thijs LG (1990) A model for the interplay of inflammatory mediators in sepsis – a study in 48 patients. Intensive Care Med 16: S187–S191

Hack CE, Hart M, Strack van Schijndel RJM et al. (1992) Interleukin-8 in sepsis: relation to shock and inflammatory mediators. Infect Immun 60: 2853–2842

Hartl WH, Günther B, Inthorn D, Heberer G (1988) Reactive hyperemia in patients with septic conditions. Surgery 103: 440–444

Hasegawa N, Husari AW, Hart WT, Kandra TG, Raffin TA (1994) Role of the coagulation system in ARDS. Chest 105: 268–277

Hebert PC, Drummond AJ, Singer J, Bernard GR, Russell JA (1993) A simple multiple system organ failure scoring system predicts mortality of patients who have sepsis syndrome. Chest 104: 230–235

Heideman M, Hugli TE (1984) Anaphylatoxin generation in multisystem organ failure. J Trauma 24: 1038–1043

Heideman M, Norder-Hansson B, Bengtsson A, Mollnes (1988) Terminal complement complexes and anaphylatoxins in septic and ischaemic patients. Arch Surg 123: 188–192

Henao FJ, Daes JE, Dennis RJ (1991) Risk factors for multiorgan failure: a case-control study. J Trauma 31: 74–80

Henson PM, Johnston RB (1987) Tissue injury in inflammation. J Clin Invest 79: 669–674

Henson PM, Larsen GL, Webster RO et al. (1982) Pulmonary vascular alteration and injury induced by complement fragments: Synergistic effect of complement activation, neutrophil sequestration, and prostaglandins. Ann N Y Acad Sci 384: 287–300

Hesselvik JF, Blombäck M, Brodin B, Maller R (1989) Coagulation, fibrinolysis and kallikrein system in sepsis: relation to outcome. Crit Care Med 17: 724–733

Hesselvik JF, Malm J, Dahlbäck B, Blombäck M (1991) Protein C, protein S and C4-binding protein in severe infection and septic shock. Thromb Haemost 65: 126–129

Hohn DC, Meyers AJ, Cherini ST, Beckmann A, Markinson RE, Churg AM (1980) Production of acute pulmonary injury by leukocytes and activated complement. Surgery 88: 48–58
Holloway WA, Reinhardt J (1984) Septic shock in the elderly. Geriatrics 39: 48–54
Hosea S, Brown E, Hammer C, Frank M (1980) Role of complement activation in a model of adult respiratory distress syndrome. J Clin Invest 66: 375–382
Hotchkiss RS, Karl IE (1992) Reevaluation of the role of cellular hypoxia and bioenergetic failure in sepsis. JAMA 267: 1503–1510
Huber AR, Kunkel SL, Tod RF III, Weiss SJ (1991) Regulation of transendothelial neutrophil migration by endogenous interleukin-8. Science 254: 99–102
Idell S, James KK, Levin EG et al. (1989) Local abnormalities in coagulation and fibrinolytic pathways predispose to alveolar fibrin deposition in the adult respiratory syndrome. J Clin Invest 84: 695–705
Jackson AC, Gilbert JJ, Young GB, Bolton CF (1985) The encephalopathy of sepsis. Can J Neurol Sci 12: 303–307
Jacob HS, Craddock PR, Hammerschmidt DE et al. (1980) Complement-induced granulocyte aggregation. An unsuspected mechanism of disease. N Engl J Med 302: 789–795
Jordan DA, Miller CF, Kubos KL, Rogers MC (1987) Evaluation of sepsis in a critically ill surgical population. Crit Care Med 15: 897–904
Kaushansky K, Broudy VC, Harlan JM, Adamson JW (1988) Tumor necrosis factor-α and tumor necrosis factor-β (lymphotoxin) stimulate the production of granulocyte-macrophage colony-stimulating factor, macrophage colony-stimulating factor, and IL-1 in vivo. J Immunol 141: 3410–3415
Kawakami M, Ishibashi S, Ogawa H, Murase T, Takaku F, Shibata S (1986) Cachectin/TNF as well as interleukin-1 induces prostacylin synthesis in cultured vascular endothelial cells. Biochem Biophys Res Commun 141: 482–487
Ketai LH, Grum CM (1986) C3a and adult respiratory distress syndrome after massive transfusion. Crit Care Med 14: 1001–1003
Kieft H, Hoepelman AIM, Zhou W, Rozenberg-Arska M, Struyvenberg A, Verhoef J (1993) The sepsis syndrome in a Dutch University Hospital: clinical observations. Arch Intern Med 153: 2241–2247
Klebanoff SJ, Vadas MA, Harlan JM et al. (1986) Stimulation of neutrophils by tumor necrosis factor. J Immunol 136: 4220–4225
Kleinknecht D (1990) Epidemiology of acute renal failure in France today. In: Bihari D, Neild G (eds) Acute renal failure in the intensive therapy unit. Springer, London, pp 13–22
Knaus WA, Draper EA, Wagner DP, Zimmerman JE (1985) Prognosis in acute organ-system failure. Ann Surg 202: 685–693
Langlois PF, Gawryl MS (1988) Accentuated formation of the terminal C5b-9 complement complex in patient plasma precedes development of the adult respiratory distress syndrome. Am Rev Respir Dis 138: 368–375
Larrick JW, Graham D, Toy K, Lin LS, Senyk G, Fendly BM (1987) Recombinant tumor necrosis factor activation of human granulocytes. Blood 69: 640–644
Le J, Vilçek J (1987) Biology of disease; tumor necrosis factor and interleukin 1: cytokines with multiple overlapping biological activities. Lab Invest 56: 234–248
Lentz SR, Tsiang M, Sadler JE (1991) Regulation of thrombomodulin by tumor necrosis factor-α: comparison of transcriptional and posttranscriptional mechanisms. Blood 77: 542–550
Lorente JA, García-Frade LJ, Landín L et al. (1993) Time course of hemostatic abnormalities in sepsis and its relation to outcome. Chest 103: 1536–1542
Lucas CE (1976) The renal response to acute injury and sepsis. Surg Clin North Am 56: 953–975
Mantovani A, Dejana E (1989) Cytokines as communication signals between leukocytes and endothelial cells. Immunol Today 10: 370–375
Marik PE (1993) Gastric intramucosal pH. A better predictor of multi organ dysfunction syndrome and death than oxygen-derived variables in patients with sepsis. Chest 104: 225–229
Marsden PA, Brenner BM (1992) Transcriptional regulation of the endothelin-1 gene by TNF-alpha. Am J Physiol 262: 854–861
Marshall JC (1994) A scoring system for multiple organ dysfunction syndrome. In: Reinhart K, Eyrich K, Sprung C (eds) Sepsis, current perspectives in pathophysiology and therapy. Springer, Berlin Heidelberg New York, pp 38–49 (Update in intensive care and emergency medicine, vol 18)
Marshall JC (1995) Multiple organ dysfunction syndrome (MODS). In: Sibbald WJ, Vincent J-L (eds) Clinical trials for the treatment of sepsis. Springer, Berlin Heidelberg New York, pp 122–138 (Update in intensive care and emergency medicine, vol 19)

Marshall JC, Christou NV, Horn R, Meakins JL (1988) The microbiology of multiple organ failure. The proximal gastrointestinal tract as an occult reservoir of pathogens. Arch Surg 123: 309–315
Martin S, Maruta K, Burkart V, Gillis S, Kolb H (1988) IL-1 and IFN-gamma increase vascular permeability. Immunology 64: 301–305
Martin C, Saux P, Mege JL, Perrin G, Papazian L, Gouin F (1994) Prognostic values of serum cytokines in septic shock. Intensive Care Med 20: 272–277
Marty C, Misset B, Tamion F, Fitting C, Carlet J, Cavaillon JM (1994) Circulating interleukin-8 concentrations in patients with multiple organ failure of septic and nonseptic origin. Crit Care Med 22: 673–679
Matuschak GM, Rinaldo JE (1988) Organ interactions in the adult respiratory distress syndrome during sepsis. Role of the liver in host defense. Chest 94: 400–406
Meakins JL, Wicklund B, Forse RA (1980) The surgical intensive care unit: current concepts in infection. Surg Clin North Am 60: 117–132
Metinko HP, Kunkel SL, Standiford TJ et al. (1992) Anoxia-hyperoxia induces monocyte-derived interleukin-8. J Clin Invest 90: 791–798
Mileski WJ, Winn RK, Vedder NB et al. (1990) Inhibition of CD18-dependent neutrophil adherence reduces organ injury hemorraghic shock in primates. Surgery 108: 206–212
Miller EJ, Cohen AB, Nagoa S et al. (1992) Elevated levels of NAP-1/interleukin-8 are present in the airspaces of patients with the adult respiratory distress syndrome and are associated with increased mortality. Am Rev Respir Dis 146: 427–432
Modig J, Samulesson T, Hallgreen R (1986) The predictive and discriminative value of biologically active products of eosinophils, neutrophils and complement in bronchoalveolar lavage and blood in patients with adult respiratory distress syndrome. Resuscitation 14: 121–128
Nachman RL, Hajjar KA, Silverstein RL, Dinarello CA (1986) Interleukin 1 induces endothelial cell synthesis of plasminogen activator inhibitor. J Exp Med 163: 1595–1600
Nawroth PP, Stern DM (1986) Modulation of endothelial cell hemostatic properties by tumor necrosis factor. J Exp Med 163: 740–745
Nawroth PP, Bank I, Handley D, Cassimeris J, Chess L, Stern D (1986) Tumor necrosis factor/cachectin interacts with endothelial cell receptors to induce release of interleukin 1. J Exp Med 163: 1363–1375
Nelson DP, Samsel RW, Wood LDH, Schumacker PT (1988) Pathological supply dependency of systemic and intestinal 0_2 uptake during endotoxemia. J Appl Physiol 64: 2410–2419
Nuijens JH, Huijbregts CCM, Eerenberg-Belmer et al. (1988) Quantification of plasma factor XIIa-C1-inhibitor and kallikrein-C1-inhibitor complexes in sepsis. Blood 72: 1841–1848
Nuijens JH, Abbink JJ, Wachtfogel YT et al. (1992) Plasma elastase α1-antitrypsin and lactoferrin in sepsis: evidence for neutrophils as mediators in fatal sepsis. J Lab Clin Med 119: 159–168
Nurnberger W, Holthausen S, Schirlau K, Michelmann I, Burdach S, Gobel U (1993) Activation of the complement and contact system during rTNF-alpha/rIFN-gamma therapy (Abstr). Mol Immunol 30: 39
Nuytinck JKS, Goris RJA, Weerts JGE, Schillings PHM, Schuurmans-Stekhoven JH (1986a) Acute generalized microvascular injury by activated complement and hypoxia: the basis of the adult respiratory distress syndrome and multiple organ failure? Br J Exp Pathol 67: 537–548
Nuytinck JKS, Goris RJA, Redl H, Schlag G, van Munster PJJ (1986b) Posttraumatic complications and inflammatory mediators. Arch Surg 121: 886–890
Ogilvie AC, Groeneveld ABJ, Straub JP, Thijs LG (1991) Plasma lipid peroxides and antioxidants in human septic shock. Intensive Care Med 17: 40–44
Ognibene FP, Martin SE, Parker MM et al. (1986) Adult respiratory distress syndrome in patients with severe neutropenia. N Engl J Med 315: 547–551
Pagani FD, Baker LS, Hsi C, Knox M, Fink MP, Visner MS (1992) Left ventricular systolic and diastolic dysfunction after infusion of tumor necrosis factor-α in conscious dogs. J Clin Invest 90: 389–398
Parker MM, Shelhamer JH, Natanson C, Alling DW, Parrillo JE (1987) Serial cardiovascular variables in survivors and nonsurvivors of human septic shock: heart rate as an early predictor of prognosis. Crit Care Med 15: 923–929
Parsons PE, Giclas PC (1990) The terminal complement complex (sC5b-9) is not specifically associated with the development of the adult respiratory distress syndrome. Am Rev Respir Dis 141: 98–103
Parsons PE, Wortehn GS, Moore EE, Tate RM, Henson PM (1989) The association of circulating endotoxin with the development of the adult respiratory distress syndrome. Am Rev Respir Dis 140: 294–301

Paterson IS, Klausner JM, Goldman G et al. (1989) Pulmonary edema after aneurysm surgery is modified by mannitol. Ann Surg 210: 796–801
Pepe PE, Potkin RT, Holtman Reus D, Hudson LD, Carrico CJ (1982) Clinical predictors of the adult respiratory distress syndrome. Am J Surg 144: 124–130
Petrak RA, Balk RA, Bone RC (1989) Prostaglandins, cyclo-oxygenase inhibitors, and thromboxane synthetase inhibitors in the pathogenesis of multiple systems organ failure. Crit Care Clin 5: 303–314
Pine RW, Wertz MJ, Lennard ES, Dellinger EP, Carrico CJ, Minshew BH (1983) Determinants of organ malfunction or death in patients with intra-abdominal sepsis. Arch Surg 118: 242–249
Pinsky MR, Vincent JL, Deviere J, Alegre M, Kahn RJ, Dupont E (1993) Serum cytokine levels in human septic shock. Relation to multiple-system organ failure and mortality. Chest 103: 565–575
Pixley RA, De La Cadena R, Page JD et al. (1993) The contact system contributes to hypotension but not disseminated intravascular coagulation in lethal bacteremia. J Clin Invest 91: 61–68
Polk HC Jr, Shields CL (1977) Remote organ failure: a valid sign of occult intra-abdominal infection. Surgery 81: 310–313
Pralong G, Calandra T, Glauser MP, Schellekens J, Verhoef J, Bachmann F, Kruithof EKO (1989) Plasminogen activator inhibitor 1: a new prognostic marker in septic shock. Thromb Haemost 612: 459–462
Rangel-Frausto MS, Pittet D, Costigan M, Hwang T, Davis CS, Wensel RP (1995) The natural history of the systemic inflammatory response syndrome (SIRS), a prospective study. JAMA 273: 117–123
Rasmussen T, Haglund U (1992) Early gut ischemia in experimental fecal peritonitis. Circ Shock 38: 122–128
Reddy VY, Pizzo SV, Weiss SJ (1989) Functional inactivation and structural disruption of human $\alpha 2$-macroglobulin by neutrophils and eosinophils. J Biol Chem 264: 13801–13809
Redl H, Schlag G, Dinges HP et al. (1991) Trauma and sepsis induced activation of granulocytes, monocytes/macrophages and endothelial cells in primates. In: Schlag G, Redl H, Siegel JH, Traber DL (eds) Shock, sepsis and organ failure. Springer, Berlin Heidelberg New York, pp 297–313 (Second Wiggers Bernard Conference)
Reilly PM, Schiller HJ, Bulkley GB (1991) Pharmacologic approach to tissue injury mediated by free radicals and other oxygen metabolities. Am J Surg 161: 488–503
Richard C, Lemonnier F, Thibault M et al. (1990) Vitamin E deficiency and lipid peroxidation during adult respiratory distress syndrome. Crit Care Med 18: 4–9
Richmond JM, Sibbald WJ, Linton AM et al. (1982) Patterns of urinary protein excretion in patients with sepsis. Nephron 31: 219–223
Robaye B, Mosselmans R, Fiers W, Dumont JE, Galand P (1991) Tumor necrosis factor induces apoptosis (programmed cell death) in normal endothelial cells in vitro. Am J Pathol 138: 447–453
Rock CS, Lowry SF (1991) Tumor necrosis factor-alpha. J Surg Res 51: 434–445
Ruokonen E, Takala J, Kari A, Alhava E (1991) Septic shock and multiple organ failure. Crit Care Med 19: 1146–1151
Ruokonen E, Takala J, Kari A et al. (1993) Regional blood flow and oxygen transport in septic shock. Crit Care Med 21: 1296–1303
Scholz W, McClurg MR, Cardenas GJ et al. (1990) C5a-mediated release of interleukin-6 by human monocytes. Clin Immunol Immunopathol 57: 297–307
Schröder J-M (1989) The monocyte-derived neutrophil activating peptide (NAP/interleukin-8) stimulates human neutrophil arachidonate-5-lipoxygenase, but not the release of cellular arachidonate. J Exp Med 170: 847–863
Schuger L, Varani J, Marks RM, Kunkel SL, Johnson KJ, Ward PA (1989) Cytotoxicity of tumor necrosis factor-alpha for human umbilical vein endothelial cells. Lab Invest 61: 62–68
Shalaby MR, Aggarwal BB, Rinderknect E, Svedersky LP, Finkle BS, Palladino MA Jr (1985) Activation of human polymorphonuclear neutrophil functions by interferon-gamma and tumor necrosis factor. J Immunol 135: 2069–2073
Shoemaker WC, Appel PL, Kram HB (1992) Role of oxygen debt in the development of organ failure sepsis, and death in high-risk surgical patients. Chest 102: 208–15
Sica A, Wang JM, Colotta F et al. (1990a) Monocyte chemotactic and activating factor gene expression induced in endothelial cells by IL-1 and tumor necrosis factor. J Immunol 144: 3034–3038
Sica A, Matsushima K, van Damme J et al. (1990b) IL-1 transcriptionally activates the neutrophil chemotactic factor IL-8 gene in endothelial cells. Immunology 69: 548–553

Silverman HJ, Tuma P (1992) Gastric tonometry in patients with sepsis. Effects of dobutamine infusions and packed red blood cell transfusion. Chest 102: 184–188

Simms HH, D'Amico R (1991) Increased PMN CD11b/CD18 expression following post traumatic ARDS. J Surg Res 50: 362–367

Skillman JJ, Bushnell LS, Goldman H et al. (1969) Respiratory failure, hypotension, sepsis and jaundice. Am J Surg 177: 523–531

Slotman GJ, Burchard KW, Yellin SA, Williams JJ (1986) Prostaglandin and complement interaction in clinical acute respiratory failure. Arch Surg 121: 271–274

Smedley LA, Tonnesen RA, Sandhaus RA, Haslett C, Guthrie LA, Johnston RB, Henson PM, Worthen GS (1986) Neutrophil-mediated injury to endothelial cells. J Clin Invest 77: 1233–1243

Stevens JH, O'Hanley P, Shapiro JM et al. (1986) Effects of anti-C5a antibodies on the adult respiratory distress syndrome in septic primates. J Clin Invest 77: 1812–1816

Stokke T, Burchardi H, Hensel I, Kostering H, Kathner T, Rahlf G (1986) Continuous intravenous infusion of elastase in normal and agranulocytic minipigs – effects on the lungs and the blood coagulation system. Resuscitation 14: 61–79

Suter PM, Suter S, Girardin E, Roux-Lombard P, Grau GE, Dayer JM (1992) High bronchoalveolar levels of tumor necrosis factor and its inhibitors, interleukin-1, interferon, and elastase in patients with adult respiratory distress syndrome after trauma, shock or sepsis. Am Rev Respir Dis 145: 1016–1022

Sznajder JL, Fraiman A, Hall JB et al. (1989) Increased hydrogen peroxide in the expired breath of patients with acute hypoxemic respiratory failure. Chest 96: 606–610

Tanaka T, Tsujinaka T, Kambayashi J et al. (1990) The effect of heparin on multiple organ failure and disseminated intravascular coagulation in a sepsis model. Thromb Res 60: 321–330

Tanaka H, Sugimoto H, Yoshiaka T, Sugimoto T (1991) Role of granulocyte elastase in tissue injury in patients with septic shock complicated by multiple organ failure. Ann Surg 213: 81–85

Taylor FB Jr, Chang A, Esmon Ct, D'Angelo A, Vigano-D'Angelo S, Blick KE (1987) Protein C prevents the coagulopathic and lethal effects of Escherichia coli infusion in the baboons. J Clin Invest 79: 918–925

Taylor FB Jr, Emerson TE Jr, Jordan R, Chang AK, Blick KE (1988) Antithrombin-III prevents the lethal effects of Escherichia coli infusion in baboons. Circ Shock 26: 227–235

Tennenberg SD, Jacobs MP, Solomkin JS (1987) Complement-mediated neutrophil activation in sepsis- and trauma-related adult respiratory distress syndrome. Arch Surg 122: 26–32

Thijs LG (1994) Regional or organ blood flow and oxygen consumption: observations in septic shock. Curr Top Intensive Care 1: 101–123

Thijs LG, Hack CE (1995) Time course of cytokine levels in sepsis. Intensive Care Med 21: S16–S21

Thijs LG, de Boer JP, de Groot MCM, Hack CE (1993) Coagulation disorders in septic shock. Intensive Care Med 19: S8–S15

Thommasen HV, Russell JA, Boyko WJ et al. (1984) Transient leucopenia associated with adult respiratory distress syndrome. Lancet 1: 809–812

Tilney NL, Bailey GL, Morgan AP (1973) Sequential system failure after rupture of abdominal aortic aneurysms: an unsolved problem in post operative care. Ann Surg 178: 117–122

Tracey KJ, Lowrey SF (1990) The role of cytokine mediators in septic shock. Adv Surg 23: 21–56

Tracey KJ, Beutler B, Lowry SF et al. (1986) Shock and tissue injury induced by recombinant human cachectin. Science 234: 470–474

Tracey KJ, Fong Y, Hesse DG et al. (1987) Anti-cachectin/TNF monoclonal antibodies prevent septic shock during lethal bacteremia. Nature 330: 662–664

Tran DD, Groeneveld ABJ, Van der Meulen J, Nauta JJP, Strack van Schijndel RJM, Thijs LG (1990) Age, chronic disease, sepsis, organ system failure and mortality in a medical intensive care unit. Crit Care Med 18: 474–479

Tsujino M, Hirata Y, Imai T, Kanno K, Eguchi S, Ito H, Marumo F (1994) Induction of nitric oxide synthase gene by interleukin-1β in cultured rat cardiocytes. Circulation 90: 375–383

Uzun O, Akalin HE, Hayran M, Ünal S (1992) Factors influencing prognosis in bacteremia due to gram-negative organisms: evaluation of 448 episodes in a Turkish University Hospital. Clin Infect Dis 15: 866–873

Van Deventer SJH, Hart M, van der Poll T, Hack CE, Aarden LA (1993) Endotoxin and tumor necrosis factor-alpha-induced interleukin-8 release in humans. J Infect Dis 167: 461–464

Van Hinsbergh VWM, Kooistra T, van den Berg EA, Princen HMG, Fiers W, Emeis JJ (1988) Tumor necrosis factor increases the production of plasminogen inhibitor in human endothelial cells in vitro and rats in vivo. Blood 72: 1467–1473

Van Hinsbergh VWM, Bauer KA, Kooistra T et al. (1990) Progress of fibrinolysis during necrosis factor infusions in humans. Concomitant increase in tissue-type plasminogen activator, plasminogen activator inhibitor type-1, and fibrin(ogen) degradation products. Blood 76: 2284–2289

Van der Poll T, Büller HR, ten Cate H et al. (1990) Activation of coagulation after administration of tumor necrosis factor to normal subjects. N Engl J Med 322: 1622–1626

Van der Poll T, Levi M, Büller HR et al. (1991) Fibrinolytic response to tumor necrosis factor in healthy subjects. J Exp Med 174: 729–732

Van der Poll T, van Deventer SJH, Hack CE et al. (1992) Effects on leukocytes after injection of tumor necrosis factor into healthy humans. Blood 79: 693–698

Van der Poll T, Jansen J, Levi M, ten Cate H, ten Cate JW, van Deventer SJH (1994) Regulation of interleukin 10 release by tumor necrosis factor in humans and chimpanzees. J Exp Med 180: 1985–1988

Varani J, Ginsberg I, Schuger L et al. (1989) Endothelial cell killing by neutrophils synergestic interaction of oxygen products and proteases. Am J Pathol 135: 435–438

Vary TC, Siegel JH, Nakatani T, Sato T, Aoyama H (1986) Effect of sepsis on activity of pyruvate dehydrogenase complex in skeletal muscle and liver. Am J Physiol 250: E634–E640

Vincent J-L, Roman A, De Backer D, Kahn RJ (1990) Oxygen uptake/supply dependency. Effects of shortterm dobutamine infusion. Am Rev Respir Dis 142: 2–7

Vito L, Dennis RC, Weisel RD et al. (1974) Sepsis presenting as acute respiratory insufficiency. Surg Gynecol Obstet 38: 896–900

Wedmore CV, Williams TJ (1981) Control of vascular permeability by polymorphonuclear leukocytes in inflammation. Nature 89: 646–650

Weinberg PF, Matthay MA, Webster RO, Roskos KV, Goldstein IM, Murray JF (1984) Biologically active products of complement and acute lung injury in patients with the sepsis syndrome. Am Rev Respir Dis 130: 791–796

Winn R, Maunder R, Chi E et al. (1987) Neutrophil depletion does not prevent lung edema after endotoxin infusion in goats. J Appl Physiol 62: 116–121

Yokoyama T, Vaca L, Rossen RD, Durante W, Hazarika P, Mann DL (1993) Cellular basis for the negative inotropic effects of tumor necrosis factor-alpha in the adult mammalian heart. J Clin Invest 92: 2303–2312

Young GB, Bolton CF, Austin TW et al. (1992) The encephalopathy associated with septic illness. Clin Invest Med 13: 297–304

Zhang H, Spapen H, Manikes P et al. (1995) Tirilazad mesylate (U 74006F) inhibits the effects of endotoxin during reductions in cardiac output in dogs. Am J Physiol (in press)

Zheutlin LM, Thonar E, Lemley-Gillespie S et al. (1987) C3a DESARG and elastase-alphal-protease inhibitor levels in lavage fluid from patients with adult respiratory distress syndrome. J Crit Care 2: 86–92

Ziegler EJ, Fisher CJ, Sprung CL et al. (1991) Treatment of gram-negative bacteremia and septic shock with HA-IA human monoclonal antibody against endotoxin. N Engl J Med 324: 429–436

Zimmerman HJ, Fang M, Utili R et al. (1979) Jaundice due to bacterial infection. Gastroenterology 77: 362–374

Involvement of Bacteria/Endotoxin Translocation in the Development of Multiple Organ Failure

S. BAHRAMI, H. REDL, Y.-M. YAO, and G. SCHLAG

1	Introduction	239
2	Hypovolemic/Traumatic Insults	240
2.1	Bacteremia/Endotoxemia and Cytokines	241
3	Burn Injury	245
4	The Role of LPS/TNF in the Development of MOF	246
5	Clinical Relevance of Translocation	248
6	Intervention Strategies Against Translocation	250
6.1	Intraluminal Approaches	250
6.2	"Maintenance" of Gut Wall	251
6.3	Intravasal Therapeutic Measures	253
7	Conclusion	254
References		254

1 Introduction

The progression of multiple organ failure (MOF) in patients with various underlying diseases continues to pose a life-threatening problem in intensive care units. Despite being the subject of an exponentially increasing number of experimental and clinical investigations during recent decades, the precise mechanisms leading to the development of MOF are not yet clearly understood. While several authors had earlier suggested that systemic sepsis due to uncontrolled infection is the leading cause of MOF (BAUE 1975; FRY et al. 1980; POLK and SHIELDS 1973), it has become increasingly clear that an infectious focus cannot be identified in all patients with MOF (CARRICO et al. 1986; FAIST et al. 1983; GORIS et al. 1985).

Although the pathogenesis of MOF is most likely multiprincipal, gut barrier failure has recently been considered to play a key role. The concept that a breakdown of the intestinal mucosa, a vital component of body's defenses against endogenous infection, may lead to the access of intraluminal pathogens into the body, contributing to morbidity and mortality of systemic sepsis, was

Ludwig Boltzmann Institute for Experimental and Clinical Traumatology, Donaueschingenstr. 13, 1200 Vienna, Austria developed by Fine in the 1950s (Fine et al. 1959). Fine's theory of endogenous infection was not generally accepted (Culbertson et al. 1959) until the 1980s, when Berg's and Deitch's groups reappraised this concept in a complementary manner. In a series of studies Deitch and coworkers demonstrated experimentally the entry of gut-associated bacteria into the lymphatic systems after burns, trauma, malnutrition, and starvation (Deitch et al. 1985, 1987a,c,d; Deitch and Bridges 1987; Deitch and Berg 1987b). These authors also showed, that endotoxin promotes translocation of bacteria from the gut (Deitch et al. 1987b). These and other similar reports (Koziol et al. 1988), renewed interest in the intestinal tract as a source of bacterial infection and led Meakins and Marshall to suggest that "the gut is the motor of multiorgan failure" (Carrico et al. 1986), a hypothesis that is still a matter of dispute. Infections with gut-associated bacteria but with no infectious focus at autopsy have been reported in patients undergoing intraoperative stress, victims of trauma, and patients developing MOF syndrome (Carrico et al. 1986; Garrison et al. 1982).

The term "translocation" was introduced in 1966 to describe the entry of endogenous micro-organisms into tissues and the circulation (Wolochow et al. 1966). This concept has been refined to include the escape of bacteria and/or bacterial products from the gastrointestinal tract (Alexander et al. 1990). Diagnostic approaches often fail to document the phenomenon of translocation. Alternative approaches, such as therapeutic measures, have therefore been employed to investigate the presence or absence of bacteria and bacterial products and their possible role in the development of systemic sepsis and the related pathophysiological alterations. Data obtained from experimental studies performed under standardized and controlled conditions provide useful information to help us better understand the entity and the phenomenon of bacterial/endotoxin translocation.

In this review we emphasize evidence and/or the significance of bacteria/endotoxin translocation in pathophysiological alterations and the development of multiple organ injury during various clinical conditions and the possible intervention strategies for them.4

2 Hypovolemic/Traumatic Insults

In general, impaired gut barrier permeability can be caused either during the shock period by decreased intestinal blood flow and reduced oxygen delivery (Haglund 1993), during reperfusion resulting in a stage of increased intestinal blood flow, or at a later stage again by reduced flow. Both experimental and clinical data suggest increasingly that one of the important initiating events in the sepsis syndrome that follows hypovolemic shock (HS) is the entry of bacteria and endotoxin into the blood of traumatized, hypotensive patients (Baker et al. 1988; Koziol et al. 1988; Rush et al. 1988; Deitch 1990b; Schlag et al. 1993b;

MARSHALL et al. 1993). Dispute continues, however, over a possible causal relationship between gut-origin endotoxemia/bacteremia and septic complications, on the one hand, and the clinical outcome of hemorrhage and trauma, on the other (MOORE et al. 1991; PEITZMAN et al. 1991; ROUMEN et al. 1993; BRATHWAITE et al. 1993).

We became interested in determining both the time course of bacterial translocation and the kinetics of lipopolysaccharide (LPS) and cytokine appearance in the circulation. Moreover, to evaluate the role of endogenous endotoxin and cytokines in HS-related pathophysiological alterations and mortality we employed anti-LPS and anti-tumor necrosis factor (TNF) measures in rats subjected to prolonged hemorrhagic insult.

2.1 Bacteremia/Endotoxemia and Cytokines

Histological examinations reveal bacterial translocation into the small intestinal wall in rats subjected to HS for 90 min as early as the end of a 30-min resuscitation period (JIANG et al. 1995). Similarly, translocation of bacteria into the small vessels of the colon has been observed in baboons subjected to a hemorrhagic traumatic shock (HTS) consisting of bleeding the animals to maintain a mean arterial pressure (MAP) of 40 mmHg for 3 h, femur fracture, and soft tissue trauma followed by a 3-h resuscitation period (SCHLAG et al. 1991). We noted positive blood cultures in HTS baboons even during the shock and/or reperfusion period (Fig. 1a). At the end of a 6-h acute experiment in baboons the highest bacterial counts were noted in the mesenteric lymph nodes (MLN) followed by the liver and spleen (SCHLAG et al. 1991). These results obtained in a subhuman primate model are in agreement with reports of bacterial translocation, mainly in rodents (DEITCH et al. 1987a; RUSH et al. 1989) and provide further evidence that the phenomenon of translocation may not be limited to lower animals.

Our experiments also document the significance of the early appearance of endotoxin in the systemic circulation following hemorrhage and/or trauma in both rodents and primates (SCHLAG et al. 1993a,b; JIANG et al. 1995; BAHRAMI et al. 1995a). Elevated plasma LPS levels were found in HTS baboons at the end of a 3-h shock period and 1 h after the beginning of reinfusion (Fig. 1b). Similarly, in a subchronic model in baboons consisting of oxygen debt controlled hemorrhage together with an infusion of zymosan-activated plasma (ZAP), to activate the complement system without trauma, the highest plasma endotoxin levels were detected at the end of the 3 to 4-h shock period and at 1 h after the start of reinfusion (SCHLAG et al. 1993b). Our rat studies reveal a marked, albeit transient increase in plasma LPS concentrations, in both the portal and the systemic circulation following HS (JIANG et al. 1995). By comparing portal and systemic endotoxin levels we found that the significant rise in the portal circulation took place 30 min before that in the systemic circulation (Fig. 2). Throughout the experiment LPS levels were higher in the portal than in the systemic circulation.

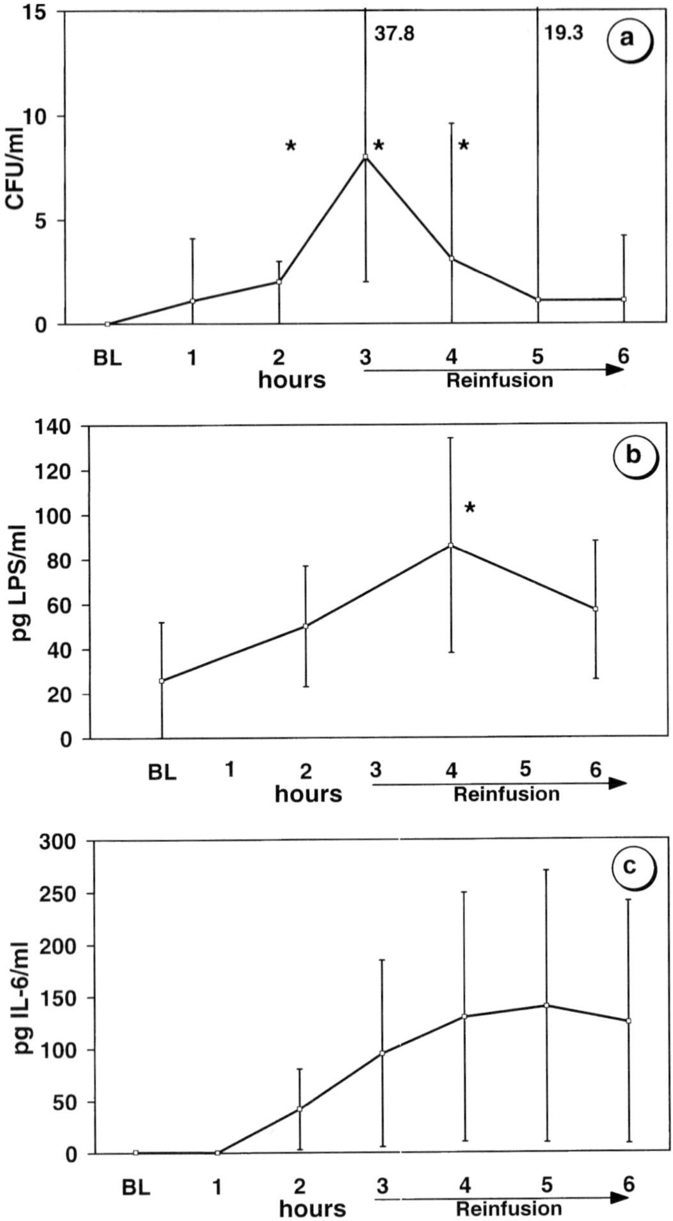

Fig. 1a-c. Measurements in baboons subjected to hypovolemic-traumatic shock. **a,** Bacterial counts in the systemic circulation **b**, Plasma endotoxin levels (mean ± SE). **c**, Plasma levels of interleukin-6 in the systemic circulation (mean ± SE). (Adapted from REDL et al. 1991)

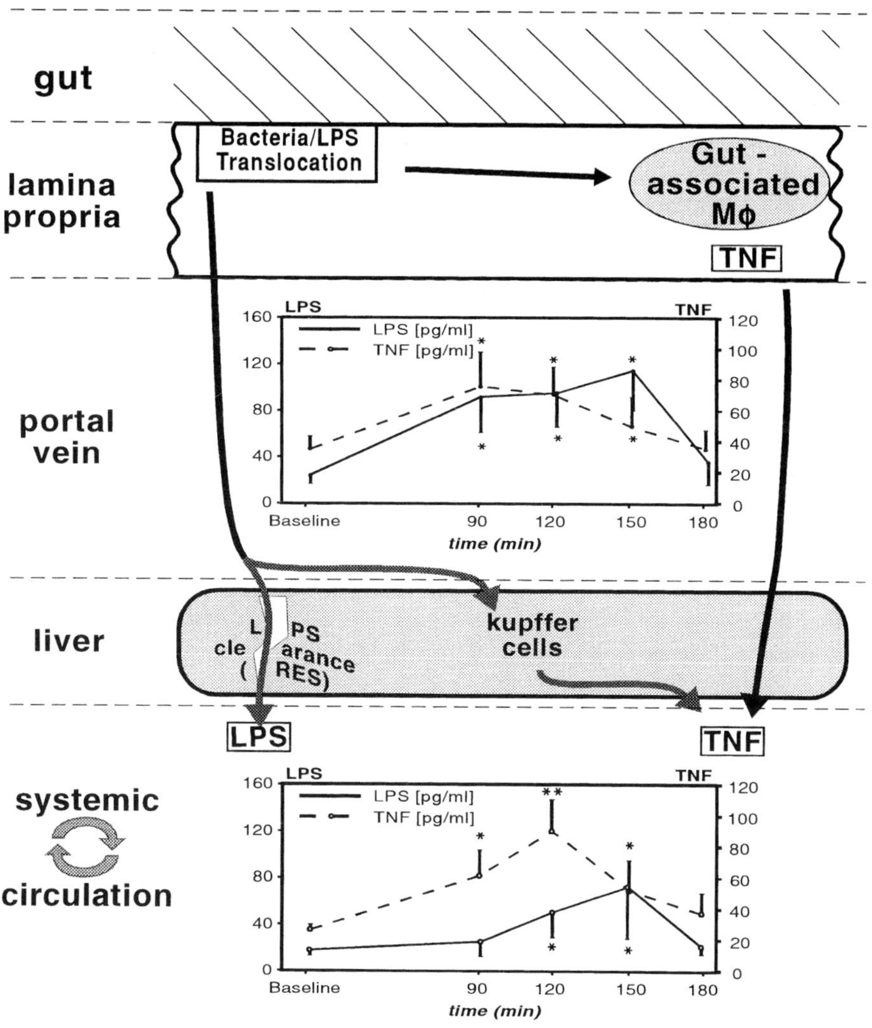

Fig. 2. Possible sequence of events following hemorrhage/trauma-induced translocation leading to cytokine formation in the presence or absence of detectable plasma endotoxin. Data of portal and systemic endotoxin and TNF plasma levels in rats subjected to 90-min hemorrhage (followed by resuscitation; mean ± SD). (Adapted from JIANG et al. 1995)

These results suggest that the development of systemic endotoxemia parallel to the endotoxin appearance in the portal vein represents spillover from elevated portal LPS levels overwhelming the liver reticuloendothelial system. In contrast to our findings, the increase in plasma endotoxin concentration has not been detected by others following hemorrhage in mice (AYALA et al. 1990) or rats (THIEMERMANN et al. 1993). Although the phenomenon of endotoxemia after shock remains controversial, our experiments provide further evidence confirming the earlier observations of Rush et al. (1988) and Baker et al. (1988), who demonstrated that there is a relationship between HS and bacterial/endotoxin translocation.

Our current data provide further evidence that HS leads to bacterial/endotoxin translocation with concomitant cytokine formation. A number of recent studies have implicated inflammatory mediators following hemorrhagic shock (Chaudry and Ayala 1993). TNF is a cytokine produced by macrophages and monocytes in response to stimulation with endotoxin/bacteria. It has been postulated that translocated endotoxin/bacteria can activate macrophages to produce various cytokines, including TNF, which may lead to the development of sepsis and MOF associated with posttraumatic hypotension (Deitch 1990b; Caty et al. 1990). The mechanism(s) of TNF formation, however, and its role in the pathophysiology of hemorrhagic shock are not clearly understood (Ayala et al. 1991; Rhee et al. 1993; Stylianos et al. 1991). According to a report by Ayala et al. (1991) using a mouse hemorrhage model, TNF was significantly increased during and up to 2 h after blood loss, despite a lack of detectable plasma endotoxin levels. Similarly, studies by Rhee et al. (1993) indicated an early appearance of TNF in rats with a fixed-volume blood loss.

Although only in a few cases have we detected plasma TNF in the HS model in baboons, the levels of soluble TNF receptor (55 kDa), however, peaked between 1 and 2 h (Bahrami et al. 1995a). It may therefore be that the lack of detectable plasma TNF levels is due to the binding of TNF to the cell or soluble receptors. Plasma IL-6 levels, however, were found to be increased in the HTS baboons and peaked after 4–5 h (Fig. 1c). The TNF concentration in the portal vein of rats subjected to HS for 90 min peaked at the end of shock (90 min) and remained at this level until 150 min after shock (Jiang et al. 1995). In the systemic circulation the TNF levels were also markedly elevated after 90 min. In contrast to portal vein TNF levels, however, systemic TNF levels reached a peak at 120 min after shock.

The significant rise in TNF concentrations in the portal blood that occurs prior to the increased concentrations in the systemic circulation indicates a gut-derived TNF formation, for which the gut-associated macrophages in the lamina propria of the intestine can be considered a potential source (Fig. 2). The parallel increase in portal and systemic TNF concentrations until the end of the shock period was followed by a continuous increase in systemic but not in portal levels. This, together with other reports of portal cytokines (Deitch et al. 1994) and studies showing that proinflammatory gene expression occurs in Peyer's patches as early as 1 h following hemorrhage (Shenkar and Abraham 1993) led us to speculate that gut-associated macrophages constitute the major source for TNF formation at the early stage of shock, and that the Kupffer cells stimulated by endotoxin transported to the liver via portal blood serve as the major cellular source for the circulating TNF at a later stage (Fig. 2). In addition, there are also several lines of evidence suggesting that changes in organ blood flow induced by hemorrhage leads to regional hypoxia, which may also contribute directly or indirectly to the induction of TNF (Ayala et al. 1990; Chaudry and Ayala 1993; Stylianos et al. 1991). Similar results have recently been found in an in vitro study (Scannell et al. 1993), showing that hypoxia may cause the release of TNF-α and its soluble receptors by human macrophages. Thus, both hypoxia and/or translocated LPS acting on local macrophages can be considered inducers of TNF formation that would not necessarily be influenced by anti-LPS treatment at the early stage of HS.

3 Burn Injury

It has been shown that burn injury can lead to acute derangements of multiple host systems, including alterations of the intestinal barrier function, which may predispose the burn victim to bacterial/endotoxin translocation from the gastrointestinal tract into the systemic circulation and to distal organs (Deitch 1990b; Maejima et al. 1984; Deitch and Berg 1987b; Alexander et al. 1990). Although the exact relationship of bacterial/endotoxin translocation to clinical sepsis and MOF is not known, extensive work performed in animal models has clearly demonstrated the importance of translocation in the development of systemic responses and, ultimately, of MOF resulting from major burns.

Using a nonlethal burn injury model, Deitch and Berg (1987b) found that 44% of the rats experiencing 40% total body surface area burn had viable, gram-negative enteric bacilli in MLN on the second postburn day. The combination of trauma and intestinal overgrowth with gram-negative enteric bacilli or *Candida* can result in the synergistic spread of bacteria from the gut to systemic organs, such as the spleen and liver (Deitch and Berg 1987b; Alexander et al. 1990). It was also shown that burn injury followed by wound infection with *Pseudomonas* causes prolonged and enhanced bacterial translocation (Jones et al. 1990b). On the other hand, using a radionuclide probe of a specific organism or endotoxin, Alexander et al. (1991) reported that both organism and endotoxin translocate very soon (within 1 h) after the burn injury. Translocation of endotoxin may have biological significance that is independent of and additional to translocation of intact bacteria. Furthermore, experimental results document that a nonlethal dose of endotoxin promotes bacterial translocation from the gut to MLN and to systemic organs, and increased mortality to 100% in burned mice, while the mortality of mice receiving only endotoxin or burn was less than 10% (Deitch and Berg 1987a). It is therefore possible that a self-sustaining cycle can be established with endotoxin itself under some circumstances because of its effects toward weakening mucosal integrity, increasing gut permeability, and promoting the translocation process (Gans and Matsumoto 1974; Garidis et al. 1973; Deitch et al. 1987b). In this way LPS can potentiate the development or progression of sepsis and result in further injury to distant organs.

In addition to small animals, bacterial translocation has been shown to take place in large animals subjected to thermal injury. MORRIS et al. (1990), using a 40% total body surface area burn in an ovine model, found significant increases in the number of bacteria cultured from the MLN, liver, and lung compared to sham animals. Decreased blood flow to the gastrointestinal tract after burn certainly plays a critical role in the pathogenesis of bacterial translocation. This phenomenon has also been seen in burned pigs (TOKYAY et al. 1990), whose gastrointestinal flora is similar to that in humans, and whose vasculature supply to the gut is nearly identical to that in humans. By nonspecifically preventing the vasoconstriction of the mesenteric artery with a smooth muscle relaxant (NAVARATNAM et al. 1989) or selectively inhibiting thromboxane A_2 induction (TOKYAY et al. 1990), gastrointestinal blood flow was maintained and translocation was reduced. Gut bacteria translocated into the MLN, spleen, and burn wound is thus probably due to gut mucosal failure after burn trauma and sepsis, thereby enhancing and generalizing the inflammatory response to thermal injury alone.

4 The Role of LPS/TNF in the Development of MOF

If invading enteric bacteria and endotoxin, however, were indeed the reason for the development of sepsis and the irreversibility of shock, animals treated with agents to eliminate bacterial endotoxin should be more resistant to hemorrhagic insult. To evaluate the role of endogenous endotoxin in pathophysiological alterations following hemorrhagic shock by neutralizing circulating LPS concentrations, we employed two different anti-LPS measures in a prolonged HS model (MAP at 30–35 mmHg for 180 min, followed by reinfusion and resuscitation over 50 min) in rats. At the onset of shock the animals were treated with an improved recombinant preparation of bactericidal/permeability increasing protein ($rBPI_{21}$), an antibiotic protein isolated from human neutrophils, which has been reported to inactivate endotoxin (YAO et al. 1995b). Furthermore, to check the origin of hemorrhage-induced endotoxemia we employed in a similar manner an anti-core LPS monoclonal antibody (WNI 222–5) which was shown to bind to all Escherichia *coli* clinical isolates and to some *Citrobacter, Enterobacter,* and *Klebsiella* isolates with a common inner core structure (DI PADOVA et al. 1993). In fact, the administration of $rBPI_{21}$ or WNI 222–5 almost neutralized the plasma endotoxin levels seen in control animals and provided noticeable protection of vital organs, particularly the lung and intestine, against hemorrhage induced injury (BAHRAMI et al. 1995a; YAO et al. 1995b). Furthermore, treatment with WNI 222–5 led to a survival rate of 71.4%, compared to 21.4% in the control group. Similarly, $rBPI_{21}$ improved 48-h survival to 68.8%, compared to 37.5% in the control group. These results suggest that systemic endotoxemia not only occurs early in the ischemic episode but also plays an important role regarding morbidity and mortality due to severe hemorrhage. Our findings are in accordance with previous studies in which Re-LPS antiserum was found to increase the survival rate in animals subjected to HS (YAO et al. 1992a,b).

The results from both the $rBPI_{21}$ and WNI 222–5 treated groups showed nevertheless that some animals still died after hemorrhage. This could be due either to our treatment protocol itself (insufficient amount of $rBPI_{21}$/WNI 222–5 and/or limited time points of treatment) or to the fact that mediators other than endotoxin may cause, at least in part, the effects on organ system dysfunction and the resulting mortality observed in this experiment.

It is interesting to note that treatment with $rBPI_{21}$ or WNI 222–5 not only neutralized systemic endotoxemia but also significantly reduced the total incidence of bacterial translocation in the intestinal wall of different segments, which supports the assumption that the vicious circle of endotoxemia-bacterial translocation can be broken by means of the inhibition of LPS activities, for example, by the action of $rBPI_{21}$ or WNI 222–5.

In our anti-LPS treatment experiments the systemic plasma TNF levels were also markedly elevated at the end of reinfusion in control animals, but surprisingly they were not significantly influenced by $rBPI_{21}$ or WNI 222–5 treatment. With regard to the potential inducers responsible for the release of TNF during hemorrhage, it would appear that the elevation in TNF found here is not due to the endotoxin derived from the gut, as neutralization of systemic endotoxemia with $rBPI_{21}$ or WNI 222–5 did not markedly influence plasma TNF concentrations. Nevertheless, the failure of anti-LPS treatment to influence the plasma TNF concentrations in our experiments, despite neutralization of circulating LPS levels, could be a result of the initial TNF formation by the LPS-stimulated gut-associated macrophages prior to the entry of LPS into the circulation and its inactivation by anti-LPS measures.

Measures that control or prohibit translocation have been evaluated in some experimental studies to assess the potential role of bacterial/endotoxin in the pathogenesis of sepsis and MOF associated with large burns. DIJKSTRA et al. (1992) reported that the incidence of bacterial translocation and endotoxemia in *P. aeruginosa* burn wound contaminated mice was significantly reduced by low-dose treatment with polymyxin B, a cationic polypeptide antimicrobial agent known to bind endotoxin via its lipid A moiety both in vitro and in vivo. This study supports the hypothesis that the vicious circle of endotoxemia-bacterial translocation can be broken by blocking the activity of endotoxin. A recent series of studies in burned rats determined that pretreatment with selective decontamination of the digestive tract (SDD) can significantly reduce the incidence of bacterial translocation and attenuate the increase of plasma endotoxin concentrations in both portalvein and systemic blood (YAO et al. 1994). Moreover, the 5-day survival rate was markedly higher with 90.0% in the SDD-treated group than the 63.3% in the control group.

To investigate the role of TNF in HS-related multiple organ injury and mortality we employed a monoclonal antibody (mab) to TNF in rats subjected to HS (MAP of 30–35 mmHg for 180 min followed by resuscitation over 50 min). Treatment of rats with a bolus dose (20 mg/kg) of TNF mab (Celltech, Berkshire,

UK) at 65 min after HS resulted in a reduction in systemic vascular resistance, increased cardiac index, significant reduction in the increase in lung wet weight 3 h after shock, and improved 48-h survival (73.3% compared to 26.7% in the control group; Bahrami et al. 1994, 1995b). Macropathological and histological evaluations also revealed a significantly lower frequency of organ injuries in the TNF mab treated animals than in the control group. These findings, taken together with observations in previous studies, lead us to suggest that hemorrhage can induce a marked release of TNF into the circulation, which may at least in part be responsible for end-organ injury in the hemorrhagic setting.

Our data suggest that systemic endotoxemia not only occurs early in the ischemic episode but also plays an important role in the development of multiple organ injury and mortality resulting from severe hemorrhage, trauma, and burns. The efficacy of mab to LPS indicates that the Enterobacteriaceae, in particular *E. coli* as the largest population among them, may serve as the main source of endotoxin following hemorrhage, at least in rats. The cytokines, in particular TNF, induced following hypovolemic insult may play a key role in the initiation of pathophysiological alterations of HS.

5 Clinical Relevance of Translocation

Although translocation of bacteria and endotoxin after hemorrhage , trauma, and burns has been well demonstrated in the above experimental models, it is difficult to extrapolate these results to humans. As a result, their clinical significance remains controversial. Rush et al. (1988) reported a significant correlation between positive blood cultures and the degree of hemorrhagic shock in patients. Patients with admission blood pressure below 80 mmHg had a 56% rate of positive blood cultures, compared to less than 10% in patients with an initial blood pressure over 110 mmHg. Among patients undergoing elective surgery, Deitch (1989) found that in only 4% did the MLN contain bacteria, compared to 59% in patients undergoing surgery for small bowel obstruction, despite the absence of necrotic bowel or positive peritoneal fluid culture. In a previous study by Cabie et al. (1993) which measured endotoxin and cytokines in systemic and portal blood of patients undergoing abdominal aortic surgery, portal LPS was detected even after bowel manipulation (36%) but was elevated especially during the reperfusion phase (71%). In contrast to endotoxin, positive blood cultures were not found in either the portal or in the systemic circulation. Endotoxin was also detected in humans in the portal vein following aortic clamping. In a previous study performed together with Borst's group we were able to detect endotoxin in the systemic circulation during cardiopulmonary bypass (Schlag et al. 1993a). We also reported the release of cytokines, especially soluble cytokine receptors such as TNF receptors (Schlag et al. 1993a).

In extensive burns early increases in plasma endotoxin have been reported by WINCHURCH et al. (1987) and DOBKE et al. (1989), indicating endotoxemia with the peak serial levels 7–12 h and 3–4 days after the burn. As the burn wound is initially "sterile", and there are no apparent infectious foci in the body, this early endotoxemia cannot be attributed to the burn wound itself. This phenomenon is thus more likely to be due to bacterial translocation and/or the passage of endotoxin across the gut barrier. Recently in a group of extensively burned patients (more than 70% TBSA) it was documented that endotoxemia was not only common in the immediate postinjury period but was also strongly associated with the development of MOF and with an adverse outcome (YAO et al. 1995c). Plasma TNF was already elevated on postburn day 1, and significantly higher values were found in the MOF group than in those without MOF throughout the observation period (SHENG and YANG 1994). In view of the above evidence it ,might be reasonable to assume that, as a result of hypoperfusion and reperfusion injury, translocated bacteria/endotoxin can act as the primary trigger in initiating an exuberant inflammatory response, leading to the clinical manifestations of the systemic inflammatory response syndrome (SIRS) and sepsis and further damage to various organs.

In addition to these studies, there have been some clinical reports that intestinal permeability may be increased shortly after injury, which correspond well to the timetable of early endotoxemia induced by burns in both animals (CARTER et al. 1990) and humans (ZIEGLER et al. 1988; DEITCH 1990a; ROUMEN et al. 1993; RYAN et al. 1992). ZIEGLER et al. (1988), using lactulose and mannitol as permeability probes, demonstrated increased bowel permeability in burned patients with infections, but most patients were not studied in the immediate post burn period (average of 15–18 days after injury). Subsequently DEITCH et al. (1991) observed increased gut permeability early after burn injury in the absence of infection. More recently, using a macromolecule as a permeability probe for the intestine, gut permeability was found to be increased at least within 72 h after the injury and in the absence of infection or other morbid complications. Furthermore, this elevated gut permeability to polythylene glycol 3350 was correlated with the extent of the injury (RYAN et al. 1992). Although limited, the existing indirect clinical evidence supports the hypothesis that bacterial/endotoxin translocation does indeed occur in humans.

In contrast to these data, a prospective clinical study did not confirm portal or systemic endotoxemia and/or bacteremia after severe trauma, despite a subsequent 30% incidence of MOF (MOORE et al. 1991). Recently ROUMEN et al. (1993) reported that intestinal permeability after severe trauma and HS was increased without any correlation to septic complications. ENDO et al. (1992) reported that in the early period after injury when no infection was present very few patients had an endotoxin level above the normal range, and that endotoxin levels were not correlated with the extent of the burn or with the prognosis. Their findings suggest that there is no marked translocation in the first week after the accident. On the other hand, in a clinical study the administration of polymyxin B successfully reduced the plasma endotoxin level during the early postburn period but did not influence the cytokine cascade, clinically manifested sepsis, or mortality rate. Therefore it has been concluded that the host inflammatory response in the early postburn state is due to mediators of injury rather than early translocated endotoxemia (MUNSTER et al. 1993). Additionally, no convincing reduction in morbidity or mortality has been shown in most clinical trials despite a consistent decrease in the infection rate as a consequence of the use of SDD (HAMMOND et al. 1992). In this respect, it seems that the clinical significance of bacterial/endotoxin translocation remains an open question. It is still unknown whether septic complications can be prevented in critically ill patients by maintaining the gut mucosal barrier and limiting the translocation process.

6 Intervention Strategies Against Translocation

When the gut is suspected of being the source of bacteria/endotoxin causing systemic infections, the first logical step is to reduce the amount of bacteria and/or endotoxin in the gut lumen and support the mucosal barrier function to minimize the chances of bacterial translocation. Next in importance, if translocation cannot be prevented, is to kill bacteria, neutralize endotoxin, and eliminate bacteria/bacterial products that enter the extraluminal compartments.

6.1 Intraluminal Approaches

Since it is almost impossible to sterilize the gut, the regimen of SDD of the gut was developed using antibiotics (STOUTENBEEK et al. 1984). This selective approach is important since one should not suppress the nonpathogenic anaerobes which are protective against bacterial translocation. Several clinical trials with SDD of the gut have been performed, with the rather consistent finding of decreased nosocomial infections; however, only one study also demonstrated clinical benefits in terms of survival (SANDERSON 1989). There are several arguments why such therapy might not be successful. Especially with regard to the trauma there is the possibility that the drugs are administered too late or are not effective because they are not transported through the gut due to decreased peristaltic movement. Another important arguement is that the host response is induced not only by bacteria but especially by bacterial toxins such as endotoxin; such toxins may in fact be more important that the bacteria, and SDD is usually not designed to neutralize endotoxin.

Endotoxin is probably one of the reasons for patients developing SIRS,but there is no clinical evidence of infection leading to MOF. Intraluminal endotoxin neutralization has been considered an important approach to minimize endotoxin translocation and induction of proinflammatory cascades in the host. Although data are available from only a few investigators, there is some evidence that bile salts are an important factor in binding and eliminating endotoxin from the gut by forming detergent like complexes which are poorly absorbed and well eliminated (BERTOK 1977; CAHILL et al. 1987). The antibacterial and anti-inflammatory properties of lactoferrin are usually explained by its ability to chelate iron. Lactoferrin was therefore used in a rat hemorrhagic shock model (NITSCHE et al. 1993). In the therapy group receiving lactoferrin a dose-dependent reduction in endotoxin transfer from the intestinal tract was measured (NITSCHE et al. 1993). Moreover, LPS binding of lactoferrin was found to decrease cytokine production (MACHNICKI et al. 1993). Another substance with antiendotoxin properties is lactulose (LIEHR et al. 1980). which could also be beneficial as an intraluminal therapeutic agent. Lactulose was administered in a rat model of hepatic failure after liver resection, and lactulose treatment reduced the level of endotoxemia and enhanced survival (VAN LEEUWEN et al. 1991).

A very simple noninvasive procedure for inhibiting translocation is to maintain peristalsis in order to prevent close contact of the bacteria and the intestinal mucosa and therefore reduce the time available for bacteria to penetrate the mucosal layer. It has been shown that in situations such as bowel obstructions bacteria adhere directly to the mucosa (PLAUT et al. 1967). SPAETH et al. (1990) demonstrated that feeding rats dietary fibers in addition to either enteral or parenteral nutrition results in a significant reduction of bacterial translocation. In addition to the direct mechanical effect, the dietary fibers were described as being important in maintaining the normal ecological balance of the micro flora; it was furthermore shown that the degradation products of fibers are trophic for enterocytes. (TOPPING and ILLMAN 1986; SAVAGE 1978)

Since results indicate that enterocytes (e.g., Caco-2 cells as in vitro models) are actively involved in bacterial translocation, receptor mechanisms have been proposed for binding of bacteria (CRUZ et al. 1994). Interfering with binding could thus pose another useful intraluminal therapeutic approach, for example, using mucin (Cruz et al., unpublished manuscript). Furthermore, secretory IgA produced by B-cells lining the mucosal surfaces binds to bacteria, preventing mucosal invasion by blocking their attachment to epithelial cells (TOMASI 1983). Therefore one might speculate that supplementary IgA may be another means to influence bacterial adherence.

6.2 "Maintenance" of Gut Wall

Probably the best strategy for maintaining the integrity of the gut wall and thus low gut permeability is immediate treatment in the posttraumatic period with both sufficient fluid support and operative treatment when necessary. In addition, a thromboxane A_2 receptor blocker could abolish the increase in mesenteric vascular resistance and subsequent bacterial translocation (TOKYAY et al. 1990).

Similar to thromboxane, angiotensin II, another potent vasoconstrictor (also of the splanchnic circulation), could be important. An angiotensin-converting enzyme inhibitor was found to reduce burn induced bacterial translocation in rodents (JONES et al. 1990a). In addition, the fact that maintenance of gastrointestinal integrity with epidermal growth factor (ZAPATA SIRVENT et al. 1993) and prevention of atrophy of the intestinal mucosa with enteral nutrition (ALEXANDER 1993) decrease the translocation of both micro-organisms and endotoxin also suggests that there is a clear link between thermal injury and septic responses.

Hemorrhage-induced mucosal damage was found to be mediated by xanthine oxidase-derived oxygen free radicals, which cause lipid peroxidation and mucosal injury (DEITCH et al. 1988). In this study the inhibition of xanthine oxidase activity by allopurinol or some other means to block xanthine oxidase significantly reduced mucosal injury and the incidence of bacterial translocation.

An excess of nitric oxide (NO) has been hypothesized as causing the hyporeactivity and decompensation following HS (THIEMERMANN et al. 1933) and contributing to peripheral vascular failure in septic shock (PETROS et al. 1991). In an intestinal ischemia/reperfusion model, PAYNE and KUBES (1993) studied the effect of exogenous NO on mucosal barrier function. NO donors such as nitroprusside, SIN-1, and L-arginine all reduced mucosal barrier dysfunction without improving intestinal flow, while NO synthase inhibitors further augmented the rise in mucosal permeability. More recently have we found that NO induced by HS in rats is an important mediator for the pathophysiological alterations associated with cardiovascular abnormalities, multiple organ injury, and lethality (BAHRAMI et al. 1995b). Our experiments demonstrate that administration of *N*-monomethyl-L-arginine, an inhibitor of both constitutive and inducible NO synthase, may be either deleterious or salutary in a dose-dependent manner. While a high dose markedly increases the organ injury seen in the control group, an intermediate dose reduces the acute injuries without affecting the histological incidence of translocation (YAO et al. 1995a). Thus the regulation of NO and the use of inhibitors may provide new advances for the treatment of HS.

Studies have suggested that the parenteral nutrition regimens currently in use do not support the gut (VAN LEEUWEN et al. 1991; WILMORE et al. 1988). There have therefore been several attempts to apply specific substrates to maintain gut mucosa integrity. Glutamine, for example, has been shown to protect intestinal mucosa even when supplied in a total parenteral nutrition regimen (BURKE et al. 1989). Others have attempted to use specific growth factors, such as insulin-like growth factor 1 (ZEIGLER et al. 1993), which may have trophic effects on the intestinal mucosa. Studies have also been performed in burn models in which specific factors were used to prevent mucosal atrophy and translocation (e.g., bombesin; HASKEL et al. 1994). Epidermal growth factor has also been shown to maintain mucosal integrity (ZEIGLER et al. 1993).

6.3 Intravasal Therapeutic Measures

If translocation cannot be prevented by the above procedures, one must deal with the bacteria/endotoxin within the vascular compartment to prevent an exaggerated host response leading to sepsis-related organ dysfunction. DONOHOE et al. (1986) reported that late mortality (72 h) in their rat hemorrhagic shock model was significantly improved when antibiotics (cefoxitin) were given to the animals in the postshock period. In a thermal injury model the increase in bacterial translocation from the gut and endotoxemia in *P. aeruginsa* burn wound colonized mice was significantly reduced by polymyxin B treatment (DIJKSTRA et al. 1992).

Anti-LPS measures such as BPI and monoclonal antibodies to LPS (WNI 222-5) prevent the elevation of systemic endotoxin levels resulting in reduction of intestinal injury and bacterial translocation, attenuation of damage to vital organs, and improvement in 48-h survival following prolonged hemorrhage (YAO et al. 1995b; BAHRAMI et al. 1995a). In experiments in which rabbits were subjected to HS after having been fed *E. coli*, bacterial and endotoxin translocation was detected. Pretreatment of these animals with an endotoxin antiserum significantly reduced the incidence of multiorgan failure and increased survival rate (YAO et al. 1992b). These results indicate that anti-LPS measures may be useful new therapeutic agents against endogenous bacteria/endotoxin-related disorders in severe HS.

There have been attempts to study the role of proinflammatory mediators in bacterial translocation. In studies by DEITCH's group blocking either platelet-activating factor with two different antagonists (DEITCH et al. 1989) or with an anti-TNF antibody (DEITCH et al. 1991) did not attenuate endotoxin-induced bacterial translocation. In contrast, other studies confirm the important role of TNF in the hemorrhagic setting. For example, it has been shown that anti-TNF antibody even with posttreatment significantly improves survival (BAHRAMI et al. 1994) over 48 h. Also in a pretreatment study anti-TNF antibody improved rat condition in an experiment lasting only 2 h (ZINGARELLI et al. 1994). Further positive effects of TNF blockade have been seen on Kupffer cell function (ERTEL et al. 1991b) and positive effects on cecal ligation and puncture sepsis after a previous hemorrhagic shock (CHAUDRY et al. 1993). Similar to anti-TNF antibodies, agents such as chloroquine which reduce TNF formation have been demonstrated to be beneficial when administered during resuscitation following shock (ERTEL et al. 1991a). Furthermore, studies by MARZI et al. (1993, 1995) have indicated that anti-TNF antibody reduces leukocyte adhesion and improves microcirculatory conditions, as seen in significantly improved negative base excess. In rabbit hemorrhagic shock anti-TNF antibody reduces postshock organ damage (FOULKES et al. 1994).

When comparing different therapeutical regimens in experimental models, it should be kept in mind that these models differ, for example, in severity of hemorrhagic shock, with a short hemorrhagic period in most of the experiments by DEITCH et al. (1987a) or CHAUDRY et al. (1993) and very long experiments (up to

250 min) in the shock models of Koziol et al. (1988), Rush (1989), and Schlag et al. (1991); our rat model (Bahrami et al. 1995a) is in between these (either for 90 or 180 min). Although bacterial translocation occurs in most of these models, the mechanisms may not be identical. The ischemia dependent tissue hypoxia is probably much more prominent in the prolonged shock models (Rush et al. 1988; Bahrami et al. 1995a) than in the 30-min model (Deitch et al. 1987a).

7 Conclusion

In all recent experimental studies the common theme is that bacterial translocation is clearly evident, not only by less reliable diagnostic techniques but especially in results of therapeutic studies. The therapeutic approach also proves the importance of the translocation event for organ dysfunction and survival, at least experimentally. The lack of convincing clinical data is due to the fact that only studies with diagnostic approaches have been performed so far. In addition to bacterial translocation, current data indicate the important role of secondary mediators such as TNF and NO in the pathogenesis of postshock/postburn sepsis.

References

Alexander JW (1993) Prevention of bacterial translocation with early enteral feeding – a feasible approach? In: Faist E, Meakins J, Schildberg FW (eds) Host defense dysfunction in trauma, shock and sepsis: mechanisms and therapeutic approaches. Springer, Berlin, Heidelberg New York, pp 903–909

Alexander JW, Boyce ST, Babcock GF, Gianotti L, Peck MD, Dunn DL, Pyles T, Childress CP, Ash SK (1990) The process of microbial translocation. Ann Surg 212: 496–510

Alexander JW, Gianotti L, Pyles T, Carey MA, Babcock GF (1991) Distribution and survival of *Escherichia coli* translocating from the intestine after thermal injury. Ann Surg 213: 558–566

Ayala A, Perrin MM, Meldrum DR, Ertel W, Chaudry IH (1990) Hemorrhage induces an increase in serum TNF which is not associated with elevated levels of endotoxins. Cytokine 2: 170–174

Ayala A, Wang P, Ba ZF, Perrin MM, Ertel W, Chaudry IH (1991) Differential alterations in plasma IL-6 and TNF levels after trauma and hemorrhage. Am J Physiol 260: R167–R171

Bahrami S, Yao YM, Leichtfried G, Redl H, Schlag G, Foulkes R (1994) Efficacy of monoclonal antibody (mab) to tumor necrosis factor (TNF) against hemorrhage-induced mortality in rats. Intensive Care Med 20: S61

Bahrami S, Schlag G, Yao YM, Redl H (1995a) Significance of translocation/endotoxin in the development of systemic sepsis following trauma and/or hemorrhage. In: Levin J, Alving CR, Munford RS et al. (eds) Bacterial endotoxins: lipopolysaccharides from genes to therapy. John Wiley, New York, 197–208

Bahrami S, Schlag G, Yao YM, Redl H (1995b) Involvement of endotoxin, tumour necrosis factor, and nitric oxide in haemorrhagic shock-related alterations. In: Schlag G, Redl H (eds) Shock, sepsis, and organ failure – nitric oxide. Springer, Berlin Heidelberg New York, pp 59–76 (Fourth Wiggers Bernard Conference 1994)

Baker JW, Deitch EA, Li M, Berg RD, Specian RD (1988) Hemorrhagic shock induces bacterial translocation from the gut. J Trauma 28: 896–906

Baue AE (1975) Multiple, progressive, or sequential systems failure: a syndrome of the 1970's. Arch Surg 110: 779–781

Bertok L (1977) Physico-chemical defense of vertebrate organisms: the role of bile acids in defense against bacterial endotoxins. Perspect Biol Med 21: 70–76

Brathwaite CEM, Ross SE, Nagele R, Mure JA, O'Malley KF, Garcia-Perez FA (1993) Bacterial translocation occurs in humans after traumatic injury: evidence using immunofluorescence. J Trauma 34: 586–589

Burke DJ, Alverdy JC, Aoys E, Moss GS (1989) Glutamine-supplemented total parenteral nutrition improves gut immune function. Arch Surg 124: 1396–1399

Cabie A, Farkas JC, Fitting C, Laurian C, Cormier JM, Carlet J, Cavaillon JM (1993) High levels of portal TNF-a during abdominal aortic surgery in man. Cytokine 5: 448–453

Cahill CJ, Pain JA, Bailey ME (1987) Bile salts, endotoxin and renal function in obstructive jaundice. Surg Gynecol Obstet 165: 519–522

Carrico CJ, Meakins JL, Marshall JC, Fry D, Maier RV (1986) Multiple-organ-failure syndrome. Arch Surg 121: 196–208

Carter EA, Tompkins RG, Schiffrin E, Burke JF (1990) Cutaneous thermal injury alters macromolecular permeability of rat small intestine. Surgery 107: 335–341

Caty MG, Guice KS, Oldham KT, Remick DG, Kunkel SI (1990) Evidence for tumor necrosis factor-induced pulmonary microvascular injury after intestinal ischemia-reperfusion injury. Ann Surg 212: 694–700

Chaudry IH, Ayala A (1993) Mechanism of increased susceptibility to infection following hemorrhage. Am J Surg 165: 59S–67S

Chaudry IH, Ertel W, Ayala A (1993) Alterations in inflammatory cytokine production following hemorrhage and resuscitation. In: Schlag G, Redl H, Traber DL (eds) Cytokine network. Springer, Berlin Heidelberg New York, pp 72–127 (Third Wiggers Bernard Conference)

Cruz N, Lu Q, Berg R, Deitch EA (1994) The Caco-2 cell monolayer system as an in vitro model for studying bacterial-enterocyte interactions and bacterial translocation. J Burn Care Rehabil 15: 207–212

Culbertson WR, Elstun W, Cole W, Altemeier WA (1959) Bacterial studies in irreversible hemorrhagic shock. Arch Surg 79: 185–189

Deitch EA (1989) Simple intestinal obstruction causes bacterial translocation in man. Arch Surg 124: 699–701

Deitch EA (1990a) Intestinal permeabilities increased in burn patients shortly after injuries. Surgery 107: 411–416

Deitch EA (1990b) The role of intestinal barrier failure and bacterial translocation in the development of systemic infection and multiple organ failure. Arch Surg 125: 403–404

Deitch EA, Berg RD (1987a) Endotoxin but not malnutrition promotes bacterial translocation of the gut flora in burned mice. J Trauma 27: 161–166

Deitch EA, Berg RD (1987b) Bacterial translocation from the gut: a mechanism of infection. J Burn Care Rehabil 8: 475–482

Deitch EA, Bridges RM (1987) Effects of stress and trauma on bacterial translocation from the gut. J Surg Res 42: 536–542

Deitch EA, Maejima K, Berg R (1985) Effect of oral antibiotics and bacterial overgrowth on the translocation of the GI tract microflora in burned rats. J Trauma 25: 385–392

Deitch EA, Baker T, Berg R, Ma L (1987a) Hemorrhagic shock promotes the systemic translocation of bacteria from the gut. J Trauma 27: 815

Deitch EA, Berg R, Specian R (1987b) Endotoxin promotes the translocation of bacteria from the gut. Arch Surg 122: 185–190

Deitch EA, Winterton J, Berg R (1987c) Effect of starvation, malnutrition, and trauma on the gastrointestinal tract flora and bacterial translocation. Arch Surg 122: 1019–1024

Deitch EA, Winterton J, Li M, Berg R (1987d) The gut as a portal of entry for bacteremia. Role of protein malnutrition. Ann Surg 205: 681–692

Deitch EA, Bridges W, Baker J, Ma JW, Ma L, Grisham MB, Granger DN, Specian RD, Berg R (1988) Hemorrhagic shock-induced bacterial translocation is reduced by xanthine oxidase inhibition or inactivation. Surgery 104: 191–198

Deitch EA, Ma L, Ma WJ, Grisham MB, Granger DN, Specian RD, Berg RD (1989) Inhibition of endotoxin-induced bacterial translocation in mice. J Clin Invest 84: 36–42

Deitch EA, Specian RD, Berg RD (1991) Endotoxin-induced bacterial translocation and mucosal permeability: role of xanthine oxidase, complement activation, and macrophage products. Crit Care Med 19: 785–791

Deitch EA, Dazhong X, Franko L, Ayala A, Chaudry IH (1994) Evidence favoring the role of the gut as a cytokine-generating organ in rats subjected to hemorrhagic shock. Shock 1: 141–146
Di Padova FE, Brade H, Barclay GR, Poxton IR, Liehl E, Schuetze E, Kocher HP, Ramsay G, Schreier MH, McClelland DB et al. (1993) A broadly cross-protective monoclonal antibody binding to *Escherichia coli* and *Salmonella* lipopolysaccharides. Infect Immun 61: 3863–3872
Dijkstra HM, Manson WL, Klasen HJ, Van der Waaij D (1992) Effect of polymyxin B on intestinal bacterial translocation in *Pseudomonas aeruginosa* wound-colonized burned mice. Eur Surg Res 24: 69–76
Dobke MK, Simoni J, Ninnemann JL, Garrett J, Harnar TJ (1989) Endotoxemia after burn injury: effect of early excision on circulating endotoxin levels. J Burn Care Rehabil 10: 107–111
Donohoe MK, Rush BFJ, Koziol JM, Machiedo GW (1986) Role of antibiotics in late survival from hemorrhagic shock. Surg Forum 27: 62–64
Endo S, Inada K, Kikuchi M, Yamada Y, Kasoi T, Kondo M, Hoshi S, Yamashita H, Suzuki M, Yoshida M (1992) Are plasma endotoxin levels related to burn size and prognosis? Burns 18: 486–489
Ertel W, Morrison MH, Ayala A, Chaudry IH (1991a) Chloroquine attenuates hemorrhagic shock-induced suppression of Kupffer cell antigen presentation and major histocompatibility complex class II antigen expression through blockade of tumor necrosis factor and prostaglandin release. Blood 78: 1781–1788
Ertel W, Morrison MH, Ayala A, Perrin MM, Chaudry IH (1991b) Anti-TNF monoclonal antibodies prevent hemorrhage-induced suppression of Kupffer cell antigen presentation and MHC class II antigen expression. Immunology 74: 290–297
Faist E, Baue AE, Dittmer H, Heberer G (1983) multiple organ failure in polytrauma patients J Trauma 23: 775–787
Fine J, Frank E, Rutenburg S, Schweinberg F (1959) The bacterial factor in traumatic shock. N Engl J Med 260: 214–220
Foulkes R, Hughes B, Kingaby R, Woodger R, Vetterlein O (1994) Anti-TNF treatment reduces severity of organ damage in a concious rabbit model of hemorrhagic/traumatic shock. Intensive Care Med 20: 72
Fry DE, Pearlstein L, Fulton RL, Polk HC (1980) Multiple system organ failure: the role of uncontrolled infection. Arch Surg 115: 136–140
Gans H, Matsumoto K (1974) The escape of endotoxin from the intestine. Surg Gynecol Obstet 139: 395–402
Garidis DT, Ischigama M, Woodruff PW, Fine J (1973) Role of intestinal flora in clearance and detoxification of circulating endotoxin. J Reticulo endothel Soc 14: 513–521
Garrison RN, Fry DE, Berberich S, Polk HC (1982) Enterococcal bacteremia: clinical implications and determinants of death. Ann Surg 196: 43–47
Goris RJ, te Boekhorst TP, Nuytinck JK, Gimbrere JS (1985) Multiple-organ failure. Generalized autodestructive inflammation? Arch Surg 120: 1109–1115
Haglund U (1993) Hypoxic damage. In: Schlag G, Redl H (eds) Pathophysiology of shock, sepsis, and organ failure. Springer, Berlin Heidelberg New York, pp 314–321
Hammond JM, Potgieter PD, Saunders GL, Forder AA (1992) Double-blind study of selective decontamination of the digestive tract in intensive care [see comments]. Lancet 340: 5–9
Haskel y, Dazhong X, Lu Q, Deitch EA (1994) Bombesin protects against bacterial translocation induced by three commercially available liquid enteral diets: a prospective, randomized, multigroup trial. Crit Care Med 22: 108–113
Jiang JX, Bahrami S, Leichtfried G, Redl H, Oehlinger W, Schlag G (1995) Kinetics of endotoxin and tumor necrosis factor appearance in portal and systemic circulation following hemorrhagic shock in rats. Ann Surg 221: 100–106
Jones WG, Minei JP, Barber AE et al. (1990a) Angiotensin converting enzyme inhibitor decreases bacterial translocation after burn injury. FASEB 9: A953
Jones WG, Minei JP, Barber AE, Rayburn JL, Fahey TJ, Shires GT, Shires GT (1990b) Bacterial translocation and intestinal atrophy after thermal injury and burn wound sepsis. Ann Surg 211: 399–405
Koziol JM, Rush BF Jr, Smith SM, Machiedo GW (1988) Occurrence of bacteremia during and after hemorrhagic shock. J Trauma 28: 10–16
Liehr H, Englisch G, Rasenack U (1980) Lactulose-a drug with anti-endotoxin effect. Hepatogastroenterology 27: 356–360
Machnicki M, Zimecki M, Zagulski T (1993) Lactoferrin regulates the release of tumor necrosis factor alpha and interleukin 6 in vivo. Int J Exp Pathol 74: 433–439

Maejima K, Deitch EA, Berg RD (1984) Bacterial translocation from the gastrointestinal tracts of rats receiving thermal injury. Infect Immun 43: 6–10

Marshall JC, Christou NV, Meakins JL (1993) The gastrointestinal tract. The "undrained abscess" of multiple organ failure [see comments]. Ann Surg 218: 111–119

Marzi I, Bauer M, Secci A, Bahrami S, Redl H, Schlag G (1995) Effect of anti-tumor necrosis factor-alpha on lekocyte adhesion in the liver after hemorrhagic shock: an intravital microscopy study in the rat. Shock 3: 27–33

Marzi M, Bauer M, Reisdorf E, Bahrami S, Redl H, Buehren V (1993) TNFa and PAF modulate leukocyte adhesion in the liver after hemorrhagic shock in vivo. Circ Shock Suppl 1: 23

Moore F, Poggetti R, McAnena O, Peterson V, Abernathy C, Parsons P (1991) Gut bacterial translocation via the portal vein: a clinical perspective with major torso trauma. J Trauma 31: 629–638

Morris SE, Navaratnam N, Herndon DN (1990) A Comparison of effects of thermal injury and smoke inhalation on bacterial translocation. J Trauma 30: 639–643

Munster AM, Smith Meek M, Dickerson C, Winchurch RA (1993) Translocation. Incidental phenomenon or true pathology? Ann Surg 218: 321–326

Navaratnam N, Morris SE, Townsend MC, Traber DL, Traber LD, Herndon DN (1989) Bacterial translocation and selective mesenteric artery perfusion with nitroprusside in an ovine model. Int J Radiat Oncol Biol Phys 21: 240

Nitsche D, Schulz C, Nebermann L (1993) Prevention of gut-derived endotoxemia by means of oral administration of lactorferrin. Circ Shock Suppl 1: 47–48

Payne D, Kubes P (1993) Nitric oxide donors reduce the rise in reperfusion-induced intestinal mucosal permeability. Am J Physiol 265: 189–195

Peitzman A, Udekwu A, Ochoa J, Smith S (1991) Bacterial translocation in trauma patients. J Trauma 31: 1083–1087

Petros A, Bennett D, Vallance P (1991) Effect of nitric oxide synthase inhibitors on hypotension in patients with septic shock [see comments]. Lancet 338: 1557–1558

Plaut AG, Goorbach SL, Nahas L, Weinstein L, Spauknebel G, Levitan R (1967) Studies of intestinal microflora. III the microbial flora of intestinal mucosa in fluids. Gastroenterology 53: 868–873

Polk HC, Shields CL (1973) Remote organ failure: a valid sign of occult intra-abdominal infection. Surgery 81: 310–313

Redl H, Schlag G, Dinges HP, Bahrami S, Buurman WA, Schade U, Ceska M (1991) Trauma and sepsis induced activation of granulocytes, monocytes/macrophages and endothelial cells in primates. In: Schlag G, Redl H, Siegel JH et al. (eds) Shock, sepsis, and organ failure Springer, Berlin Heidelberg New York, pp 297–313 (Second Wiggers Bernard Conference)

Rhee P, Waxman K, Clark L, Kaupke CJ, Vaziri ND, Tominaga G, Scannell G (1993) Tumor necrosis factor and monocytes are released during hemorrhagic shock. Resuscitation 25: 249–255

Roumen RM, Hendriks T, Wevers RA, Goris JA (1993) Intestinal permeability after severe trauma and hemorrhagic shock is increased without relation to septic complications. Arch Surg 128: 453–457

Rush BFJ (1989) Irreversibility in hemorrhagic shock is caused by sepsis. Ann Surg 55: 204–208

Rush BFJ, Sori AJ, Murphy TF, Smith S, Flanagan JJ, Machiedo GW (1988) Endotoxemia and bacteremia during hemorrhagic shock: the link between trauma and sepsis? Ann Surg 207: 549–554

Rush BFJ, Redan JA, Flanagan JJ, Heneghan JB, Hsieh J, Murphy TF, Smith S, Machiedo GW (1989) Does the bacteremia observed in hemorrhagic shock have clinical significance? A study in germfree animals. Ann Surg 210: 342–347

Ryan CM, Yarmush ML, Burke JF, Tompkins RG (1992) Increased gut permeability early after burns correlates with the extent of burn injury. Crit Care Med 20: 1508–1512

Sanderson PJ (1989) Selective decontamination of the digestive tract. Br Med J 299: 1413–1414

Savage DC (1978) Factors involved in the colonization of the gut epithelial surface. Am J Clin Nutr 31: 131–135

Scannell G, Waxman K, Kaml GJ, Ioli G, Gatanaga T, Yamamoto R, Granger GA (1993) Hypoxia induces a human macrophage cell line to release tumor necrosis factor-alpha and its soluble receptors in vitro. J Surg Res 54: 281–285

Schlag G, Redl H, Dinges HP, Radmore K (1991) Bacterial translocation in a baboon model of hypovolemic-traumatic shock. In: Schlag G, Redl H, Siegel JH et al. (eds) shock, sepsis and organ failure. Springer, Berlin Heidelberg New York, pp 53–90 (Second Wiggers Bernard Conference)

Schlag G, Redl H, Bahrami S, Davies J, Smuts P, Marzi I (1993a) Trauma and cytokines. In: Schlag G, Redl H, Traber DL (eds) Shock, sepsis organ failure. Springer, Berlin Heidelberg New York pp 128–155

Schlag G, Redl H, Khakpour Z, Davies J, Pretorius J (1993b) Hypovolemic-traumatic shock models in baboons. In: Schlag G, Redl H (eds) Pathophysiology of shock, sepsis, and organ failure. Springer, Berlin Heidelberg New York pp 384–402

Sheng ZY, Yang HM (1994) The concept and the diagnosis of multiple systems organ failure Chin Med J 107: 563–569

Shenkar R, Abraham E (1993) Effects of hemorrhage on cytokine gene transcription Lymphokine Cytokine Res 12: 237–247

Spaeth G, Specian RD, Berg RD, Deitch EA (1990) Food without fiber promotes bacterial translocation from the gut. Surgery 108: 240–247

Stoutenbeek CP, Saene van HKF, Miranda DR et al. (1984) The effect of selective contamination of the digestive tract on colonization and infection rate in multiple trauma patients. Intensive Care Med 10: 185–192

Stylianos S, Wakabayashi G, Gelfand JA, Harris BH (1991) Experimental hemorrhage and blunt trauma do not increase circulating tumor necrosis factor. J Trauma 31: 1063–1067

Thiemermann C, Szabo C, Mitchell JA, Vane JR (1993) Vascular hyporeactivity to vasoconstrictor agents and hemodynamic decompensation in hemorrhagic shock is mediated by nitric oxide. Proc Natl Acad Sci USA 90: 267–271

Tokyay R, Traber DL, Herndon DN (1990) Thromboxane synthese inhibition prevents the increased mesenteric vascular resistance seen after major thermal injury in a chronic porcine model. Surg Forum 41: 56–59

Tomasi TB (1983) Mechanisms of immune regulation at mucosal surfaces.Rev Infect Dis 5 Suppl 4: S784–S792

Topping DL, Illman RJ (1986) Bacterial fermentation in the human large bowel. Time to change from the roughage model of dietary fibre? Med J Aust 144: 307–309

Van Leeuwen PAM, Hong RW, Rounds JD, Roderick ML, Wilmore D (1991) Hepatic failure and coma after liver resection is reversed by manipulation of gut contents: the role of endotoxin. Surgery 110: 169–175

Wilmore DW, Smith RJ, O'Dwyer ST, Jacobs DO, Ziegler TR, Wang XD (1988) The gut: a central organ after surgical stress. Surgery 104: 917–923

Winchurch RA, Thupari JN, Munster AM (1987) Endotoxemia in burn patients: levels of circulating endotoxins are related to burn size. Surgery 102: 808–812

Wolochow H, Hildebrand GJ, Lammanna C (1966) Translocation of microorganisms across the intestinal wall in rats: effect of microbial size and concentration. J Infect Dis 116: 523–528

Yao Y, Tian H, Sheng Z, Wang Y, Shi Z, Lan F (1992a) The role of endotoxin in the pathogenesis of experimental multiple system organ failure a preliminary report. Chin Med Sci J 7 : 161–165

Yao Y, Tian H, Wang Y, Sheng Z, Xu S (1992b) Protective effect of re-LPS antiserum on experimental multiple system organ failure. Chin Med J 105: 833-838

Yao YM, Yu Y, Wang YP, Lu LR, Tian HM, Sheng ZY (1994) Selective decontamination of the digestive tract (SDD) prevents bacterial/endotoxin translocation and mortality in thermal-injured rats. In: Mutz NJ, Koller W, Benzer H (eds) 7th European congress on intensive care medicine. Monduzzi, Bologna, pp 1097–1102

Yao YM, Bahrami S, Leichtfried G, Redl H, Schlag G (1995a) Effects of NG-monomethyl-L-arginine (L-NMMA) on hemodynamics and mortality in rats following hemorrhagic shock. Shock 3 Suppl: 71

Yao YM, Bahrami S, Redl H, Schlag G (1995b) Pathogenesis of hemorrhage/induced endotoxemia in rats: effects of recombinant bactericidal/increasing protein ($rBPI_{21}$). Ann Surg 221: 398–405

Yao YM, Sheng ZY, Tian HM, Yu Y, Wang YP, Yang HM, Guo ZR, Gao WY (1995c) The association of circulating endotoxemia with the development of multiple organ failure in burned patients. Burns (in press)

Zapata Sirvent RL, Hansbrough JF, Wolf P, Grayson LS, Nicolson M (1993) Epidermal growth factor limits structural alterations in gastronintestinal tissues and decreases bacterial translocation in burned mice. Surgery 113: 564–573

Zeigler ST, Traber DL, Herndon D (1993) Bacterial translocation in burns. In: Schlag G, Redl H (eds) Pathophysiology of shock, sepsis, and organ failure. Springer, Berlin Heidelberg New York, pp 300–313

Ziegler TR, Smith RJ, O'Dwyer ST, Demling RH, Wilmore DW (1988) Increased intestinal permeability associated with infection in burn patients. Arch Surg 123: 1313–1319

Zingarelli B, Squadrito F, Altavilla D, Calapai G, Di Rosa M, Caputi Ap (1994) Role of tumor necrosis factor-a in acute hypovolemic hemorrhagic shock in rats. Am J Physiol 266: H1512–H1515

The Mechanisms of Host Defense Dysfunction Following Shock and Trauma

E. Faist

1 Introduction 259

2 Immunological Derangements of Mononuclear Leukocytes Following Injury 261
2.1 T-Cells 261
2.2 B-Cells 263
2.3 Monocytes 263

3 Downregulation of the Monocytic Inflammatory Potential 264

4 Trauma-Induced Impairment of Nonspecific Immune Functions 267

5 Is There a Need for Posttraumatic Therapeutic Immunointervention? 269

References 271

1 Introduction

Major injury is a profound threat to the individual. Both the extent of injuries and the preinjury health of the patient greatly influence the outcome. In our experience with individuals having an injury severity score (ISS) of 30 points or higher who survive the immediate injury phase, only 50% begin convalescence without inflammatory or infectious complications (Faist et al. 1993). Thus in every second patient the metabolic and neuroendocrine stress reactions do not shut down and return to normal regulation and homeostasis. The causes of this phenomenon in multisystem injuries are extensive tissue necrosis, prolonged hemorrhagic shock, and perhaps immediate invasive endotoxemia by way of gut translocation.

Severe injury results in a profound dysfunction of host defense mechanisms. This we see as the dire consequence of a systemic, nondiscriminant, and excessive inflammatory response, together with a dramatic paralysis of cell-mediated immune function. We have learned that parts of the immune system are stimulated while others are depressed in a complex order of events that is as yet not well understood. On the one hand, the nonspecific immune system is

Ludwig-Maximilian University of Munich, Department of Surgery, Marchioninistr. 15, 81377 München, Germany activated as white blood cells and macrophages (Mϕ) mobilize to the site of injury, with the activation of complement and opsonins. Interleukin (IL)-1 should stimulate IL-2, which activates T-helper cells. Acute-phase proteins are produced as a part of the immune response with IL-6 production. On the other hand, shock depresses the activity of white blood cells and lung and liver Mϕ). Catabolism with protein-calorie malnutrition alters immunoglobulin (Ig) production and depresses both T- and B-cell function. Cortisol decreases immune responses, as do some anesthetic agents. Circulating immunosuppressive factors, including suppressor T-cells, appear because of tissue necrosis. With tissue injury, protein denaturation may occur, which could be interpreted as the presence of nonself protein. With a burn injury neutrophils (PMN) are activated systematically, and chemotaxis is depressed. All these factors contribute to increased susceptibility to infection. The immune system can cope with a modest injury. However, when the injury is severe and overwhelming, with extensive tissue necrosis, the immune system is activated systematically and may become self-destructive (Baue 1990). A response which should be local in the area of the wound and critical for survival may become generalized and produce remote organ damage (Baue 1992).

Infection following an operation may result in further insult to the delicate balance of cytokine regulation. Since trauma induces immunosuppression as well as hyperactivation, immunomodulation should not be limited to restoration of depressed immune responses but should also include the regulation of exaggerated responses where appropriate. Support of host resistance is becoming very important (Fig. 1). An incentive for the development of therapeutic regimens comes from the recognition that infection is a major contributor to late morbidity and mortality after injury (second-phase multiple-organ failure, MOF; Faist et al. 1983). Such therapy is based on studies of the downregulatory mechanisms responsible for injury-related immune problems.

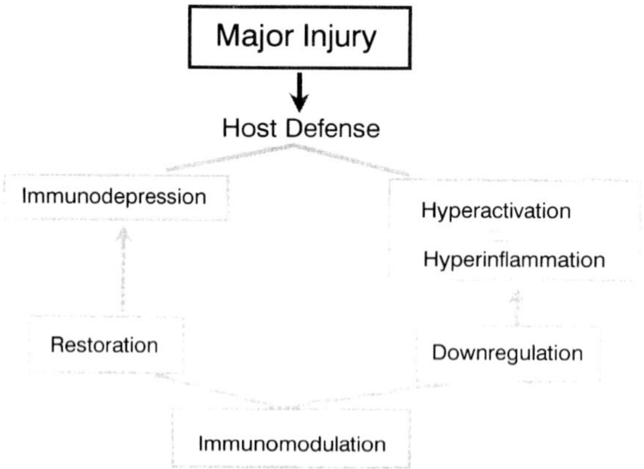

Fig. 1. Components of posttraumatic host defense dysfunction and counterregulatory immunomodulation

2 Immunological Derangements of Mononuclear Leukocytes Following Injury

There is a wide array of immunological derangements which occur after injury, shock, trauma, burns, and extensive surgical procedures. These are induced mainly by Mϕ activation, a significant inhibition of the bone marrow, and changes in the specific and nonspecific immune system (Chaudry et al. 1990; Munster 1984; Ninnemann 1987; Mannick 1993). The specific immune functions which are altered following hemorrhage and tissue trauma are:

Myelodepression
Lymphopenia
CD4/CD8 ratio <1
T- and B-cell proliferation ↓
Release of lymphokines (IL-2, IL-3, γ-IFN) ↓
IL-2 receptor expression ↓
NK cell activity ↓
Alteration of Mϕ/T-cell interaction
HLA-DR antigen expression ↓
DTH skin test reactivity ↓
Inhibition of B-lymphocyte differentiation
Plasma cells: IgM-synthesis and secretion ↓
Shift from IgM ⟶ IgG Synthesis

2.1 T-Cells

Following major injury one often observes absolute lymphopenia ($CD3^+$ lymphocytes) and simultaneous monocytosis ($CD14^+$ cells). Analysis of T-lymphocyte subpopulations demonstrates a massive decline in $CD4^+$ T-helper cells while IL-2 receptor expression on the surface of lymphocytes is also reduced (Antonacci et al. 1984, O'Mahoney et al. 1984, Faist et al. 1987, Ertel and Faist 1989). The $CD8^+$ cytotoxic and suppressor-active T-cells that play a central role within the posttraumatic immunosuppression are found to be unchanged or elevated. This leads to a conversion of the CD4/CD8 ratio to a value below 1. These alterations have been observed until day 21 after injury (Faist et al. 1986). Shock, trauma, and burns also produce massive inhibition of lymphocyte proliferation from mitogenic stimulation. Immunomodulatory in vitro studies clearly indicate that excessive release of immunoreactive prostaglandin E_2(PGE_2) is a major lymphoproliferative inhibitory factor (Ertel et al. 1992; Rodrick et al. 1992). Faist et al. (1986) demonstrated in 11 patients with major trauma that lymphoblastogenesis was impaired in more than 70%, and 8 patients had a major infectious episode. Keane et al. (1983) reported a correlation between the degree and duration of inhibition of lymphoproliferation following trauma and the incidence of a "sepsis syndrome." Comparable relationships have been described for lymphokine synthesis. Impairment of IL-2 biosynthesis is the most striking and persistent abnormality in the wide spectrum of defective immune responses. Inhibition of IL-2 production occurs instantaneously after injury and persists for a long time (Faist et al. 1986). Wood et al. (1984), investigating 23 patients with major burns, observed in 9 patients with major sepsis that the invitro mitogen-induced IL-2 release was significantly lower than in patients with an uneventful clinical course. Similar results have also been reported for patients in hemorrhagic shock and following major surgical procedures (Stephan et al. 1986). In addition to the impairment of IL-2 secretion, the suppression of IL-3 (Meldrum et al. 1991) and γ-interferon (γIFN; Faist et al. 1988; Livingston et al. 1988) synthesis from lymphocytes (splenocytes) has been described. Quattrocchi et al. (1991) showed that isolated brain injury results in massive inhibition of lymphoblastogenesis and IL-2 receptor expression.

In addition to the inhibition of IL-2 receptor expression, there is a striking decrease in expression of human leukocyte antigen (HLA) DR, a class II major histocompatibility complex antigen on Mϕ following injury (Ayala et al. 1990; Cheadle et al. 1989; Gibbons et al. 1989). HLA-DR was depressed most markedly in patients who developed sepsis or died a septic death compared with patients who had uneventful recoveries from trauma (Polk et al. 1986). When γIFN was added to peripheral mononuclear leukocyte cultures obtained from these patients, there was an increase in HLA-DR antigen density of expression on Mϕ. The increase and density of HLA-DR antigen expression did return to the normal range in controls, but there was a positive response to the γIFN treatment (Hershman et al. 1989).

An effective means of global assessment of the host defense response is the delayed-type hypersensitivity (DTH) response to ubiquitous antigens. This is an easily performed and objective measurement in trauma patients. The response depends on prior sensitization where a nonimmune lymphocyte having antigen presented on the surface of Mϕ becomes an immune lymphocyte. The second step involves recognition, where an immune lymphocyte encounters the antigen injected into the skin and then sets up a cytokine and lymphokine response of local nonspecific inflammation with detection of a "bump" in the skin. Major credit to the development of the DTH skin test response method and its clinical application belongs to the group of MacLean (Christou et al. 1980). These investigators demonstrated that the incidence of anergy following trauma increases with the severity of the injury. Patients with an ISS higher than 25 are usually anergic upon admission, and this can last up to 40 days depending on the degree of injury. Anergy varied from 3 to 4 days with an ISS of 10 or less, to more than 40 days with an ISS of 40 or greater. The clinical implication of anergy is that infection occurs more readily in anergic patients than in reactive patients after trauma. Also, a higher percentage of infected anergic patients die than infected reactive patients. Christou and Tellado (1993) postulate that patients with an intact DTH skin response after trauma respond appropriately to their infection, do not develop multiple organ failure, and survive, whereas the anergic patients either fail to control their infections, have dissemination of the infection and death, or "overreact" to their infection or the signals triggered by their infection and develop multiple-organ failure and die.

2.2 B-Cells

The effects of trauma on the B-lymphocyte system have also been investigated extensively. Nohr et al. (1983, 1984) studied the in vitro and in vivo antibody response to tetanus toxoid following major surgery and found a deficit of IgG production which was correlated with DTH skin test responses. In contrast, Shorr et al. (1984) found a normal in vivo antibody response to a rechallenge with tetanus toxoid in burned patients. However, lymphocyte in vitro cultures showed that burned patients had an elevated spontaneous Ig production with depressed antigen response. Wood et al. (1986) and Teodorczyk-Injeyan et al. (1986) demonstrated in burn patients a reduced in vivo IgG production following tetanus immunization and increased levels of in vitro IgG production following pokeweed mitogen stimulation. In patients with major trauma we found massive suppression of B-cell differentiation and IgM synthesis in peripheral blood mononuclear cell cultures following pokeweed mitogen stimulation (Ertel and Faist 1989; Faist et al. 1989). The degree of suppression was highly correlated with the amount of Mϕ suppression in the peripheral blood mononuclear cell cultures while IgG and IgA synthesis were within the normal range or elevated. Further studies in purified B-cell cultures confirmed these data, demonstrating a persistent IgM/IgG shift. Investigations of the influence of recombinant lymphokines on Ig synthesis showed only a partial restoration of depressed B-cell function by IL-2. Very recently we showed that an anti-μ-induced B-cell proliferation per se is not suppressed by mechanical trauma after elimination of suppressor active extracellular factors (Ertel et al. 1990). Trauma caused a highly elevated antigen-induced proliferation, but the capacity of IL-2, IL-4, or T-cell supernatants to increase proliferation was depressed. We concluded that mechanical trauma causes a high preactivation of B-lymphocytes but impairs the responsiveness to most lymphokines except γIFN.

2.3 Monocytes

There is compelling evidence that altered behavior patterns of activated Mϕ play a key role in the abnormal inflammatory and cell-mediated immune responses. Shock, trauma, and burns similar to endotoxemia and bacteremia induce profound activation of monocyte/Mϕ with elevated synthesis and secretion of inflammatory mediators. Hemorrhagic shock without significant tissue trauma (Ertel et al. 1991) results in a steep elevation of tumor necrosis factor-α (TNFα) plasma levels similar to those detected during septic states (Marks et al. 1990; Michie et al. 1988). Ertel et al. (1991) demonstrated enhanced secretion of TNFα and IL-6 from peritoneal Mφ as well as from Kupffer cells TAKAYAMA et al. (1990) described an accelerated release of these mediators in Mφ cultures from patients following major surgery. The amount of TNFα secretion (>30 ng/10^6 Mφ) was correlated with the onset of septic complications. IL-6 plasma levels in patients with major burns were shown by GUO et al. (1989) to be correlated with posttraumatic infections and outcome. In contrast to many of these studies, we found different results by careful study of Mφ behavior in patients following major trauma and during the sepsis syndrome. In many patients there is an initially marked Mφ suppression to synthesize and secrete proinflammatory cytokine to endotoxin stimuli. There ia a profound deficit of IL-1 and IL-8 production for several days after injury (FAIST et al. 1992; ZIMMER et al. 1994). This deficit may well result in the immunodeficiency of traumatized and septic patients since these cytokines in low concentrations are involved in the upregulation of all essential humoral and cellular immune functions. Inadequate IL-1 synthesis appears to be a substantial component of defective IL-2 synthesis following trauma (FAIST et al. 1987; MILLER-GRAZIANO et al. 1988). A crucial immunomechanistic consequence of Mφ PGE_2 release after injury is its downregulation of IL-2 synthesis (RAPPAPORT and DODGE 1982).

3 Downregulation of the Monocytic Inflammatory Potential

Within the complex regulatory mechanisms of cytokine synthesis, as far as they are presently understood, a number of naturally occurring negative feedback mechanisms exist that appear able to downregulate an activated Mφ system. Modified mimicking of these regulatory loops via exogenous delivery of such a broad spectrum mediator might represent a new avenue to regain early control over injury-induced host defense catastrophes. While endotoxin, TNFα, and γIFN are most potent inducers of Mφ activation with subsequent release of TNFα, IL-1, and PGE_2; PGE_2 turn is capable of reducing TNFα secretion via the activated Mφ so as to form a counter-regulatory system (RENZ et al. 1988). Transforming growth factors-beta 1 and 2 are interesting molecules capable of deactivating Mφ, mainly by inhibiting its H_2O_2 production; however, it has been found that the crucial characteristics of two other cytokines. IL-4 and IL-10, are their ability to inhibit the Mφ production of numerous inflammatory mediators (TSUNAWAKI et al. 1988; ERTEL et al. 1993, DE WAAL MALEFYT et al. 1991; HOWARD and O'GARRA 1992; SPITS and DE WAAL HALEFYT et al. 1992; OSWALD et al. 1992). Another lymphokine, IL-13, has been described to have potential anti-inflammatory activity, as judged by its capacity to suppress the production of proinflammatory mediators such as TNFα and IL-6 and to upregulate granulocyte and Mf IL-1ra production (MINTY et al. 1993; DE WAAL MALEFYT et al. 1993; MUZIO et al. 1994). IL-13 is produced from the activated T helper 2 cell subset which is also able to produce IL-4, IL-5, and IL-6 cytokines which enable T-cells to provide help to B lymphocytes for immunoglobulin synthesis in humoral immune responses (DEFRANCE et al. 1994; MCKENZIE et al. 1992). IL-13 has several biological activities in common with IL-4, and there is accumulating evidence that IL-13 and IL-4 receptors share components which are important for signal transduction (ZURAWSKI and DE VRIES 1994).

In a most recent study we examined the potential of IL-13 to modify in vitro LPS induced Mϕ activity in human cells from individuals that had sustained either major mechanical or burn injury, and whether there is a difference in IL-13 impact on Mϕ that have been preactivated under traumatic stress versus monocytic cells from a more homeostatic environment, as exists in healthy volunteers. Peripheral Mϕ from 20 controls (6 women, 14 men 29 ± 2 years) and 16 patients (2 women, 14 men; 43 ± 4 years) following major burn and mechanical trauma (average ISS 47 ± 3) were separated on postinjury days 1, 3, 5, and 7 and incubated with LPS (1 μg/ml) in the presence or absence of IL-13 (10 ng/ml) over a period of 4 h and for 20 h. Thereafter we examined from the culture supernatants the concentration of inducible NO and its metabolites NO_2^- NO_3^-, a most important mediator of inflammation, and the production of the immunoregulatory proteins TNFα, IL-1, IL-6, and IL-8. Ex vivo LPS activated Mϕ, compared to control cells, displayed a considerably enhanced inflammatory activity during the immediate posttraumatic course with a substantial and consistent elevation in TNFα and IL-6 (peak values 9- and 16-fold above controls, respectively). The addition of hrIL-13 to the Mϕ cultures resulted in an effective downregulation of the synthesis of TNFα (–33%), IL-1β (–35%), IL-6(–35%), and IL-8 (–28%), showing an average reduction in mediator production to two-thirds of the value found in corresponding, solely LPS-stimulated cultures (Fig. 2). The impact of hrIL-13 with respect to inflammatory cytokine production on control monocytes was more or less identical for IL-6 (–35%) and IL-1β (–40%), slightly lower for IL-8 (–20%), and nonexistent for TNFα compared to the values obtained for in vivo traumatic stress preactivated monocytes. While we observed no consistent elevation in monocytic NO release during the posttraumatic course, in a number of individual determinations the addition of hrIL-13 resulted in a striking reduction in peak NO output toward baseline levels.

We observed in coexperiments (results not shown here), when comparing the impact of hrIL-13 on IL-1β and IL-8 synthesis with that achieved by the addition of hrIL-10 to Mϕ cultures, that IL-10 is significantly more potent in terms of its downregulatory capacity for the Mϕ inflammatory potential. In LPS-stimulated Mϕ cultures from controls, burned and polytrauma patients, and patients undergoing cardiac surgery, the administration of hrIL-10 (10 ng/ml) resulted in an average of 80% suppression of inflammatory cytokine release compared to solely LPS-stimulated cultures. These data together with preexisting reports about the biological properties of IL-10 indicate that this cytokine, also unlike IL-13, acts as a major suppressor factor for T-cell performance (proliferation, IL-2 synthesis; DE WAAL MALEFYT et al. 1992; TAGA and TOSATO 1992; AYALA et al. 1994), downregulates Mϕ antigen presenting capacity (FIORENTINO et al.

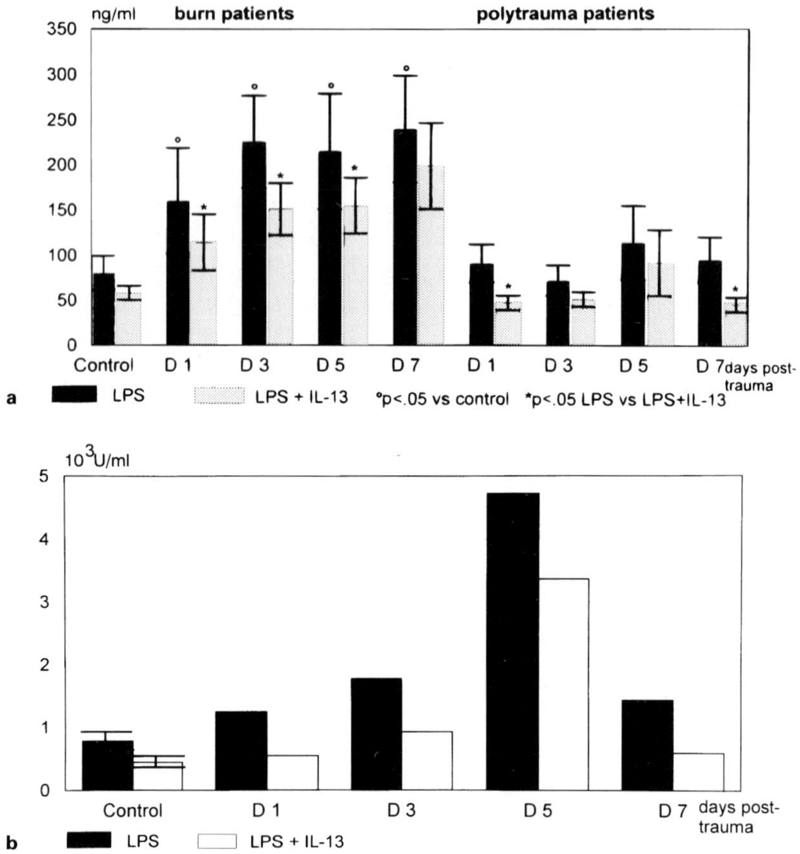

Fig. 2. a Changes in the in vitro capacity of adherent Mϕ to release IL-8 in response to LPS (1 μg/ml) over a period of 20 h in the presence or absence of hrIL-13 (10 ng/ml) after they were harvested on consecutive days (*D*) after trauma from patients who had sustained major burn (*n*=12) or mechanical (*n*=4) injuries and from 20 healthy controls. °*p* 0.05 vs control; **p* 0.05 vs. LPS+IL-13. **b** Changes in the in vitro capacity of adherent Mϕ to release IL-6 in response to LPS (1 μg/ml) over a period of 20 h in the presence or absence of hrIL-13 (10 ng/ml) after they were harvested on consecutive days (*D*) after trauma from an individual patient (man, 24 years old, ISS 66)

1991), prompts suppression of γIFN synthesis (D'Andrea et al. 1993) and B-cell function, resulting in an overall immunosuppressive capacity. These facts created doubts about the utility of IL-10 as a biological response modifier for states of posttraumatic SIRS and sepsis, as suggested by others (Napolitano et al. 1994; Vincent 1995).

It is our conviction that the trauma-induced coexistence of two immunomechanistic entities – hyperinflammation and depression of cell-mediated immune responses – requires an integrated and concerted response in terms of a target-seeking counterregulation. Therefore we suggest that hrIL-13, rather than hrIL-10, possesses the essential characteristics which warrant testing this agent as a modifier of posttraumatic states of deficient host defenses. We observed that hrIL-13 promises moderate reduction rather than complete shut-off of the inflammatory component of immunity; it supports, to a certain degree, cell-mediated immune responses rather than inducing overall immunosuppression; and it possesses a forward regulatory functionality towards a variety of important cellular targets. From this study and other findings we concluded that its biological properties recommend hrIL-13 for testing as a biological response modifier for acute states of trauma-induced host defense deficiency. It is our hypothesis that two recently popular agents granulocyte colony-stimulating factor (HARTUNG et al. 1995) and pentoxifylline (HOFFMANN et al. 1995; SCHADE and ZABEL 1994) which have been found to possess enormous potential for the deactivation of Mϕ predominantly via the downregulation of TNFα synthesis may very well prove useful agents together with hrIL-13 in a combined-drug broad-spectrum therapeutic intervention to control SIRS.

4 Trauma-Induced Impairment of Nonspecific Immune Functions

Nonspecific immune functions which are altered following hemorrhage and tissue injury include:

Monocytosis
Plasma levels of TNF-α and IL-6 ↑
Mϕ TNF-α and IL-6 secretion accelerated
Mϕ IL-1 secretion ↓
Mϕ PGE_2 synthesis ↑
PGE_2 plasma levels ↑
PMN Chemotaxis ↓
  Phagocytic capacity ↓
  β-integrin expression ↓
  Leukotriene B4 synthesis ↓
  Release of O_2 radicals ↓↑
  Release of elastate ↑
Acute-phase protein synthesis ↑
C3a plasma levels ↑
Cathepsin, lactoferrin, myeloperoxidase levels ↑
Neopterin plasma levels ↑
Depletion of fibronectin, opsonins and anglotensin III plasma levels ↑

Abnormalities in PMN function after serious injury have been interpreted by some as signs of unresponsiveness, whereas others have demonstrated evidence for PMN hyperactivation. Several laboratories have found a significant correlation between the degree of PMN dysfunction and morbidity of the patient

(Babcock et al. 1993). A decrease in PMN phagocytic activity (Grogan 1976; Heck et al. 1980), reduction in intracellular bacterial killing (Bjerknes 1989), and large number of cellular parameters were found to be altered. This included a decrease in glucose oxidation and oxygen consumption, reduced H_2O_2 production, alterations in phagolysosomal acidification, and a loss of or defect in lysosomal enzymes (Heck et al. 1975; Duque et al. 1985; Gadd and Hansbrough 1989; Bjerknes et al. 1989; Gallin 1985). Koller et al. (1989) reported lower leukotriene, including B_4, generation from PMNs of severely burn patients, which could account for some of the chemotactic and chemokinetic alterations reported by others. Rothe et al. (1990) used flow cytometry to distinguish the effect on PMN phagocytosis of trauma from that of trauma plus sepsis. They found that severe trauma causes hypoeregic phagocytosis while sepsis following trauma causes hypoergic phagocytosis. Recently, Babcock et al (1994) and Faist and Babbock (1994) found PMN dysfunction in circulating blood within 72 h after trauma; 70% of all PMNs had downregulated or nonexistent β-integrins. These cells did not phagocytose, had reduced respiratory burst, and appeared to be completely degranulated. Other reports, however, stress increased expression of $CR3^+$ receptors (CD11, CD18 complex) on circulating PMNs, increasing the ability of these cells to adhere to intercellular adhesion molecules and thus contributing to pulmonary capillary leakage.

Repair and healing or perpetuation of acute inflammation is characterized by a complex network of interacting cellular and humoral defense mechanisms. Of the numerous inflammatory mediators, the proteolytic lysosomal enzyme PMN elastase is highly destructive when released extracellularly, thus contributing to organ dysfunction after injury. Dittmer et al. (1986) and Nast-Kolb et al. (1992) have described significant increases in plasma elastase following major trauma and demonstrated a correlation between elevated elastase levels and the onset of organ failure. Pacher et al. (1989) predicted the onset of organ failure with a sensitivity of 88% and a specificity of 83% by elevated elastase levels after injury. When neopterin levels were elevated reflecting Mϕ activation, sensitivity was increased to 91% and specificity to 99% (Huber et al. 1987) Brandel et al. (1989) and Strohmaier et al. (1987) observed that increased neopterin in injured patients correlated with lethal outcome, compared to patients with an uneventful clinical course and low neopterin levels. Recently Gaitzsch et al. (1994) studied 40 patients with multiple trauma (mean age, 42 years; average ISS, 28) and found the acute-phase protein C-reactive protein together with IL-6 and TNFα, when elevated, to be most sensitive predictors of septic complications 24 h before the onset of sepsis. Persisting decreases in plasma anglotensin III were also correlated with the onset of multiple-organ failure (Risberg et al. 1986).

5 Is There a Need for Posttraumatic Therapeutic Immunointervention?

It is our understanding that the endogenous provisions of the organism to survive following overwhelming trauma are insufficient and require major exogenous support. It must be the principal clinical goal of modern immunotherapy to prevent a state of systemic inflammatory response in an immunocompromised host from converting into a state of bacterial sepsis. Several strategic approaches to prevent the development of late multiple-organ dysfunction syndrome/multiple-organ failure appear to be feasible (Fig. 3). As of today no clinical trials in septic patients with gram-negative bacterial infections employing therapeutic tools such as anti-LPS monoclonal antibodies, anti-TNF antibodies, soluble TNF receptors, and IL-1 receptor antagonists have shown an overall valid, clinically important, reproducible, and statistically significant treatment benefit. This bitter lesson has convinced us that the earliest possible blockade of the initiation of host responses following tissue destruction with or without a bacterial component may prove to be the most efficacious or perhaps even the only approach for the prevention of life-threatening bacterial infections.

Ideally, the immunoaugmentative intervention should prevent the posttraumatic systemic inflammatory response from turning into a nonreversible autodestructive inflammation with or without infection. This intervention must be employed in a calculated preventive fashion as early as possible following trauma. It should protect multiple cellular targets (e.g., lymphocytes, Mϕ, granulocytes, and endothelial cells), and its mode of action should protect the host from cell hyperactivation and cell exhaustion. It is also most likely from our present immunomechanistic understanding that only a combination of several

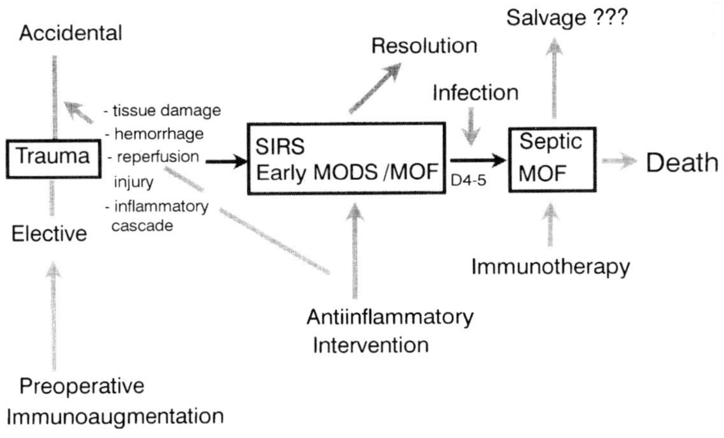

Fig. 3. Potential strategies for a therapeutic anti-inflammatory approach to intercept the development of nonreversible bacterial sepsis

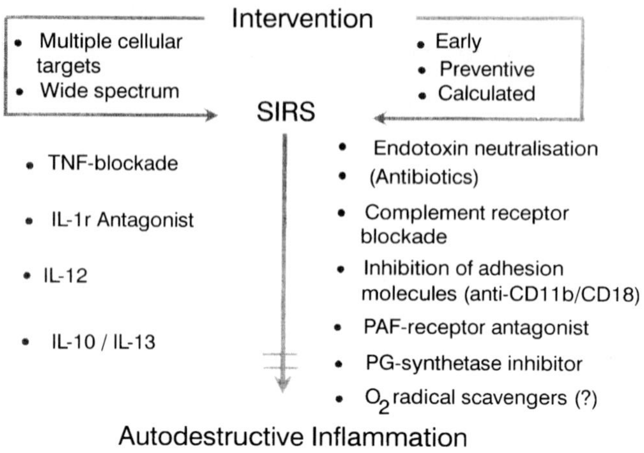

Fig. 4. Schematic depiction of the possibilities for dierentiated therapeutic intervention to prevent or alleviate posttraumatic multiple-organ dysfunction syndrome/multiple-organ failure

drugs will be effective in controlling the posttraumatic dyshomeostasis of the various cell systems. Figure 4 presents a variety of potentially useful pharmacological regimens for preventing autodestructive inflammation. Crucial issues within the complex field of preventive immunomodulation for the control of SIRS include patient selection, timing of administration, the difficulty in avoiding a complete shut-off of inflammatory responses, and cost of therapy.

It is clear that the high susceptibility to infection following trauma is correlated with injury severity and thus with the degree of uncontrolled inflammation. We do know, however, that the onset of infection most frequently occurs in a delayed fashion and appears to be based on a secondary phase of dyshomeostasis rather than on a primary event (Fig. 3). Nevertheless it appears that only primary states of dyshomeostasis can be corrected via immediate cell-directed intervention to prevent a malignant nonreversible systemic inflammatory response. It is our hypothesis that a combined therapeutic strategy should include (a) a global short-term ($\leq$72 h) downregulation of inflammatory Mϕ and PMN activity as previously outlined, (b) the prevention of excessive Mϕ stimulation via neutralization of circulating endo-/exotoxins with high doses of polyvalent immunoglobulins and soluble complement receptors, and (c) cell-mediated specific immune performance upregulation to overcome posttraumatic paralysis via the administration of thymomimetic hormones, γIFN, and gramulocyte colony-stimulating factor.

Inspite of all the progress that has been made in understanding globally the mechanisms of host defense dysfunction in trauma, shock, and sepsis the precondition to employ immunotherapeutic interventions effectively in surgical patients in the near future will depend on (a) our ability to measure the activation state of host defenses accurately, (b) a clear comprehension of the interactions among the various components of the immune system during health and disease, and (c) a much more rapid and precise identification of pathogens and microbial toxins (Dunn 1944).

References

Antonacci AC, Reaves LE, Calvano SE, Amand R, De Riesthal HF, Shires GT (1984) Flow cytometric analysis of lymphocyte subpopulations after thermal injury in human beings. Surg Gynecol Obstet 159: 1

Ayala A, Perrin MM, Chaudry IH (1990) Defective macrophage antigen presentation following hemorrhage is associated with the loss of the MHC class II (IA) antigens. Immunology 70: 33

Ayala A, Lehman DL, Herndon CD, Chaudry IH (1994) Mechanism of enhanced susceptibility to sepsis following hemorrhage. Interleukin-10 suppression of T-cell response is mediated by eicosanoid-induced interleukin-4 release. Arch Surg 129(11): 1172–1178

Babcock GF, White-Owen CL, Alexander JW, Warden GD (1993) Characteristics of neutrophill dysfunction. In: Faist E, Meakins J, Schildberg FW (eds) Host defense dysfunction in trauma, shock and sepsis. Springer, Berlin Heidelberg New York, p 95

Babcock GF, Rodeberg DA, White-Owen C (1994) Changes in neutrophil function following major trauma or thermal injury. J Intensive Care Med 20(1): 172

Baue AE (1990) Multiple organ failure: patient care and prevention. Mosby, St Louis

Baue AE (1992) The horror autotoxicus and multiple organ failure. Arch Surg 127: 1451–1461

Bjerknes R, Vindenes H (1989) Neutrophil dysfunction after thermal injury: alteration of phagolysosomal acidification in patients with large burns. Burns 15: 77–81

Bjerknes R, Vindenes H, Pitkanen J, Winneman J, Laerum OD, Abyholm F (1989) Altered polymorphonuclear neutrophilic granulocyte functions in patients with large burns. J Trauma 29: 847–855

Brandel M, Pscheidl E, Amann W, Barjasic A, Pasch T (1989) Biochemical and hormonal parameters in patiens with multiple trauma. In: Schlag R, Redl H (eds) Second Vienna Shock Forum. Liss, New York, p 743

Chaudry IH, Ayala A, Ertel W, Stephen RN (1990) Editorial review. Hemorrhage and resuscitation: immunological aspects. Am J Physiol 259:663

Cheadle WG, Hershman MJ, Wellhausen SR, Polk HC (1989) Role of monocyte HLA-DR expression following trauma in predicting clinical outcome. In: Faist E, Ninnemann J, Green D (eds) Immune consequences of trauma, shock and sepsis. Springer, Berlin Heidelberg New York, p 119

Christou NV, Tellado JM (1993) The impact of preexisting disease conditions for host defense integrity in traumatized and critically ill patients. In: Faist E, Meakins J, Schildberg FW (eds) Host defense dysfunction in trauma, shock and sepsis. Springer, Berlin Heidelberg New York, p 73

Christou NV, McLean APH, Meakins JL (1980) Host defense in blunt trauma: interrelationships of kinetics of anergy and depressed neutrophil function, nutritional status, and sepsis. J Trauma 20: 833–841

D'Andrea A, Aste-Amezaga M, Valiante N, Ma X, Kubin M, Trinchieri G (1993) Interleukin 10 (IL-10) inhibits human lymphocyte interferon γ-production by suppressing natural killer cell stimulatory factor/IL-12 synthesis in accessory cells. J Exp Med 178: 1041–1048

Defrance T, Carayon P, Billian G et al. (1994) Interleukin 13 is a B cell stimulating factor. J Exp Med 179 (1): 135–143

De Waal Malefyt R, Abrams J, Bennett B, Figdor CG, de Vries J (1991) Interleukin 10 (IL-10) inhibits cytokine synthesis by human monocytes: an autoregulatory role of IL-10 produced by monocytes. J Exp Med 174 (5): 1209–1220

De Waal Malefyt R, Yssel H, Roncaralo MG, Spits H, de Vries J (1992) Interleukin-10. Curr Opin Immunol 4: 314–320

De Waal Malefyt R, Figdor CG, Huijbens R et al. (1993) Effects of IL-13 on phenotype, cytokine production, and cytoxic function of human monocytes. J Immunol 151: 6370–6381

Dittmer H, Jochum M, Fritz H (1986) Freisetzung von granulozytrer Elastase und Plasmaprotein und Plasmaproteinvernderungen nach traumatisch-hmorrhagischem Schock. Unfallchirurg 89: 160

Dunn D (1994) Immunomodulation. In: Meakins JL (ed) Surgical infections. Scientific American, New York, p 475

Duque R, Phan S, Hudson J, Till G, Ward P (1985) Functional defects in phagocytic cells following thermal injury. Am J Pathol 118: 116–127

Ertel W, Faist E (1989) The influence of mechanical trauma on the B-cell system: phenotypes terminal B-cell maturation, immunologlobulin synthesis and influence of lymphokines. In: Faist E, Ninnemann J, Green D (eds) Immune consequences of trauma, shock and sepsis. Springer, Berlin Heidelberg New York, p 143

Ertel W, Faist E, Salmen B, Huber P, Heberer G (1989) Influence of cyclooxygenase inhibition on interleukin-2 receptor (IL-2R) expression after major trauma. Surg Res Commun 5: 17

Ertel W, Faist E, Nestle C, Storck M, Schildberg FW (1990) Changes in B-lymphocyte proliferative response following major trauma: the influence of recombinant lymphokines. Surg Res Commun 8: 253–261

Ertel W, Morrison MH, Ayala A, Chaudry IH (1991a) Chloroquine attenuates hemorrhagic shock induced suppression of Kupffer cell antigen presentation and MHC class II antigen expression through blockade of tumor necrosis factor and prostaglandin release. Blood 78: 1781

Ertel W, Morrison MH, Ayala A, Perrin MM, Chaudry IH (1991b) Anti-cachectin/TNF monoclonal antibodies prevent suppression of Kupffer cell antigen presentation and MHC class II antigen expression Immunology 74: 290

Ertel W, Morrison MH, Meldrum DR, Ayala A, Chaudry IH (1992) Ibuprofen restores immunity and decreases susceptibility to sepsis following hemorrhage. J Surg Res 53: 55

Ertel W, Karres I, Kremer JP, Kenney JS, Faist E, Schildberg FW (1993) Transforming growth factor type β1 (TGF-β1) is a strong anti-inflammatory cytokine that inhibits transcription, translation, and secretion of inflammatory mediators on different levels. Surg Forum 44: 105–108

Faist E, Babcock GF (1994) Host defense dysfunction following trauma, shock and sepsis. J Intensive Care Med 20(1): 106

Faist E, Baue AE, Dittmer H, Heberer G (1983) Multiple organ failure in polytrauma patients. J Trauma 23: 775–787

Faist E, Kupper TS, Baker CC, Chaudry IH Dwyer J, Baue AE (1986) Depression of cellular immunity after major injury. Arch Surg 121: 1000

Faist E, Mewes A, Baker CC, Strasser T, Alkan SS, Rieber P, Heberer G (1987) Prostaglandin E_2 dependent suppression of interleukin-2 production in patients with major trauma. J Trauma 28: 837

Faist E, Mewes A, Strasser T, Walz A, Alkan S, Baker C, Ertel W, Heberer G (1988) Alterations of monocyte function following major injury. Arch Surg 123: 287

Faist E, Ertel W, Baker CC et al. (1989) Terminal B-cell maturation and immunoglobulin (Ig) synthesis in vitro in patients with major injury. J Trauma 29: 2–9

Faist E, Storck M, Hueltner L, Redl H, Ertel W, Walz A, Schildberg FW (1992) Functional analysis of monocyte (Mϕ) activity via synthesis patterns of interleukin 1,6,8 (IL-1, IL-6, IL-8) and neopterin (NPT) in surgical intensive care patients. Surgery 9: 562–572

Faist E, Schinkel C, Zimmer S, Kremer JP, Donnersmarck GH, Schildberg FW (1993) Inadequate interleukin-2 synthesis and interleukin-2 messenger expression following thermal and mechanical trauma in human is caused by defective transmembrane signalling. J Trauma 36(6) 1–9

Fiorentino DF, Zlotnik A, Mosmann TR, Howard M, O'Garra A (1991) IL-10 inhibits cytokine production by activated macrophages. J Immunol 147(11): 3815–3822

Gadd M, Hansbrough J (1989) The effect of thermal injury on murine neutrophil oxidative metabolism. J Burn Care Rehabil 10: 125–130

Gaitzsch A, Lefering R, Weber D, Sauerland S, Dangel ML, Fink PC (1994) Early recognition of sepsis using IL-6, TNF, elastase and CRP. J Intensive Care Med 20(1): 109

Gallin J (1985) Neutrophil specific granule deficiency. Annu Rev Med 36: 263–274

Gibbons RAS, Martinez OM, Lim V, Lim R, Horn J, Garovoy MR (1989) Early alterations in HLA class II expression and response to tetanus toxoid by peripheral blood monocytes from patient with injury from burns or trauma. In: Faist E, Ninnemann J, Green D (eds) Immune consequences of trauma, shock and sepsis. Springer, Berlin Heidelberg New York, p 123

Grogan J (1976) Altered neutrophil phagocytic function in burn patiets. J Trauma 16: 734–738

Guo Y, Wiener S, Winchurch RA, Adler WH, Munster AM (1989) Interleukin-6, a potent immunoregulatory cytokine in burn patient serum. Surg Forum 40: 58

Hartung T, Döcke WD, Gantner F et al. (1995) Effect of granulocyte colony-stimulating factor treatment on ex-vivo blood cytokins response in human volunteers. Blood (in press)

Heck E, Browne L, Curreri PW, Baxter C (1975) Evaluation of leukocyte function in burned individuals by in vitro oxygen consumption. J Trauma 15: 486–488
Heck E, Edgar M, Hunt J, Baxter C (1980) A comparison of leukocyte function and burn mortality. J Trauma 20: 75–77
Hershman MJ, Appel SH, Wellhausen SR, Sonnenfeld G, Polk HC (1989) Interferon-gamma treatment increases HLA-DR expression on monocytes in severely injured patients. Clin Exp immunol 77: 67–70
Hoffmann H, Reichert K, Markewitz A, Jochum M, Faist E (1995) Supplemental treatment with pentoxifylline in patients at risk for sepsis: pilot study. 15 Annual Meeting of the Surgical Infection Society, p 14
Howard M, O'Garra A (1992) Biological properties of interleukin 10. Immunol Today 13 (6): 198–200
Huber C, Troppmair J, Rokos H, Curtius HC (1987) Neopterin heute. Dtsch Med Wochenschr 11: 107
Keane MR, Birmingham W, Shatney CM, Winchurch RA, Munster AM (1983) Prediction of sepsis in the multitraumatic patient by assays of lymphocyte respsonsiveness. Surg Gynecol Obstet 156: 163
Koller M, Konig W, Brom J, Erbs G, Muller F (1989) Studies on the mechanisms of granulocyte dysfunctions in severely burned patients – evidence for altered leukotriene generation. J Trauma 29: 435–445
Livingston DH, Appel SH, Wellhausen SR, Sonnenfeld G, Polk HC (1988) Depressed interferon-μ production and monocyte HLA-DR expression after severe injury. Arch Surg 123: 1309
Mannick JA (1993) Trauma, sepsis, and immune defects. In: Faist E, Meakins J, Schildberg FW (eds) Host defense dysfunction in trauma, shock and sepsis. Springer, Berlin Heidelberg New York, p 15
Marks JD, Marks CB, Luce JM, Montgomery AB, Turner J, Metz CA, Murray JF (1990) Plasma tumor necrosis factor in patients with septic shock. Am Rev Respir Dis 141: 94
McKenzie ANJ, Culpepper JA, de Waal Malefyt R et al. (1992) Interleukin 13, a T-cell derived cytokine that regulates human monocyte and B-cell function. Proc Natl Acad Sci USA 90: 3735–3739
Meldrum DR, Ayala A, Wang P, Ertel W, Chaudry IH (1991) Association between decreased splenic ATP levels and immunodepression: Ameliroration with ATP-$MgCl_2$. Am J Physiol 261–351
Michie HR, Manogue KR, Spriggs DR, Revhaug A, O'Dwyer S, Dinarello CA, Cerami A, Wolff SM, Wilmore DW (1988) Detection of circulating tumor necrosis factor after endotoxin administration. N Engl J Med 318: 1481
Miller-Graziano CL, Fink M, Wu JY, Szabo G, Kodys K (1988) Mechanisms of altered monocyte prostaglandin E_2 production in severely injured patients. Arch Surg 123: 293
Minty A, Chalon P, Derocq JM et al. (1993) Interleukin-13 is a new human lymphokine regulating inflammatory and immune responses. Nature 362(6417): 248–250
Munster AM (1984) Immunological response of trauma and burns. An overview. Am J Med 76: 142
Muzio M, Re F, Sironi M et al. (1994) Interleukin-13 induces the production of interleukin-1 receptor antagonist (IL-1ra) and the expression of the mRNA for the intracellular (keratinocyte) form of IL-1ra in human myelomonocytic cells. Blood 83(7): 1738–1743
Napolitano LM, Cara L, Campbell BS (1994) Inhibition of nitric oxide (NO) synthesis normalizes splenocyte interleukin-10 (IL-10) synthesis in murin thermal injury. J Intensive Care Med 20(1): 59
Nast-Kolb D, Waydhas C, Jochum M, Duswald KH, Machleidt, W, Spannagl M, Schramm W, Fritz H, Schweiberer L (1992) Biochemische Faktoren als objektive Parameter zur Prognoseabschtzung beim Polytrauma. Unfallchirurg 95: 59
Ninnemann JL (1987) Trauma, sepsis, and the immune response. Burn Care Rehabil 8: 462
Nohr CW, Christou NV, Broadhead M et al. (1983) Failure of humoral immunity in surgical patients. Surg Forum 34: 127–129
Nohr CW, Christou NV, Rode H et al. (1984) In vivo and in vitro humoral immunity in surgical patients. Ann Surg 200: 373–381
O'Mahoney JB, Palder SB, Wood JJ, McIrvine A, Rodrick ML, Demling RH, Mannick JA (1984) Depression of cellular immunity after multiple trauma in the absence of sepsis. J Trauma 24: 869
Oswald IP, Gazzinelli RT, Sher A, James SL (1992) IL-10 synergizes with IL-4 and transforming growth factor-b to inhibit macrophage cytotoxic activity. J IMMUNOL 148: 3578
Pacher R, Redl H, Frass M, Petzl DH, Schuster E, Woloszczuk W (1989) Relationship between neopterin and granulocyte elastase plasma levels and the severity of multiple organ failure. Crit Care Med 17: 221
Polk HC, George CD, Wellhausen SR, Cost K, Davidson PR, Regan MP, Borzotta AP (1986) A systematic study of host defense processes in badly injured patients. Ann Surg 204: 282

Quattrocchi KB, Frank EH, Miller CH, Dull ST, Howard RR, Wagner FC (1991) Severe head injury: effect upon cellular immune function. Neurol Res 13: 13

Rappaport RS, Dodge GR (1982) Prostaglandin E inhibits the production of human interleukin 2. J Exp Med 155: 943

Renz H, Gong JH, Schmidt A et al. (1988) Release of tumor necrosis factor-alpha from macrophages. Enhancement and suppression are dose-dependently regulated by prostaglandin E2 and cyclic nucleotides. J Immunol 141: 2388–2393

Risberg B, Medegard A, Heideman M, Gyzander E, Bundsen P, Oden M, Teger-Nilsson AC (1986) Early activation of humoral proteolytic systems in patients with multiple trauma. Crit Care Med 14: 917

Rodrick ML, Moss NM, Grbic JT, Revhaug A, O'Dwyer ST, Michie HR, Gough DB, Dubravec D, Manson JM, Wilmore DW, Mannick JA (1992) Effects of in vivo endotoxin on in vitro cellular immune responses in humans. J Clin Immunol 12: 440–450

Rothe G, Kellerman W, Valet G (1990) Flow cytometric parameters of neutrophil function as early indicators of sepsis- or trauma-related pulmonary or cardiovascular organ failure. J Lab Clin Med 115: 52–61

Schade FU, Zabel P (1964) Modulating effects of pentoxifylline on cytokine release syndromes. In: Reinhart K, Eyrich K, Sprung C (eds) Sepsis – current perspectives in pathophysiology and therapy, Springer, Berlin Heidelberg New York, p 371

Shorr RM, Ersher WB, Gamelli RL (1984) Immunoglobulin production in burned patients. J Trauma 24: 319–322

Spits H, de Waal Malefyt R (1992) Functional characterization of human IL-10. Int Arch Immunol 99: 8–15

Stephan RN, Conrad PJ, Janeway CA, Geha AS, Baue AE, Chaudry IH (1986) Decreased interleukin-2 production following simple hemorrhage. Surg Forum 37: 73

Strohmaier W, Redl H Schlag G, Inthorn D (1987) D-Erythroneopterin plasma level in intensive care patients with and without septic complications. Crit Care Med 15: 757

Taga K, Tosato G (1992) IL-10 inhibits human T-cell proliferation and IL-2 production. J Immunol 148(4): 1143–1148

Takayama TK, Miller C, Szabo G (1990) Elevated tumor necrosis factor production concomitant to elevated prostaglandin E_2 production by trauma patients' monocytes. Arch Surg 125: 29

Teodorczyk-Injeyan JA, Sparkes BG, Falk RE et al. (1986) Polyclonal immunoglobulin production in burned patients: kinetics and correlation with T-cell activity. J Trauma 26: 834–839

Tsunawaki S, Sporn M, Ding A et al. (1988) Deactivation of macrophages by transforming growth factor-beta 334: 260–262

Vincent JL (1995) The role of tumour necrosis factor-α (TNF). Sepsis Newslett 1(1): 1–8

Wood JJ, O'Mahoney JB, Rodrick ML, Mannick JA (1984) Immunoglobulin production and suppressor T-cells after thermal injury. Surg Forum 35: 619

Wood JJ, O'Mahony JB, Rodrick ML et al. (1986) Abnormalities of antibody production after thermal injury. Arch Surg 121: 108–116

Zimmer S, Schinkel C, Kremer JP, Walz A, Alkan S, Henckel von Donnersmarck, Faist E (1994) Decreased in-vitro cytokine production following major mechanical trauma. J Intensive Care Med 20(1): 79

Zurawski G, de Vries J (1994) Interleukin 13, an interleukin 4-like cytokine that acts on monocytes and B cells, but not on T cells. Immunol Today 15(1): 19–26

Endotoxin-Based Molecular Strategies for the Prevention and Treatment of Gram-Negative Sepsis and Septic Shock

M. Pollack and C.A. Ohl

1 Introduction 275

2 Lipopolysaccharide Antibodies 276

3 BPI and Other Granulocyte-Derived (CAP) Proteins 279

4 Anti-LPS Factors from Horseshoe Crabs 281

5 Polymyxin B and PmB-Related Synthetic Peptides 282

6 Interruption of LPS-Induced, LBP- and CD14-Mediated Cellular Activation Pathways 282

7 Serum Lipoproteins 285

8 Natural and Synthetic LPS and Lipid A Analogs or Partial Structures That Antagonize or Induce Tolerance to LPS 286

9 Plasmapheresis, Exchange Transfusion, and Extracorporal LPS Adsorption 288

References 289

1 Introduction

As documented in the other chapters of this volume, there is a rapidly growing appreciation that gram-negative sepsis is both an infectious and a systemic inflammatory disease. Extraordinary progress has been achieved in describing at least some of the molecular mechanisms which produce sepsis, especially as they relate to the causative role of bacterial lipopolysaccharides (LPS) or endotoxin.

Recent research has elucidated structure-function relationships of LPS and has begun to unravel some of the mysteries surrounding molecular events involved in mammalian responses to this complex macromolecule (Rietschel et al. 1993, 1994). Our expanding understanding of LPS-induced biological responses has been stimulated by the characterization of CD14, a putative endotoxin receptor (Lee et al. 1993; Pugin et al. 1993a,b; Tobias and Ulevitch 1993; Ulevitch and Tobias 1994), and LPS-binding protein (LBP), a serum protein that catalyzes

Department of Medicine, Uniformed Services University of the Health Sciences, F. Edward Hébert School of Medicine and Department of Internal Medicine, National Naval Medical Center, Bethesda, MD 20814-4799, USA the CD14-mediated recognition of LPS-LBP complexes by eukaryotic cells (SCHUMANN et al. 1990; TOBIAS et al. 1992; MATHISON et al. 1993; HAILMAN et al. 1994).

There is accumulating evidence that the recognition of LPS by target cells of the infected host, a process mediated through membrane-bound or soluble forms of CD14 (FREY et al. 1992; PUGIN et al. 1993a,b) and facilitated by LBP, is a major triggering event in endotoxin-induced inflammation associated with the development of sepsis. Current research seeks to discover how CD14 and other putative LPS receptors are functionally linked, through signal transduction and transcriptional regulation, to the initiation of intracellular events associated with LPS-induced cell activation (WEINSTEIN et al. 1993; ULEVITCH and TOBIAS 1994).

Meanwhile, much has been learned about host-derived proteins which, through their interactions with LPS, attenuate its proinflammatory activities. These include granulocyte-derived cationic antimicrobial proteins (CAP) such as CAP18 (LARRICK et al. 1995) and bactericidal/permeability-increasing protein (BPI; TOBIAS et al. 1988; HEUMANN et al. 1993; WILDE et al. 1994), the latter a natural homolog of LBP; serum transport proteins such as low-density lipoprotein (LDL), high-density lipoprotein (HDL), and very low-density lipoprotein (VLDL)/chylomicrons (HARRIS et al. 1990; FLEGEL et al. 1993); and antibodies directed against various structural elements of the LPS macromolecule, including the O polysaccharide, core oligosaccharide, and biologically active lipid A moiety (POLLACK 1992a; POLLACK et al. 1995).

This chapter evaluates therapeutic strategies that target the LPS molecule itself or host-derived factors through which LPS acts to produce a pathogenic proinflammatory response. The former include mammalian proteins with endotoxin-neutralizing activity such as leukocyte-derived BPI and CAP proteins, serum transport proteins, anti-LPS antibodies, nonmammalian proteins such as anti-LPS factors from horseshoe crabs (WARREN et al. 1992b; ROTH and TOBIAS 1993), and polymyxin B (PmB) or PmB-related synthetic peptides (MORRISON and JACOBS 1976; RUSTICI et al. 1993). The latter include antibodies to CD14 and LBP, soluble CD14, and natural or synthetic lipid A analogs which act as lipid A antagonists or agonists with anti-inflammatory or tolerizing activities. Also discussed are therapeutic measures which involve the replacement or extracorporal purification of a septic patient's blood or plasma to remove endotoxin (and inflammatory mediators).

2 Lipopolysaccharide Antibodies

LPS antibodies mediate natural and acquired immunity to gram-negative bacteria. They are ubiquitous in human serum and other extracellular fluids, appearing as a function of age and cumulative antigenic exposure related to clinical and subclinical infection or simple colonization.

Naturally acquired antibodies are directed toward epitopes in all three structural domains of LPS (POLLACK 1992a). In general, antibodies to the phylogenetically heterogeneous O polysaccharide are species and serotype specific, while antibodies which recognize more conserved epitopes on the core oligosaccharide or lipid A tend to react more broadly among phylogenetically diverse LPS. Differences in the submolecular specificity of LPS antibodies may be associated with other functional disparities as well. For example, antibodies that recognize relatively exposed epitopes associated with repeating oligosaccharide subunits of the LPS O-side chain may exhibit more extensive binding and express higher functional affinity toward whole LPS compared with antibodies directed toward less well exposed and less numerous epitopes associated with the relatively cryptic, nonrepeating LPS core structure (POLLACK et al. 1990; POLLACK 1992a).

LPS antibodies mediate antibacterial as well as antiendotoxic functions. Antibodies specific for the O-polysaccharide of smooth LPS or core component of phenotypically rough LPS (so-called lipooligosaccharide) may deposit complement on the bacterial cell surface, leading to production of membrane attack complexes, pore formation, and subsequent cell lysis (OISHI et al. 1992). Alternatively, such antibodies may mediate complement-dependent opsonophagocytic bacterial killing in conjunction with professional phagocytes (PIER et al. 1994).

LPS antibodies may also promote the complement-dependent phagocytic uptake of free LPS (KRIEGER et al. 1993), while inhibiting its uptake by the more proinflammatory CD14-mediated pathway (POLLACK et al. 1995). Moreover, in humans LPS antibodies may mediate the clearance of LPS-containing immune complexes through CR1 receptors on erythrocytes which are thought to transport and deliver the immune complexes to fixed tissue macrophages of the reticuloendothelial system (KATSIKIS et al. 1993; KRIEGER et al. 1993; TONOLI et al. 1993).

LPS antibodies may thus mediate bacterial killing, facilitate the sequestration or removal of LPS from sites of production, participate in the clearance of circulating LPS, and prevent its inflammation-provoking recognition by monocytes/macrophages.

There is little question that LPS antibodies serve an important protective function against gram-negative bacterial disease. This has been more easy to document experimentally; however, in the case of O-side chain-specific antibodies expressing antibacterial activities (KIRKLAND and ZIEGLER 1984; STOLL et al. 1986; OISHI et al. 1992; HOFFMAN et al. 1994; PIER et al. 1994) than in the case of core- or lipid-A specific antibodies exhibiting antiendotoxic properties (BAUMGARTNER et al. 1990; APPELMELK and COHEN 1992; POLLACK 1992a). However, more therapeutic interest has focused in recent years on the core region of LPS, and upon the core hypothesis which predicts that antibodies directed toward phylogenetically conserved, biologically active epitopes in the core/lipid A region express cross-reactive, endotoxin-neutralizing, and cross-protective functions (APPELMELK and COHEN 1992; POLLACK 1992a). The popularity of the core hypothesis has been driven by its clear theoretical rationale, a substantial body of suggestive if not definitive experimental evidence, and the widely perceived practical need for a single antiendotoxin reagent with broad therapeutic application in gram-negative sepsis of diverse etiology.

Two putative lipid A-specific IgM monoclonal antibodies (MAbs), one murine, E5 (Wood et al. 1992), and the other human, HA-1A (Teng et al. 1985; Bogard et al. 1993), have undergone extensive preclinical and clinical evaluation. Some evidence suggests that under appropriate conditions both MAbs are able to react with phenotypically smooth as well as rough LPS and both are capable of cross-reacting with LPS produced by gram-negative organisms representing different bacterial species, genera, and families (Parent et al. 1992; Wood et al. 1992; Bogard et al. 1993; Fujihara et al. 1993a,b; Mascelli et al. 1993; Siegel et al. 1993). The apparent cross-reactivity of the MAbs may be attributable in part to their low intrinsic but higher functional affinity, characteristics which often typify polyvalent antibodies of the IgM class and render them more cross-reactive than IgG antibodies of similar specificity (Pollack 1992a).

The putative endotoxin-neutralizing properties of the E5 and HA-1A MAbs have been supported by certain in vitro and in vivo experiments (Teng et al. 1985; Chen et al. 1993; Katsikis et al. 1993; Paleolog et al. 1993; Ono et al. 1994) but not by others (Baumgartner et al. 1990; Warren et al. 1993). Similarly conflicting preclinical data have been produced with regard to in vivo protection against live challenge and cross-protection against heterologous challenge (Teng et al. 1985; Baumgartner et al. 1990; Quezado et al. 1993; Romulo et al. 1993; Rogy et al. 1994). Some of the negative or nonconfirmatory data collected in these experiments have been explained or rationalized on the basis of inadequate animal models of septic shock.

Clinical trial results obtained with the E5 and HA-1A MAbs have also been controversial. In each case initial phase III trials in gram-negative sepsis demonstrated significant MAb-associated reductions in mortality (Greenman et al. 1991; Ziegler et al. 1991; Wortel et al. 1992), albeit only in subgroups of the entire study population (i.e., patients *not* in shock in the case of E5 and patients with documented gram-negative bacteremia with or without shock in the case of HA-1A). Unfortunately, neither of these results was reproduced in second trials designed to confirm the results of the earlier trials (Wenzel et al. 1991; McCloskey et al. 1994). It is plausible that the lack of reproducibility of the E5 and HA-1A clinical trials reflects more the complexity of the disease process, the heterogeneity of the patient populations studied, and the intricacies of study design than the inherent biological limitations of the MAbs themselves. Further phase III trials of both MAbs are still in progress.

It has been argued that the E5 and HA-1A clinical trials were premature in the absence of definitive preclinical data and an adequate understanding of the MAbs' mechanism of action (Warren et al. 1992a). The main counterargument has been that inadequate in vitro and in vivo models existed for testing the core hypothesis prior to human trials of core or lipid A reactive antibodies, and that the decision to conduct such trials was therefore a necessary and courageous one

(WENZEL 1992; ZIEGLER and SMITH 1992). In any case the E5 and HA-1A MAbs were among the earlier examples of such reagents available for testing but not necessarily optimal therapeutic candidates. It is theoretically possible that genetic reengineering of HA-1A or E5 or the development of new MAbs with improved binding and endotoxin-neutralizing characteristics might yield therapeutic reagents superior to the original E5 and HA-1A MAbs.

It has been suggested on the basis of animal data that core-specific MAbs may offer greater therapeutic potential than lipid A-specific antibodies. Moreover, core-reactive MAbs with a narrower spectrum of activity against a single bacterial species, genus, or closely related group of genera (TERASHIMA et al. 1991; DI PADOVA et al. 1993) may prove more therapeutically active against these more restricted targets than more broadly cross-reactive antibodies. Presumably, these more narrowly targeted core-reactive MAbs could be administered in combination, if necessary, to achieve broader coverage.

There has been considerable recent interest in the exploitation of possible therapeutic synergy between antibiotics which disrupt the bacterial cell wall, releasing LPS into the surrounding environment, and LPS antibodies which neutralize or facilitate the clearance of LPS thus released (COLLINS et al. 1989; KRIEGER et al. 1993; ROMULO et al. 1993; SIEGEL et al. 1993). There is also keen interest in combining LPS antibodies with anti-inflammatory agents, such as anticytokine antibodies (CROSS et al. 1993; CROSS and OPAL 1994), in order to interrupt the proinflammatory cascade at its point of origin as well as further downstream in the hope that such a multifocal approach will provide additive or synergistic therapeutic benefits.

Thus, although there is currently a general disaffection with antiendotoxin MAbs as therapeutics in gram-negative sepsis based, in part, upon their uneven performance in clinical trials, the core hypothesis survives intact if somewhat tarnished and a new generation of therapeutically active MAbs directed toward core- or lipid-A associated epitopes appears to be at least a theoretical possibility (CROSS 1994; CROSS and OPAL 1994). LPS antibodies remain an attractive alternative to other adjunctive therapies in gram-negative sepsis because of their potentially beneficial activities against both LPS and bacteria and the fact that they target a critical virulence factor produced by the invading pathogen rather than the delicate homeostasis-maintaining apparatus of the infected host.

3 BPI and Other Granulocyte-Derived (CAP) Proteins

Another therapeutic strategy utilizes a variety of native or recombinant proteins that bind LPS and block its biological functions. Mammalian polymorphonuclear leukocyte granules contain several cationic proteins and peptides with bactericidal and LPS-neutralizing properties (TOBIAS et al. 1988; ELSBACH et al. 1994).

One such protein, BPI (CAP57), is a 55- to 60-kDa azurophilic granule protein with 45% sequence homology with LBP (Tobias et al. 1988). BPI and LBP both bind with high affinity through the 25-kDa N-terminal portion of their respective molecules to lipid A, smooth and rough LPS, and intact gram-negative bacteria (Gazzano-Santoro et al. 1992; Appelmelk et al. 1994b; Theofan et al. 1994). Although LBP binds to LPS and enhances various LPS-induced cellular responses, BPI binds with greater avidity and antagonizes the biological activities of LPS (Heumann et al. 1993; Gazzano-Santoro et al. 1994). In vitro, BPI inhibits LPS uptake by monocytes and neutrophils, LPS-induced cytokine and nitric oxide release by mononuclear cells, LPS priming of neutrophils, and LPS-mediated endothelial cell activation and injury (Ooi et al. 1991; Dentener et al. 1993; Appelmelk et al. 1994b; Arditi et al. 1994; Kohn and Kung 1995). BPI also inhibits bacterial proliferation and exhibits extracellular cytotoxic activity against a broad range of gram-negative bacteria related to its high affinity for outer membrane bound LPS (Weiss et al. 1992; Elsbach et al. 1994).

It is suspected but unproven that granulocyte-derived, extracellular forms of BPI play a protective role in gram-negative infections distinct from the strictly bactericidal functions of BPI. BPI is found in higher concentrations than LBP in certain closed space infections, while LBP predominates in plasma from septic patients. This suggests that measured levels of endogenous BPI are sufficient to counteract the proinflammatory effects of LBP at local tissue sites but not in plasma (Calvano et al. 1994; Opal et al. 1994).

Recombinant BPI (rBPI) and its 23-kDa N-terminal portion (rBPI-23) retain the biological activities of the natural protein or exceed it in potency (Ooi et al. 1991; Gazzano-Santoro et al. 1992, 1994; Weiss et al. 1992; Dentener et al. 1993; Theofan et al. 1994). The evaluation of rBPI-23 in animals revealed a reduction in LPS-induced proinflammatory cytokine release; normalization of LPS-associated hemodynamic, pulmonary and metabolic dysfunction; and protection against lethal endotoxemia (Kohn et al. 1993; Fisher et al. 1994; Kung et al. 1994; Parent 1994; Evans et al. 1995). In recent experiments rBPI-23 reduced mortality in mice and rabbits subjected to otherwise lethal intravenous, intraperitoneal, or intratracheal challenge with a variety of gram-negative bacteria and in most cases reduced levels of bacteremia (Kelly et al. 1993; Ammons et al. 1994; Parent 1994; Evans et al. 1995). The administration of rBPI-23 to human volunteers subsequently challenged with intravenous endotoxin reduced fever and leukopenia, blocked cytokine release, and prevented endotoxin-induced fibrinolysis (van Deventer et al. 1994).

Although BPI and its recombinant products have demonstrated therapeutic promise in preclinical studies, reduced efficacy has been documented in association with delays in treatment. Moreover, intravenously administered rBPI-23 failed to achieve therapeutic concentrations in the cerebrospinal fluid or to reduce endotoxin-induced meningeal inflammation in a rabbit model of bacterial meningitis (Kartalija et al. 1995). These studies suggest that the clinical utility of BPI may be limited by a stringent requirement for early administration relative to the onset of infection and suboptimal pharmacokinetic properties.

Both rBPI and rBPI-23 have been given safely to animals and human volunteers. Since these proteins have a short serum half-life, they must be administered by repeated bolus injection or continuous intravenous infusion. Fusion chimeras comprised of amino acid sequences derived from both BPI and LBP have been constructed and exhibit a longer circulating half-life while retaining the ability to neutralize endotoxin and protect animals against endotoxin-induced lethality (MARRA et al. 1994).

The recombinant product rBPI-21 has undergone human phase I safety and pharmacokinetic evaluation, and separate safety and clinical efficacy trials have recently begun in patients with meningococcemia and hemorrhagic trauma (the latter assessing prevention of secondary infections, organ damage, or death).

Several other cationic proteins and peptides with bactericidal and endotoxin-neutralizing properties have been isolated from neutrophil granules. The 18-kDa protein designated CAP18, originally isolated from rabbit granulocytes, was recently identified in normal human granulocytes and cloned (LARRICK et al. 1995). Rabbit and human CAP18, and the active C-terminal fragment derived from the parent molecules (CAP7 and $CAP18^{104-135}$, respectively), bind with high affinity to LPS, exhibit LPS-neutralizing properties, inhibit LPS-related proinflammatory activities, and protect mice from LPS-induced lethality (HIRATA et al. 1994a; LARRICK et al. 1995).

A number of natural and synthetic peptides have been identified with molecular and biological characteristics similar to those of CAP18. Most share homology with the N-terminal region of cathelin, a potent cysteine protease inhibitor (LEVY et al. 1993; GENNARO et al. 1994; HIRATA et al. 1994b; LARRICK et al. 1995). Other granulocyte-derived antimicrobial proteins with antiendotoxin activities include the defensins (LEHRER et al. 1993), lactoferrin (APPELMELK et al. 1994a), and azurocidin, also known as CAP37 (PEREIRA et al. 1993, 1994).

4 Anti-LPS Factors from Horseshoe Crabs

Two distinct basic proteins which share partial sequence homology and inhibit LPS-induced amebocyte lysate coagulation have been isolated from the horseshoe crabs *Tachypleus tridentatus* and *Limulus polyphemus* (TANAKA et al. 1982). These anti-LPS factors of *T. tridentatus* (TALF) and *L. polyphemus* (LALF), inhibit the growth of gram-negative bacteria, bind phylogenetically diverse endotoxins with high affinity, and inhibit a variety of LPS-associated biological activities in vitro and in vivo (MORITA et al. 1985; DESCH et al. 1989; ALPERT et al. 1992; WARREN et al. 1992b). Recombinant LALF (also known as endotoxin-neutralizing protein or ENP) and synthetic fragments of TALF exhibit endotoxin-neutralizing properties similar to those of the native proteins, including the ability to protect animals against experimental endotoxemia and sepsis (FLETCHER et al. 1993; KLOCZEWIAK et al. 1994; KUPPERMANN et al. 1994). The administration of ENP to rats or rabbits up to 60 min after *Escherichia coli* challenge appeared to lower circulating endotoxin and tumor necrosis factor (TNF) levels, and reduce mortality (Saladino et al. 1994; Nelson et al. 1995).

5 Polymyxin B and PmB-Related Synthetic Peptides

PmB is a cationic cyclic decapeptide that binds avidly to hydrophilic and hydrophobic elements of LPS and lipid A and has the capacity to disorganize the outer membrane of gram-negative bacteria (Morrison and Jacobs 1976; Storm et al. 1977). In addition to its bactericidal activity, PmB inhibits many of the in vitro and in vivo biological activities of endotoxin and prevents LPS-induced lethality (Corrigan and Bell 1971; Walterspiel et al. 1986; Flynn et al. 1987; Baldwin et al. 1991; Coyne and Fenwick 1993; Durando et al. 1994). The endotoxin-neutralizing activity of PmB has been attributed to ionic interactions between diaminobutyric acid residues of the cationic antibiotic and mono- or diphosphate groups of lipid A. The resulting LPS-PmB complex has a reduced capacity for association with LBP and/or recognition by membrane-bound CD14 (Coyne and Fenwick 1993; Gallay et al. 1993b; Helander et al. 1994). Recently designed synthetic peptides mimic primary and secondary structural motifs of PmB and further define molecular structures required for LPS binding and neutralization (Danner et al. 1989; Rustici et al. 1993). Although devoid of antibiotic activity, some of these synthetic peptides exhibit endotoxin-neutralizing properties equivalent or superior to those of PmB (Rustici et al. 1993). Toxicity limits the use of PmB as a systemic antibiotic, but endotoxin-neutralizing activity is expressed at relatively nontoxic serum concentrations (Durando et al. 1994). The low toxicity and intact antiendotoxic activity of PmB-related synthetic peptides make them particularly attractive therapeutic candidates.

6 Interruption of LPS-Induced, LBP- and CD14-Mediated Cellular Activation Pathways

Cell-bound and soluble forms of the 50- to 55-kDa glycoprotein CD14 and the 58- to 60-kDa serum glycoprotein LBP are the major mammalian host factors implicated in the cellular recognition of LPS leading to transmembrane signaling and cell activation (Ulevitch and Tobias 1994). Accordingly, these proteins are prime targets for interventional therapy in gram-negative sepsis.

The cell-associated form of CD14 (mCD14) is a glycosylphosphatidyl inositol-anchored membrane glycoprotein expressed on monocytes, macrophages, and polymorphonuclear leukocytes (PMNs). Recent studies indicate that mCD14 is probably one part of multicomponent LPS receptor functionally linked to the initiation of intracellular events associated with LPS-induced cell activation (ULEVITCH and TOBIAS 1994).

The soluble form of CD14 (sCD14), which lacks the glycosylphosphatidyl inositol anchor of mCD14, is found in normal human serum. High circulating levels of sCD14 are associated with increased mortality in gram-negative septic shock (LANDMANN et al. 1995). Endothelial and epithelial cells, which lack mCD14, are able to recognize and respond to LPS/sCD14 complexes through an as yet unidentified membrane-bound element(s) (FREY et al. 1992; ARDITI et al. 1993; PUGIN et al. 1993a,b; GOLDBLUM et al. 1994; ULEVITCH and TOBIAS 1994).

A normal constituent of plasma, LBP increases in concentration in association with the acute-phase response and appears to mediate the lethal effects of endotoxemia (GALLAY et al. 1993a). LBP controls cellular responses to LPS through the formation of high-affinity LPS-LBP complexes. These complexes are recognized by mCD14 on myeloid cells, facilitating LPS-induced cell activation at LPS concentrations encountered in sepsis (SCHUMANN et al. 1990; WRIGHT et al. 1990; MATHISON et al. 1992; TOBIAS and ULEVITCH 1994). Recent evidence (HAILMAN et al. 1994) suggests that LBP functions catalytically as a lipid transfer protein, accelerating the binding of LPS to both mCD14 and sCD14. LBP may also transfer LPS to serum lipoproteins, thus acting as a cofactor for an activity which results in the neutralization of LPS (WURFEL et al. 1994).

The recognition of LPS-LBP complexes by mCD14 on monocytes of macrophages initiates LPS-induced protein tyrosine phosphorylation (WEINSTEIN et al. 1993), NF-κB activation (LEE et al. 1993), the release of proinflammatory cytokines such as TNF-α (WRIGHT et al. 1990), and enhanced tissue factor expression which is causally related to sepsis-induced coagulopathy (STEINEMANN et al. 1994; YANG et al. 1994). These events are all important in the pathogenesis of gram-negative sepsis and septic shock, and all are inhibited by antibodies to LBP or CD14 (WRIGHT et al. 1990; TOBIAS et al. 1992; GALLAY et al. 1993b; LEE et al. 1993; TOBIAS and ULEVITCH 1993; GALLAY et al. 1994; MARTIN et al. 1994; STEINEMANN et al. 1994; YANG et al. 1994). Similarly, LPS-induced, serum-dependent adherence of PMNs to endothelium (WORTHEN et al. 1992), LPS priming of PMNs for enhanced release of superoxide in response to stimulation with fMet-Leu-Phe (YASUI et al. 1994), and LPS-mediated upregulation of CR1 receptors on PMNs (WEINGARTEN et al. 1993) are all blocked by anti-CD14 or anti-LBP antibodies. LPS-induced endothelial cell activation and injury are also inhibited by CD14-specific antibodies, whether these sepsis-related events are produced directly through the action of LPS-sCD14 complexes in plasma or indirectly through LPS-mCD14 complexes on monocytes in whole blood, (FREY et al. 1992; ARDITI et al. 1993; PUGIN et al. 1993a,b; GOLDBLUM et al. 1994).

Despite extensive in vitro evaluation of the LPS-neutralizing properties of LBP-and CD14-specific antibodies there are few published data concerning the in vivo effects of these reagents in animal models. One study (GALLAY et al. 1993a) documented that an anti-LBP antibody reduced circulating TNF levels and protected mice against the lethal effect of LPS when given at the same time as LPS challenge, but failed to protect mice when delayed 15 min after LPS challenge. Another more recent study (LETURCQ et al. 1994) evaluated two CD14-specific MAbs in a primate model of endotoxin-induced shock, and found that these MAbs reduced systemic hypotension, abnormalities in lung permeability, and elevations in circulating proinflammatory cytokines. The authors concluded that "strategies targeting the CD14 receptor will have utility in diseases such as septic shock or ARDS [adult respiratory distress syndrome] where LPS plays a central role in initiating injury."

One hypothesis holds that sCD14, which is shed from myeloid cells in tissue culture and is found in normal plasma, might be a natural inhibitor of endotoxin (WRIGHT 1991). This hypothesis predicts that sCD14 would "neutralize" LPS-LBP complexes by binding the complexes in competition with mCD14 receptors on macrophages, thereby preventing endotoxin-induced cell activation and TNF release. Several lines of evidence support the endotoxin-neutralizing capacity of sCD14. One study, for example, demonstrated that the addition of human sCD14 to calf serum abrogated the binding of fluorescein isothiocyanate-conjugated LPS to bovine monocytes and protected the cells from LPS-induced activation (GRUNWALD et al. 1993). The dose-dependent reduction by sCD14 of LPS-inducible reactive oxygen species in human monocytes was documented in a second study (SCHÜTT et al. 1992). Confirmatory data came from another group which found that recombinant sCD14 (rsCD14) antagonized the response of human and murine mononuclear cells to LPS and inhibited the release of TNF-α by monocytes/macrophages (HAZIOT and GOYERT 1994). It was further discovered that administration of rsCD14 following an injection of LPS protected mice from LPS-induced lethality. The authors concluded than even though rsCD14 enhances endothelial (and epithelial) cell responses to LPS in vitro, these effects are overshadowed in vivo by the inhibitory effect of rsCD14 on monocyte/macrophage responses to LPS which are apparently more relevant with respect to LPS-induced mortality. On this basis, the authors predicted that rsCD14 "might provide a new therapeutic for endotoxin shock in man" (HAZIOT and GOYERT 1994).

In order to characterize structure-function relationships within LBP investigators have created and expressed truncated forms of human LBP (HAN et al. 1994; THEOFAN et al. 1994). One group has characterized a truncated form of the molecule (NH-LBP) comprising amino acid residues 1–197 of the parent molecule (HAN et al. 1994). NH-LBP was found to bind LPS efficiently but not transfer the LPS to either mCD14 or sCD14. The LBP partial structure inhibited LPS binding to LBP, LBP-enhanced binding of LPS to CD14, and LBP-dependent activation of rabbit peritoneal exudate macrophages. On the basis of these findings the authors concluded that the LPS-binding site of LBP resides in the amino-terminal half of LBP, and that the CD14 interaction site resides in the carboxyl-terminal half of the parent molecule. It was further concluded that "appropriate modifications of the primary sequence of LBP might provide a novel protein that would bind LPS with high affinity, but prevent its interaction with either mCD14 or sCD14 . . . [and that] such entities may have therapeutic potential in diseases where LPS plays a prominent role'' (HAN et al. 1994).

Despite their apparent therapeutic potential, no LBP- or CD14-related therapeutic products are presently in clinical development.

7 Serum Lipoproteins

In contrast with LBP, a serum glycoprotein which facilitates the CD14-mediated, proinflammatory recognition and uptake of LPS by myeloid cells, serum lipoproteins bind and neutralize LPS, suggesting a possible anti-inflammatory role for these complex macromolecules in gram-negative sepsis (ULEVITCH and JOHNSTON 1978; ULEVITCH et al. 1979; CAVAILLON et al. 1990; EMANCIPATOR et al. 1992; PARKER et al. 1995). In addition, serum lipoproteins appear to enhance the clearance of circulating endotoxins through hepatobiliary excretion (READ et al. 1993).

Recent data (WURFEL et al. 1994) actually suggest that serum LBP, previously characterized as a lipid transport enzyme, is capable of transferring LPS to HDL as well as to soluble or membrane-bound forms of CD14. Moreover, much of the LBP found in human blood is closely associated with the apolipoprotein A-I (ApoA-I) component of native HDL. It thus appears that ApoA-I and LBP-containing HDL particles are naturally equipped to either bind and neutralize LPS or participate in its transfer to CD14-mediated proinflammatory pathways.

Incubation of LPS with plasma, but not delipidated plasma, reduces the proinflammatory activity of endotoxin in animals (ULEVITCH and JOHNSTON 1978; ULEVITCH et al. 1979). Chylomicrons, VLDL, LDL, HDL, ApoA-I, reconstituted HDL (R-HDL), and synthetic lipid emulsions may all bind and inactivate LPS, as documented experimentally by decrements in LPS-induced *Limulus* amebocyte lysate-gelating activity, cytokine release, physiologic responses, or lethality (HARRIS et al. 1990; EMANCIPATOR et al. 1992; FLEGEL et al. 1993; HARRIS et al. 1993; HUBSCH et al. 1993; LEVINE et al. 1993; PARKER et al. 1995; READ et al. 1995).

A single published study (QUEZADO et al. 1995) employed a canine model of *E. coli* peritonitis and septic shock to evaluate the ability of an infused lipoprotein to protect against a live bacterial challenge. Although the intravenous administration of R-HDL reduced endotoxemia, associated leukopenia, and cytokine release, treated dogs showed signs of liver toxicity and some developed seizures, contributing to decreased survival in the treated group compared with controls. The apparent toxicity of R-HDL exhibited in this canine experiment had not been noted in previous studies using small animals.

8 Natural and Synthetic LPS and Lipid A Analogs or Partial Structures That Antagonize or Induce Tolerance to LPS

With detailed descriptions of LPS and lipid A structures, biosynthetic pathways, and structure-function relationships has come the discovery of relatively nontoxic natural forms of LPS and lipid A or lipid A precursors. These discoveries, in turn, have facilitated the laboratory preparation or synthesis of a variety of nontoxic lipid A analogs or partial structures capable of competing with LPS for binding sites on LPS target cells, inducing endotoxin tolerance, stimulating anti-inflammatory pathways, and recruiting nonspecific mechanisms of host resistance. Such lipid A-like molecules may thus have prophylactic or therapeutic applications in gram-negative sepsis and septic shock, and some are currently undergoing preclinical and early clinical evaluation.

Natural lipid A precursors like lipid X and lipid IV_A, synthetic analogs of these molecules such as 3-aza-lipid X-4-phosphate (SDZ880.431) and phosphonooxyethyl derivatives of lipid IV_A, and the non-toxic *Rhodobacter sphaeroides* diphosphoryl lipid A (rsDPLA) all bind to LPS receptors but express markedly reduced endotoxinlike (agonist) activities. These lipid A-like molecules function as LPS antagonists. They compete in a dose-dependent manner with native LPS for occupancy of LPS receptors, prevent the cellular uptake of LPS, block LPS-induced cell activation, and protect against pathophysiological responses to LPS in vivo (Takayama et al. 1989; Kovach et al. 1990; Golenbock et al. 1991, 1994; Lynn et al. 1991; Qureshi et al. 1991; Wang et al. 1991, 1992; Kitchens et al. 1992; Ulmer et al. 1992, 1994; Lei et al. 1993, Manthey et al. 1993; Heine et al. 1994).

A particularly promising LPS receptor antagonist is so-called E5531, a novel synthetic derivative of non-toxic *Rhodobacter capsulatus* lipid A (Christ et al. 1995). In vitro, E5531 demonstrated potent antagonism of cellular activation induced by phylogenetically diverse LPS in a variety of systems. It was reported to be 260 000 times more effective than lipid X, the monosaccharide lipid A precursor, at inhibiting TNF release while displaying no agonistic activity (Kawata et al. 1992). In vivo, E5531 protected mice from LPS-induced lethality, and, in combination with an antibiotic, protected mice against lethal *E. coli* infections (Christ et al. 1995). In healthy volunteers E5531 produced a dose-dependent decrease in LPS-induced fever, chills, headache, nausea, and myalgias (Bunnell et al. 1995a); reduced TNF and interleukin (IL)-6 responses; and blocked the hyperdynamic vascular and myocardial depressant effects of LPS (Bunnel et al. 1995b).

The ability of certain lipid A analogs to induce tolerance is distinct from the ability to competitively inhibit or antagonize LPS-induced cellular activation and corresponding LPS-induced in vivo events (Carpati et al. 1993). Endotoxin tolerance represents a temporary endotoxin-specific hyporesponsive state induced by prior LPS exposure in relation to subsequent reexposure. It is both an in vitro and in vivo phenomenon. In vitro exposure of macrophages to LPS or certain less toxic lipid A analogs results in a refractoriness to subsequent LPS stimulation marked by a reduction in LPS-induced TNF synthesis and release (MATHISON et al. 1990). Repeated administration of LPS, or lipid A analogs, to experimental animals is associated with a corresponding attenuation of LPS-induced fever and cytokine release, supporting the hypothesis that a critical component of endotoxin tolerance is the reduced production and release of cytokines by monocytes and macrophages upon reexposure to LPS (ROTH et al. 1994). Moreover, recent clinical evidence suggests that a tolerance-like state occurs in septic patients wherein peripheral blood monocytes exhibit markedly diminished proinflammatory cytokine responses to LPS stimulation ex vivo (MUNOZ et al. 1991; VAN DEUREN et al. 1994).

Monophosphoryl lipid A (MPL), a weak LPS agonist, induces tolerance to a subsequent endotoxin challenge in both normal and D-galactosamine sensitized mice. A 4300-fold MPL-induced reduction in the sensitivity of galactosamine-treated mice to endotoxin appears to be mediated by MPL itself as well as by TNF released in response to MPL (RUDBACH et al. 1994). MPL-induced tolerance, as that induced by LPS, entails reduced production of proinflammatory cytokines, including TNF, γ-interferon, colony-stimulating factor, and IL-6 (VOGEL and HENRICSON 1990). However, more MPL than LPS is required to induce the same degree of early endotoxin tolerance, which may be due to the relative inefficiency or low affinity with which MPL binds to LPS cell receptors (VOGEL and HENRICSON 1990). While the results of one study suggested that LPS and MPL are significantly less effective in protecting mice against lethality from peritonitis than against endotoxemia (ASTIZ et al. 1993), another report indicated that MPL given 1 or 2 days, or as little as 6 h, before live *E. coli* challenge produced complete protection (RUDBACH et al. 1994).

In studies reminiscent of those carried out with MPL, rsDPLA induced tolerance to endotoxic shock in rats (ZUCKERMAN and QURESHI 1992). The administration of rsDPLA 48 h before endotoxin challenge reduced the severity of hemodynamic sequelae and mortality associated with endotoxin challenge. This observed tolerance appeared at least in part related to macrophage hyporesponsiveness, as suggested by a decrease in circulating TNF levels in rsDPLA pretreated rats (ZUCKERMAN and QURESHI 1992).

Lipid A analogs may mediate antiendotoxic, anti-inflammatory, or protective activities by mechanisms other than classical LPS antagonism or tolerance. The monosaccharide lipid A analog, SDZ MRL 953, for example, induces resistance to bacterial and fungal infections in part because of its ability to stimulate high levels of secretion of granulocyte colony-stimulating factor and IL-6, and low levels of proinflammatory cytokines such as TNF (ENGELHARDT et al. 1994). This lipid A analog is also capable of inducing high circulating levels of IL-1 receptor antagonist in rhesus monkeys (ENGELHARDT et al. 1994). This activity, together with the development of hyporesponsiveness to endotoxin, may provide a basis for observed protection against infection.

It has been postulated that the in vivo protective effect of rsDPLA against endotoxin challenge in mice actually involves three major mechanisms: antagonism of LPS receptors, induction of tolerance, and ability to elicit an anti-inflammatory response (Zuckerman and Qureshi 1992; Hofman et al. 1994). The latter is supported by an observed stimulatory effect of rsDPLA on the hypothalamic-pituitary-adrenal axis, resulting in enhanced secretion of adrenocorticotrophic hormone and corticosteroids. Elevated corticosterone levels, in turn, inhibit TNF and IL-1 biosynthesis at the pretranslational level, providing a basis for the anti-inflammatory effect of rsDPLA. The inability of rsDPLA to inhibit LPS-induced increases of serum TNF in adrenalectomized mice suggests that the in vivo protective effect of rsDPLA is mediated largely through corticosterone induction and not LPS antagonism or tolerance (Zuckerman and Qureshi 1992).

In addition to its tolerance-inducing activity, MPL induces nonspecific resistance to a variety of infectious agents, including gram-positive bacteria, and is capable of modulating the inflammatory response to host-generated proinflammatory mediators as well as to microbial products (Rudbach et al. 1994).

MPL has been administered safely to healthy volunteers and tolerized them to subsequently administered LPS (Astiz et al. 1995). Plans are underway to evaluate the ability of MPL to prevent infectious and febrile episodes in neutropenic leukemia patients, and consideration is being given to the possible use of MPL as an immunomodulator in human immunodeficiency virus infections and tuberculosis. The intended clinical uses of MPL are thus far broader than those suggested by its endotoxin-tolerizing activities alone.

9 Plasmapheresis, Exchange Transfusion, and Extracorporal LPS Adsorption

An alternative to the systemic administration of specific endotoxin-neutralizing agents in the treatment of gram-negative sepsis is the replacement or purification of a septic patient's blood or plasma in order to remove endotoxin and inflammatory mediators (Pollack 1992b). A number of reports have suggested clinical responses to plasmapheresis or exchange transfusion in septic patients. Most of these studies, however, have been anecdotal, uncontrolled, and/or nonrandomized, and the majority have been limited to patients with meningococcemia (Togari et al. 1983; Drapkin et al. 1989; Westendorp et al. 1990, van Deuren et al. 1992). In contrast, a controlled trial of plasmapheresis in an antibiotic-treated canine septic shock model revealed increased mortality and hemodynamic instability in animals undergoing this treatment (Natanson et al. 1989).

A more specific therapeutic approach to endotoxemia involves extracorporal removal of circulating endotoxin by plasma filtration or selective adsorption (COHEN et al. 1987; HANASAWA et al. 1988; BYSANI et al. 1990). An evaluation, in animal models, of extracorporal adsorption of endotoxin-containing plasma with PmB immobilized on insoluble fibers demonstrated effective endotoxin removal, reduced LPS-induced toxicity, and enhanced survival (HANASAWA et al. 1988). This method of extracorporal adsorption has undergone evaluation in septic patients, and is currently in clinical use in Japan.

Acknowledgements. This work was supported by Naval Medical Research and Development Command Grant 63706N.M00095.001.9218.

References

Alpert G, Baldwin G, Thompson C, Wainwright N, Novitsky TJ, Gillis Z, Parsonnet J, Fleisher GR, Siber GR (1992) Limulus antilipopolysaccharide factor protects rabbits from meningococcal endotoxin shock. J Infect Dis 165: 494–500

Ammons WS, Kohn FR, Kung AHC (1994) Protective effects of an N-terminal fragment of bactericidal/permeability-increasing protein in rodent models of gram-negative sepsis: role of bactericidal properties. J Infect Dis 170: 1473–1482

Appelmelk BJ, Cohen J (1992) The protective role of antibodies to the lipopolysaccharide core region. In: Ryan JL, Morrison DC (eds) Immunopharmacology and pathophysiology. CRC Press, Boca Raton, pp 375–410 (Bacterial endotoxic lipopolysaccharides, vol 2)

Appelmelk BJ, An Y-Q, Geerts M, Thijs BG, de Boer HA, MacLaren DM, de Graaff J, Nuijens JH (1994a) Lactoferrin is a lipid A-binding protein. Infect Immun 62: 2628–2632

Appelmelk BJ, An Y-Q, Thijs BG, MacLaren DM, de Graaff J (1994b) Recombinant human bactericidal/permeability-increasing protein ($rBPI_{23}$) is a universal lipopolysaccharide-binding ligand. Infect Immun 62: 3564–3567

Arditi M, Zhou J, Dorio R, Rong GW, Goyert SM, Kim KS (1993) Endotoxin-mediated endothelial cell injury and activation: role of soluble CD14. Infect Immun 61: 3149–3156

Arditi M, Zhou J, Huang SH, Luckett PM, Marra MN, Kim KS (1994) Bactericidal/permeability-increasing protein protects vascular endothelial cells from lipopolysaccharide-induced activation and injury. Infect Immun 62: 3930–3936

Astiz ME, Saha DC, Brooks K, Carpati CM, Rackow EC (1993) Comparison of the induction of endotoxin tolerance in endotoxemia and peritonitis by monophosphoryl lipid A and lipopolysaccharide. Circ Shock 39: 194–198

Astiz ME, Rackow EC, Still JG, Howell ST, Cato A, Von Eschen KB, Ulrich JT, Rudbach JA, McMahon G, Vargas R, Stern W (1995) Pretreatment of normal humans with monophosphoryl lipid A induces tolerance to endotoxin: a prospective, double-blind, randomized, controlled trial. Crit Care Med 23: 9–17

Baldwin G, Alpert G, Caputo GL, Baskin M, Parsonnet J, Gillis ZA, Thompson C, Siber GR, Fleisher GR (1991) Effect of polymyxin B on experimental shock from meningococcal and *Escherichia coli* endotoxins. J Infect Dis 164: 542–549

Baumgartner JD, Heumann D, Gerain J, Weinbreck P, Grau GE, Glauser MP (1990) Association between protective efficacy of anti-lipopolysaccharide (LPS) antibodies and suppression of LPS-induced tumor necrosis factor α and interleukin 6. J Exp Med 171: 889–896

Bogard WC Jr, Siegel SA, Leone AO, Damiano E, Shealy DJ, Ely TM, Frederick B, Mascelli MA, Siegel RC, Machielse B, Naveh D, Kaplan PM, Daddona PE (1993) Human monoclonal antibody HA-1A binds to endotoxin via an epitope in the lipid A domain of lipopolysaccharide. J Immunol 150: 4438–4449

Bunnell E, Lynn M, Parrillo JE, Habet K, Friedhoff LT, Rogers SL (1995a) Effect of E5531 on systemic responses to endotoxin in healthy volunteers. Crit Care Med 23: A147

Bunnell E, Neumann A, Lynn M, Friedhoff LT, Rogers SL, Habet K, Parrillo JE (1995b) E5531, an endotoxin antagonist, blocks the hyperdynamic and depressant cardiovascular effects of endotoxin in healthy subjects. Crit Care Med 23: A151

Bysani GK, Shenep JL, Hildner WK, Stidham GL, Roberson PK (1990) Detoxification of plasma containing lipopolysaccharide by adsorption. Crit Care Med 18: 67–71

Calvano SE, Thompson WA, Marra MN, Coyle SM, de Riesthal HF, Trousdale RK, Barie PS, Scott RW, Moldawer LL, Lowry SF (1994) Changes in polymorphonuclear leukocytes surface and plasma bactericidal/permeability-increasing protein and plasma lipopolysaccharide binding protein during endotoxemia or sepsis. Arch Surg 129: 220–226

Carpati CM, Astiz ME, Saha DC, Rackow EC (1993) Diphosphoryl lipid A from *Rhodopseudomonas sphaerodies* induces tolerance to endotoxic shock in the rat. Crit Care Med 21: 753–758

Cavaillon JM, Fitting C, Haeffner-Cavaillon N, Kirsch SJ, Warren HS (1990) Cytokine response by monocytes and macrophages to free and lipoprotein-bound lipopolysaccharide. Infect Immun 58: 2375–2382

Chen T-Y, Zapol WM, Greene E, Robinson DR, Rubin RH (1993) Protective effects of E5, an antiendotoxin monoclonal antibody, in the ovine pulmonary circulation. J Appl Physiol 75: 233–239

Christ WJ, Asano O, Robidoux ALC, Perez M, Wang Y, Dubuc GR, Gavin WE, Hawkins LD, McGuinness PD, Mullarkey MA, Lewis MD, Kishi Y, Kawata T, Bristol JR, Rose JR, Rossignol DP, Kobayashi S, Hishinuma I, Kimura A, Asakawa N, Katayama K, Yamatsu I (1995) E5531, a pure endotoxin antagonist of high potency. Science 268: 80–83

Cohen J, Aslam M, Pusey CD, Ryan CJ (1987) Protection from endotoxemia: a rat model of plasmapheresis and specific adsorption with polymyxin B. J Infect Dis 155: 690–695

Collins HH, Cross AS, Dobek A, Opal SM, McClain JB, Sadoff JC (1989) Oral ciprofloxacin and a monoclonal antibody to lipopolysaccharide protect leukopenic rats from lethal infection with *Pseudomonas aeruginosa*. J Infect Dis 159: 1073–1082

Corrigan JJ Jr, Bell BM (1971) Endotoxin-induced intravascular coagulation: prevention with polymyxin B sulfate. J Lab Clin Med 77: 802–810

Coyne CP, Fenwick BW (1993) Inhibition of lipopolysaccharide-induced macrophage tumor necrosis factor-α synthesis by polymyxin B sulfate. Am J Vet Res 54: 305–314

Cross AS (1994) Antiendotoxin antibodies: a dead end? Ann Intern Med 121: 58–60

Cross AS, Opal S (1994) Therapeutic intervention in sepsis with antibody to endotoxin: is there a future? J Endotoxin Res 1: 57–69

Cross AS, Opal SM, Palardy JE, Bodmer MW, Sadoff JC (1993) The efficacy of combination immunotherapy in experimental *Pseudomonas* sepsis. J Infect Dis 167: 112–118

Danner RL, Joiner KA, Rubin M, Patterson WH, Johnson N, Ayers K, Parrillo JE (1989) Purification, toxicity and antiendotoxin activity of polymyxin B nonapeptide. Antimicrob Agents Chemother 33: 1428–1434

Dentener MA, Von EJU, Francot GJM, Marra MN, Buurman WA (1993) Antagonistic effects of lipopolysaccharide binding protein and bactericidal/permeability-increasing protein on lipopolysaccharide-induced cytokine release by mononuclear phagocytes. Competition for binding to lipopolysaccharide. J Immunol 151: 4258–4265

Desch CE, O'Hara P, Harlan JM (1989) Antilipopolysaccharide factor from horseshoe crab, *Tachypleus tridentatus,* inhibits lipopolysaccharide activation of cultured human endothelial cells. Infect Immun 57: 1612–1614

Di Padova FE, Brade H, Barclay GR, Poxton IR, Liehl E, Schuetze E, Kocher HP, Ramsay G, Schreier MH, McClelland DB, Rietschel ET (1993) A broadly cross-protective monoclonal antibody binding to *Escherichia coli* and *Salmonella* lipopolysaccharide. Infect Immun 61: 3863–3872

Drapkin MS, Wisch JS, Gelfand JA, Cannon JG, Dinarello CA (1989) Plasmapheresis for fulminant meningococcemia. Pediatr Infect Dis J 8: 399–400

Durando MM, MacKay RJ, Linda S, Skelley LA (1994) Effects of polymyxin B and *Salmonella typhimurium* antiserum on horses given endotoxin intravenously. Am J Vet Res 55: 921–927

Elsbach P, Weiss J, Doerfler M, Shu C, Kohn C, Ammons WS, Kung AHC, Meszaros K, Parent JB, Gazzano-Santoro H, Huang K, Fishwild D (1994) The bactericidal/permeability increasing protein of neutrophils is a potent antibacterial and anti-endotoxin agent in vitro and in vivo. Prog Clin Biol Res 388: 41–51

Emancipator K, Csako G, Elin RJ (1992) In vitro inactivation of bacterial endotoxin by human lipoproteins and apolipoproteins. Infect Immun 60: 596–601

Engelhardt R, Foster C, Golenbock D, Hildebrandt J, Landmann R, Mayer P, Stern AC, Stütz P (1994) SDZ MRL 953, a lipid A analog as selective cytokine inducer. J Endotoxin Res 1 Suppl 1: 23

Evans TJ, Carpenter A, Moyes D, Martin R, Cohen J (1995) Protective effects of a recombinant amino-terminal fragment of human bactericidal/permeability-increasing protein in an animal model of gram-negative sepsis. J Infect Dis 171: 153–160

Fisher CJ Jr, Marra MN, Palardy JE, Marchbanks CR, Scott RW, Opal SM (1994) Human neutrophil bactericidal/permeability-increasing protein reduces mortality rate from endotoxin challenge: a placebo-controlled study. Crit Care Med 22: 553–558

Flegel WA, Baumstark MW, Weinstock C, Berg A, Northoff H (1993) Prevention of endotoxin-induced monokine release by human low- and high-density lipoproteins and by apolipoprotein A-I. Infect Immun 61: 5140–5146

Fletcher MA, McKenna TM, Quance JL, Wainwright NR, Williams TJ (1993) Lipopolysaccharide detoxification by endotoxin neutralizing protein. J Surg Res 55: 147–154

Flynn PM, Shenep JL, Stokes DC, Fairclough D, Hildner WK (1987) Polymyxin B moderates acidosis and hypotension in established, experimental gram-negative septicemia. J Infect Dis 156: 706–712

Frey EA, Miller DS, Jahr TG, Sundan A, Bazil V, Espevik T, Finlay BB, Wright SD (1992) Soluble CD14 participates in the response of cells to lipopolysaccharide. J Exp Med 176: 1665–1671

Fujihara Y, Bogard WC, Lei M-G, Daddona PE, Morrison DC (1993a) Monoclonal anti-lipid A IgM antibodies HA-1A and E-5 recognize distinct epitopes on lipopolysaccharide and lipid A. J Infect Dis 168: 1429–1435

Fujihara Y, Lei M-G, Morrison DC (1993b) Characterization of specific binding of a human immunoglobulin M monoclonal antibody to lipopolysaccharide and its lipid A domain. Infect Immun 61: 910–918

Gallay P, Heumann D, Le Roy D, Barras C, Glauser MP (1993a) Lipolysaccharide-binding protein as a major plasma protein responsible for endotoxemic shock. Proc Natl Acad Sci USA 90: 9935–9938

Gallay P, Jongeneel CV, Barras C, Burnier M, Baumgartner J-D, Glauser MP, Heumann D (1993b) Short time exposure to lipopolysaccharide is sufficient to activate human monocytes. J Immunol 150: 5086-5093

Gallay P, Barras C, Tobias PS, Calandra T, Glauser MP, Heumann D (1994) Lipopolysaccharide (LPS)-binding protein in human serum determines the tumor necrosis factor responses of monocytes to LPS. J Infect Dis 170: 1319–1322

Gazzano-Santoro H, Parent JB, Grinna L, Horwtiz A, Parsons T, Theofan G, Elsbach P, Weiss J, Conlon PJ (1992) High-affinity binding of the bactericidal/permeability-increasing protein and a recombinant amino-terminal fargment to the lipid A region of lipopolysaccharide. Infect Immun 60: 4754–4761

Gazzano-Santoro H, Mészáros K, Birr C, Carroll SF, Theofan G, Horwitz AH, Lim E, Aberle S, Kasler H, Parent JB (1994) Competition between $rBPI_{23}$, a recombinant fragment of bactericidal/permeability-increasing protein, and lipopolysaccharide (LPS)-binding protein for binding to LPS and gram-negative bacteria. Infect Immun 62: 1185–1191

Gennaro R, Scocchi M, Storichi P, Romeo D, Tossi A, Zanetti M (1994) Identification of active domain in the LPS-binding CAP18 and in two homologous proteins from pig myeloid cells. J Endotoxin Res 1 Suppl 1: 92

Goldblum SE, Brann TW, Ding X, Pugin J, Tobias PS (1994) Lipopolysaccharide (LPS)-binding protein and soluble CD14 function as accessory molecules for LPS-induced changes in endothelial barrier function, in vitro. J Clin Invest 93: 692–702

Golenbock DT, Hampton RY, Qureshi N, Takayama K, Raetz CRH (1991) Lipid A-like molecules that antagonized the effects of endotoxins on human monocytes. J Biol Chem 266: 19490–19498

Golenbock DT, Delude RL, Christ W, Rossigno D, Millham F (1994) Relationship of LPS antagonists to CD14. J Endotoxin Res 1 Supp 1: 21

Greenman RL, Schein RMH, Martin MA, Wenzel RP, MacIntyre NR, Emmanuel G, Chmel H, Kohler RB, McCarthy M, Plouffe J, Russel JA, the XOMA Sepsis Study Group (1991) A controlled clinical trial of E5 murine monoclonal IgM antibody to endotoxin in the treatment of gram-negative sepsis. JAMA 266: 1097–1102

Grunwald U, Krüger C, Schütt C (1993) Endotoxin-neutralizing capacity of soluble CD14 is a highly conserved specific function. Circ Shock 39: 220–225

Hailman E, Lichenstein HS, Wurfel MM, Miller DS, Johnson DA, Kelley M, Busse LA, Zukowski MM, Wright SD (1994) Lipopolysaccharide (LPS)-binding protein accelerates the binding of LPS to CD14. J Exp Med 179: 269–277

Han J, Mathison JC, Ulevitch RJ, Tobias PS (1994) Lipopolysaccharide (LPS) binding protein, truncated at Ile-197, binds LPS but does not transfer LPS to CD14. J Biol Chem 269: 8172–8175

Hanasawa K, Tani T, Oka T, Yoshioka T, Aoki H, Endo Y, Kodama M (1988) Selective removal of endotoxin from the blood by extracoporeal hemoperfusion with polymyxin B immobilized fiber. Prog Clin Biol Res 264: 337–341

Harris HW, Grunfeld C, Feingold KR, Rapp JH (1990) Human very low density lipoproteins and chylomicrons can protect against endotoxin-induced death in mice. J Clin Invest 86: 696–702

Harris HW, Grunfeld C, Reingold KR, Read TE, Kane JP, Jones AL, Eichbaum EB, Bland GF, Rapp JH (1993) Chylomicrons alter the fate of endotoxin, decreasing tumor necrosis factor release and preventing death. J Clin Invest 91: 1028–1034

Haziot A, Goyert SM (1994) Recombinant soluble CD14 inhibits LPS-induced mortality in a murine model. J Endotoxin Res 1 Suppl 1: 20

Heine H, Brade H, Kusumoto S, Kusama T, Rietschel ET, Flad H, Ulmer AJ (1994) Inhibition of LPS binding on human monocytes by phosphonooxyethyl analogs of lipid A. J Endotoxin Res 1: 14–20

Helander IM, Nummila K, Kilpeläinen I, Vaara M (1994) Increased substitution of phosphate groups in lipopolysaccharides and lipid A of polymyxin-resistant mutants of *Escherichia coli* and *Salmonella typhimurium*. J Endotoxin Res 1 Suppl 1: 1

Heumann D, Gallay P, Betz-Corradin S, Barras C, Baumgartner J-D, Glauser MP (1993) Competition between bactericidal/permeability-increasing protein and lipopolysaccharide-binding protein for lipopolysaccharide binding to monocytes. J Infect Dis 167: 1351–1357

Hirata M, Shimomura Y, Yoshida M, Morgan JG, Palings I, Wilson D, Yen MH, Wright SC, Larrick JW (1994a) Characterization of a rabbit cationic protein (CAP18) with lipopolysaccharide-inhibitory activity. Infect Immun 62: 1421–1426

Hirata M, Shimomura Y, Yoshida M, Wright SC, Larrick JW (1994b) Endotoxin-binding synthetic peptides with endotoxin-neutralizing, antibacterial and anticoagulant activities. Prog Clin Biol Res 388: 147–159

Hoffman WD, Pollack M, Banks SM, Koev LA, Solomon MA, Danner RL, Koles N, Guelde G, Yatsiv I, Mouginis T, Elin RJ, Hosseini JM, Bacher J, Porter JC, Natanson C (1994) Distinct functional activities in canine septic shock of monoclonal antibodies specific for the O polysaccharide and core regions of *Escherichia coli* lipopolysaccharide. J Infect Dis 169: 553–561

Hofman J, Qureshi N, Takayama K, Kalin N, Manthey CL, Vogel SN, Lei M, Morrison DC (1994) Lipopolysaccharide-antagonizing effects of diphosphoryl lipid A from *Rhodobacter sphaeroides* (Rs-DPLA). Prog Clin Biol Res 388: 95–106

Hubsch AP, Powell FS, Lerch PG, Doran JE (1993) A reconstituted, apolipoprotein A-1 containing lipoprotein reduces tumor necrosis factor release and attenuates shock in endotoxemic rabbits. Circ Shock 40: 14–23

Kartalija M, Kim Y, White ML, Nau R, Tureen JH, Täuber MG (1995) Effect of a recombinant N-terminal fragment of bactericidal/permeability-increasing protein ($rBPI_{23}$) on cerebrospinal fluid inflammation induced by endotoxin. J Infect Dis 171: 948–953

Katsikis P, Harris G, Page T, Paleolog E, Feldman M, Dalesandro MR, Kinney CS, Siegel SA, Woody JN, Daddona PE (1993) Antilipid A monoclonal antibody HA-1A: immune complex clearance of endotoxin reduces TNF-α, IL-1β and IL-6 production. Cytokine 5: 348–353

Kawata T, Bristol J, McGuigan L, Rossignol D, Christ W, Robidoux A, Perez M, Asano O, Dubuc G, Hawkins L, Wang Y, Kishi Y, Kobayashi S, Katayama K, Yamatsu I (1992) Anti-endotoxin activities of E5531, a novel synthetic derivative of lipid A. 32nd Interscience Conference on Antimicrobial Agents and Chemotherapy American Society for Microbiology, Washington, DC, p 337

Kelly CJ, Cech AC, Argenteanu M, Gallagher H, Shou J, Minnard E, Daly JM (1993) Role of bactericidal permeability-increasing protein in the treatment of gram-negative pneumonia. Surgery 114: 140–146

Kirkland TN, Ziegler EJ (1984) An immunoprotective monoclonal antibody to lipopolysaccharide. J Immunol 132: 2590–2592

Kitchens RL, Ulevitch RJ, Munford RS (1992) Lipopolysaccharide (LPS) partial structures inhibit responses to LPS in a human macrophage cell line without inhibiting LPS uptake by a CD14-mediated pathway. J Exp Med 176: 485–494

Kloczewiak M, Black KM, Loiselle P, Cavaillon J-M, Wainwright N, Warren HS (1994) Synthetic peptides that mimic the binding site of horseshoe crab antilipopolysaccharide factor. J Infect Dis 170: 1490–1497

Kohn FR, Kung AHC (1995) Role of endotoxin in acute inflammation induced by gram-negative bacteria: specific inhibition of lipopolysaccharide-mediated responses with an amino-terminal fragment of bactericidal/permeability-increasing protein. Infect Immun 63: 333–339

Kohn FR, Ammons WS, Horwitz A, Grinna L, Theofan G, Weickmann J, Kung AHC (1993) Protective effect of a recombinant amino-terminal fragment of bactericidal/permeability-increasing protein in experimental endotoxemia. J Infect Dis 168: 1307–1310

Kovach NL, Yee E, Munford RS, Raetz CRH, Harlan JM (1990) Lipid IVA inhibits synthesis and release of tumor necrosis factor induced by lipopolysaccharide in human whole blood ex vivo. J Exp Med 172: 77–84

Krieger JI, Fletcher RC, Siegel SA, Fearon DT, Neblock DS, Boutin RH, Taylor RP, Daddona PE (1993) Human anti-endotoxin antibody HA-1A mediates complement-dependent binding of *Escherichia coli* J5 lipopolysaccharide to complement receptor type 1 of human erythrocytes and neutrophils. J Infect Dis 167: 865–875

Kung AHC, Ammons WS, Lin Y, Kohn FR (1994) Efficacy of a recombinant amino terminal fragment of bactericidal/permeability increasing protein in rodents challenged with LPS or *E. coli* bacteria. Prog Clin Biol Res 388: 255–263

Kuppermann N, Nelson DS, Saladino RA, Thompson CM, Sattler F, Novitsky TJ, Fleisher GR, Siber GR (1994) Comparison of a recombinant endotoxin-neutralizing protein with a human monoclonal antibody to endotoxin for the treatment of *Escherichia coli* sepsis in rats. J Infect Dis 170: 630–635

Landmann R, Zimmerli W, Sansano S, Link S, Hahn A, Glauser MP, Calandra T (1995) Increased circulating soluble CD14 is associated with high mortality in gram-negative septic shock. J Infect Dis 171: 639–644

Larrick JW, Hirata M, Balint RF, Lee J, Zhong J, Wright SC (1995) Human CAP18: a novel antimicrobial lipopolysaccharide-binding protein. Infect Immun 63: 1291–1297

Lee J, Kravchenko V, Kirkland TN, Han J, Mackman N, Moriarty A, Leturcq D, Tobias PS, Ulevitch RJ (1993) Glycosyl-phosphatidylinositol-anchored or integral membrane forms of CD14 mediate indentical cellular responses to endotoxin. Proc Natl Acad Sci USA 90: 9930–9934

Lehrer RI, Lichtenstein AK, Ganz T (1993) Defensins: antimicrobial and cytotoxic peptides of mammalian cells. Annu Rev Immunol 11: 105–128

Lei M-G, Qureshi N, Morrison DC (1993) Lipopolysaccharide (LPS) binding to 73-kDa and 38-kDa surface proteins on lymphoreticular cells: preferential inhibition of LPS binding to the former by *Rhodopseudomonas sphaeroides* lipid A. Immunol Lett 36: 245–250

Leturcq DJ, Moriarty AM, Winn RK, Talbott G, Martin TR, Ulevitch R (1994) Therapeutic strategies to block LPS activation through the CD14 receptor. J Endotoxin Res 1 Suppl 1: 20

Levine DM, Parker TS, Donnelly TM, Walsh A, Rubin AL (1993) In vivo protection against endotoxin by plasma high density lipoprotein. Proc Natl Acad Sci USA 90: 12040–12044

Levy O, Weiss J, Zarember K, Ooi CE, Elsbach P (1993) Antibacterial 15-kDa protein isoforms (p15s) are members of a novel family of leukocyte proteins. J Biol Chem 268: 6058–6063

Lynn WA, Raetz CRH, Qureshi N, Golenbock DT (1991) Lipopolysaccharide-induced stimulation of CD11b/CD18 expression on neutrophils. Evidence of specific receptor-based response and inhibition by lipid A-based antagonists. J Immunol 147: 3072–3079

Manthey CL, Perera PY, Qureshi N, Stütz PL, Hamilton TA, Vogel SN (1993) Modulation of lipopolysaccharide-induced macrophage gene expression by *Rhodobacter sphaeroides* lipid A and SDZ 880.431. Infect Immun 61: 3518–3526

Marra MN, Au-Young J, Lin L, Lane JC, Snable JL, Thornton MB, McKelligon BM, Seilhamer JJ, Scott RW (1994) Identification of bactericidal/permeability increasing protein (BPI) and lipopolysaccharide binding protein (LBP) domains which are important for LPS neutralizing or transducing activity and circulating half-life. J Endotoxin Res 1 Suppl 1: 18

Martin TR, Mongovin SM, Tobias PS, Mathison JC, Moriarty AM, Leturcq DJ, Ulevitch RJ (1994) The CD14 differentiation antigen mediates the development of endotoxin responsiveness during differentiation of mononuclear phagocytes. J Leukocyte Biol 56: 1–9

Mascelli MA, Frederick B, Ely T, Neblock DS, Shealy DJ, Pak KY, Daddona PE (1993) Reactivity of the human antiendotoxin immunoglobulin M monoclonal antibody HA-1A with lipopolysaccharides from rough and smooth gram-negative organisms. Infect Immun 61: 1756–1763

Mathison JC, Virca GD, Wolfson E, Tobias PS, Glaser K, Ulevitch RJ (1990) Adaptation to bacterial lipopolysaccharide controls lipopolysaccharide-induced tumor necrosis factor production in rabbit macrophages. J Clin Invest 85: 1108–1118

Mathison JC, Tobias PS, Wolfson E, Ulevitch RJ (1992) Plasma lipopolysaccharide (LPS)-binding protein. A key component in macrophage recognition of gram-negative LPS. J Immunol 149: 200–206

Mathison J, Wolfson E, Steinemann S, Tobias P, Ulevitch R (1993) Lipopolysaccharide (LPS) recognition in macrophages. Participation of LPS-binding protein and CD14 in LPS-induced adaptation in rabbit peritoneal exudate macrophages. J Clin Invest 92: 2053–2059

McCloskey RV, Straube RC, Sanders C, Smith SM, Smith CR, the CHESS Trial Study Group (1994) Treatment of septic shock with human monoclonal antibody HA-1A. A randomized, double-blind, placebo-controlled trial. Ann Intern Med 121: 1–5

Morita T, Ohtsubo S, Nakamura T, Tanaka S, Iwanaga S, Ohashi K, Niwa M (1985) Isolation and biological activities of limulus anticoagulant (anti-LPS factor) which interacts with lipopolysaccharide (LPS). J Biochem 97: 1611–1620

Morrison DC, Jacobs DM (1976) Binding of polymyxin B to the lipid A portion of bacterial lipopolysaccharides. Immunochemistry 13: 813–818

Munoz C, Carlet J, Fitting C, Misset B, Blériot J-P, Cavaillon J-M (1991) Dysregulation of in vitro cytokine production by monocytes during sepsis. J Clin Invest 88: 1747–1754

Natanson C, Hoffman WD, Danner RL, Koev LL, Banks SM, Walker LD, Parrillo JE (1989) A controlled trial of plasmapheresis (P) fails to improve outcome in an antibiotic treated canine model of human septic shock (SS). Clin Res 37: 346A

Nelson D, Kuppermann N, Fleisher GR, Hammer BK, Thompson CM, Garcia CT, Novitsky TJ, Parsonnett J, Onderdonk A, Siber GR, Saladino RA (1995) Recombinant endotoxin neutralizing protein improves survival from *Escherichia coli* sepsis in rats. Crit Care Med 23: 92–98

Oishi K, Koles NL, Guelde G, Pollack M (1992) Antibacterial and protective properties of monoclonal antibodies reactive with *Escherichia coli* O111: B4 lipopolysaccharide: relation to antibody isotype and complement-fixing activity. J Infect Dis 165: 34–45

Ono Y, Tokumura Y, Miyashita T, Kobayashi K, Kunii O (1994) Anti-endotoxic activity of murine monoclonal antibody (E5) assessed by inhibition of priming of human phagocytes by endotoxin. Kansenshogaku Zasshi 68: 346–352

Ooi CE, Weiss J, Doerfler ME, Elsbach P (1991) Endotoxin-neutralizing properties of the 25 kD N-terminal fragment and a newly isolated 30 kD C-terminal fragment of the 55–60 kD bactericidal/permeability-increasing protein of human neutrophils. J Exp Med 174: 649–655

Opal SM, Palardy JE, Marra MN, Fisher CJ Jr, McKelligon BM, Scott RW (1994) Relative concentrations of endotoxin-binding proteins in body fluids during infection. Lancet 344: 429–431

Paleolog EM, Katsikis P, Harris G, Daddona P, Dalesandro MR, Kinney CS, Woody JN, Feldmann M (1993) Antilipid A monoclonal antibody HA-1A decreases the capacity of bacterial lipopolysaccharide to activate human vascular endothelial cells by an immune adherence mechanism. Cytokine 5: 570–577

Parent JB (1994) Recombinant bactericidal/permeability increasing protein (rBPI23) has potent antibacterial and antiendotoxin activity in vitro and in vivo. J Endotoxin Res 1 Suppl 1: 18

Parent JB, Gazzano-Santoro H, Wood DM, Lim E, Pruyne PT (1992) Reactivity of monoclonal antibody E5 with endotoxin. II. Binding to short-and long-chain smooth lipopolysaccharides. Circ Shock 38: 63–73

Parker TS, Levine DM, Chang JCC, Laxer J, Coffin CC, Rubin AL (1995) Reconstituted high-density lipoprotein neutralizes gram-negative bacterial lipopolysaccharides in human whole blood. Infect Immun 63: 253–258

Pereira HA, Lacquement M, Spitznagel JK (1994) CAP37, a multifunctional neutrophil protein with lipopolysaccharide (LPS) binding activity. J Endotoxin Res 1 Suppl 1: 93

Pereira HA, Erdem I, Pohl J, Spitznagel JK (1993) Synthetic bactericidal peptide based on CAP37: a 37-kDa human neutrophil granule-associated cationic antimicrobial protein chemotactic for monocytes. Proc Natl Acad Sci USA 90: 4733–4737

Pier GB, Koles NL, Meluleni G, Hatano K, Pollack M (1994) Specificity and funciton of murine monoclonal antibodies and immunization-induced human polyclonal antibodies to lipopolysaccharide subtypes of *Pseudomonas aeruginosa* serogroup 06. Infect Immun 62: 1137–1143

Pollack M (1992a) Specificity and function of lipopolysaccharide antibodies. In: Ryan JL, Morrison DC (eds) Immunopharmacology and pathophysiology. CRC Press, Boca Raton, pp 347–374 (Bacterial endotoxic lipopolysaccharides vol 2)

Pollack M (1992b) Blood exchange and plasmapheresis in sepsis and septic shock. Clin Infect Dis 15: 431–433

Pollack M, Oishi K, Chia J, Evans ME, Guelde G, Koles NL (1990) Specificity and function of monoclonal antibodies reactive with discrete structural elements of bacterial lipopolysaccharide. Adv Exp Med Biol 256: 331–340

Pollack M, Espinoza AM, Guelde G, Koles NL, Wahl LM, Ohl CA (1995) Lipopolysaccharide (LPS)-specific monoclonal antibodies regulate LPS uptake and LPS-induced tumor necrosis factor-α responses by human monocytes. J Infect Dis 172: 794–804

Pugin J, Schürer-Maly C-C, Leturcq D, Moriarty A, Ulevitch RJ, Tobias PS (1993a) Lipopolysaccharide activation of human endothelial and epithelial cells is mediated by lipopolysaccharide-binding protein and soluble CD14. Proc Natl Acad Sci USA 90: 2744–2748

Pugin J, Ulevitch RJ, Tobias PS (1993b) A critical role for monocytes and CD14 in endotoxin-induced endothelial cell activation. J Exp Med 178: 2193–2200

Quezado ZMN, Natanson C, Alling DW, Banks SM, Koev CA, Elin RJ, Hosseini JM, Bacher JD, Danner RL, Hoffman WD (1993) A controlled trial of HA-1A in a canine model of gram-negative septic shock. JAMA 269: 2221–2227

Quezado ZMN, Natanson C, Banks SM, Alling DW, Koev CA, Danner RL, Elin RJ, Hosseini JM, Parker TS, Levine DM, Rubin AL, Hoffman WD (1995) Therapeutic trial of reconstituted human high-density lipoprotein in an canine model of gram-negative septic shock. J Pharmacol Exp Ther 272: 604–611

Qureshi N, Takayama K, Kurtz R (1991) Diphosphoryl lipid obtained from the nontoxic lipopolysaccharide of *Rhodopseudomonas sphaeroides* is an endotoxin antagonist in mice. Infect Immun 59: 441–444

Read TE, Harris HW, Grunfeld C, Feingold KR, Calhoun MC, Kane JP, Rapp JH (1993) Chylomicrons enhance endotoxin excretion in bile. Infect Immun 61: 3496–3502

Read TE, Grunfled C, Kumwenda Z, Calhoun MC, Kane JP, Feingold KR, Rapp JH (1995) Triglyceride-rich lipoproteins improve survival when given after endotoxin in rats. Surgery 117: 62–67

Rietshcel ET, Kirikae T, Schade FU, Ulmer AJ, Holst O, Brade H, Schmidt G, Mamat U, Grimmecke H-D, Kusumoto S, Zähringer U (1993) The chemical structure of bacterial endotoxin in relation to bioactivity. Immunobiology 187: 169–190

Riestchel ET, Kirikae T, Schade FU, Mamat U, Schmidt G, Loppnow H, Ulmer AJ, Zähringer U, Seydel U, Di Padova F, Schreier M, Brade H (1994) Bacterial endotoxin: molecular relationships of structure to activity and function. FASEB J 8: 217–225

Rogy MA, Moldawer LL, Oldenburg HSA, Thompson WA, Montegut WJ, Stackpole SA, Kuman A, Palladino MA, Marra MN, Lowry SF (1994) Anti-endotoxin therapy in primate bacteremia with HA-1A and BPI. Ann Surg 220: 77–85

Romulo RL, Palardy JE, Opal SM (1993) Efficacy of anti-endotoxin monoclonal antibody E5 alone or in combination with ciprofloxacin in neutropenic rats with *Pseudomonas* sepsis. J Infect Dis 167: 126–130

Roth J, McClellan JL, Kluger MJ, Zeisberger E (1994) Attenuation of fever and release of cytokines after repeated injections of lipopolysaccharide in guinea-pigs. J Physiol (Lond) 477(1): 177–185

Roth RI, Tobias PS (1993) Lipopolysaccharide-binding proteins of *Limulus* amebocyte lysate. Infect Immun 61: 1033–1039

Rudbach JA, Myers KR, Rechtman DJ, Ulrich JT (1994) Prophylactic use of monophosphoryl lipid A in patients at risk for sepsis. Prog Clin Biol Res 388: 107–124

Rustici A, Velucchi M, Faggioni R, Sironi M, Ghezzi P, Quataert S, Green B, Porro M (1993) Molecular mapping and detoxification of the lipid A binding site by synthetic peptides. Science 259: 361–365

Saladino R, Garcia C, Thompson C, Hammer B, Parsonnet J, Novitsky T, Siber G, Fleisher G (1994) Efficacy of a recombinant endotoxin neutralizing protein in rabbits with *Escherichia coli* sepsis. Circ Shock 42: 104–110

Schumann RR, Leong SR, Flaggs GW, Gray PW, Wright SD, Mathison JC, Tobias PS, Ulevitch RJ (1990) Structure and function of lipopolysaccharide binding protein. Science 249: 1429–1431

Schütt C, Schilling T, Grunwald U, Schönfeld W, Krüger C (1992) Endotoxin-neutralizing capacity of soluble CD14. Res Immunol 143: 71–78

Siegel SA, Evans ME, Pollack M, Leone AO, Kinney CS, Tam SH, Daddona PE (1993) Antibiotics enhance binding by the human lipid A-reactive monoclonal antibody HA-1A to smooth gram-negative bacteria. Infect Immun 61: 512–519

Steinemann S, Ulevitch RJ, Mackman N (1994) Role of the lipopolysaccharide (LPS)-binding protein/CD14 pathway in LPS induction of tissue factor expression in monocytic cells. Arteriosclerosis Thromb 14: 1202–1209

Stoll BJ, Pollack M, Young LS, Koles NL, Gascon R, Pier GB (1986) Functionally active monoclonal antibody that recognizes an epitope on the O side chain of *Pseudomonas aeruginosa* immunotype-1 lipopolysaccharide. Infect Immun 53: 656–662

Storm DR, Rosenthal KS, Swanson PE (1977) Polymyxin and related peptide antibiotics. Annu Rev Biochem 46: 723–763

Takayama K, Qureshi N, Beutler B, Kirkland TN (1989) Diphosphoryl lipid A from *Rhodopseudomonas sphaeroides* ATCC 17023 blocks induction of cachectin in macrophages by lipopolysaccharide. Infect Immun 57: 1336–1338

Tanaka S, Nakamura T, Morita T, Iwanaga S (1982) Limulus anti-LPS factor: an anticoagulant which inhibits the endotoxin-mediated activation of limulus coagulation system. Biochem Biophys Res Commun 105: 717–723

Teng NNH, Kaplan HS, Hebert JM, Moore C, Douglas H, Wunderlich A, Braude AI (1985) Protection against gram-negative bacteremia and endotoxemia with human monoclonal IgM antibodies. Proc Natl Acad Sci USA 82: 1790–1794

Terashima M, Uezumi I, Tomio T, Kato M, Irie K, Okuda T, Yokota S, Noguchi H (1991) A protective human monoclonal antibody directed to the outer core region of *Pseudomonas aeruginosa* lipopolysaccharide. Infect Immum 59: 1–6

Theofan G, Horwitz AH, Williams RE, Liu P-S, Chan I, Birr C, Carroll SF, Mészáros K, Parent JB, Kasler H, Aberle S, Trown PW, Gazzano-Santoro H (1994) An amino-terminal fragment of human lipopolysaccharide-binding protein retains lipid A binding but not CD14-stimulatory activity. J Immunol 152: 3623–3629

Tobias PS, Ulevitch RJ (1993) Lipopolysaccharide binding protein and CD14 in LPS dependent macrophage activation. Immunobiology 187: 227–232

Tobias PS, Ulevitch RJ (1994) Lipopolysaccharide-binding protein and CD14 in the lipopolysaccharide-dependent activation of cells. Chest 105 Suppl 3: 48S–50S

Tobias PS, Mathison JC, Ulevitch RJ (1988) A family of lipopolysaccharide binding proteins involved in responses to gram-negative sepsis. J Biol Chem 263: 13479–13481

Tobias PS, Mathison J, Mintz D, Lee JD, Kravchenko V, Kato K, Pugin J, Ulevitch RJ (1992) Participation of lipopolysaccharide-binding protein in lipopolysaccharide-dependent macrophage activation. Am J Respir Cell Mol Biol 7: 239–245

Togari H, Mikawa M, Iwanaga T, Matsumoto N, Kawase A, Hagisawa M, Ogino T, Goto R, Watanabe I, Kito H, Ogawa Y, Wada Y (1983) Endotoxin clearance by exchange blood transfusion in septic shock neonates. Acta Paediatr Scand 72: 87–91

Tonoli M, Davies KA, Norsworthy PJ, Cohen J, Walport MJ (1993) The anti-lipid A antibody HA-1A binds to rough gram-negative bacteria, fixes complement and facilitates binding to erythrocyte CR1 (CD35). Clin Exp Immunol 92: 232–238

Ulevitch RJ, Johnston AR (1978) The modification of biophysical and endotoxic properties of bacterial lipopolysaccharides by serum. J Clin Invest 62: 1313–1324

Ulevitch RJ, Tobias PS (1994) Recognition of endotoxin by cells leading to transmembrane signaling. Curr Opin Immunol 6: 125–130

Ulevitch RJ, Johnston AR, Weinstein DB (1979) New function for high density lipoproteins. Their participation in intravascular reaction of bacterial lipopolysaccharides. J Clin Invest 64: 1516–1524

Ulmer AJ, Feist W, Heine H, Kirikae T, Kirikae F, Kusumoto S, Kusama T, Brade H, Schade U, Rietschel ET, Flad H-D (1992) Modulation of endotoxin-induced monokine release in human monocytes by lipid A partial structures that inhibit binding of ^{125}I-lipopolysaccharide. Infect Immun 60: 5145–5152

Ulmer AJ, Heine H, Feist W, Wang M-H, Loppnow H, Kirikae T, Kirikae F, Kusumoto S, Kusama T, Brade H, Schade U, Rietschel ET, Flad H-D (1994) Biological activity of lipid A partial structures. Prog Clin Biol Res 388: 71–83

Van Deuren M, Santman FW, van Dalen R, Sauerwein RW, Span LFR, van der Meer JWM (1992) Plasma and whole blood exchange in meningococcal sepsis. Clin Infect Dis 15: 424–430

Van Deuren M, van der Ven-Jongekrijg J, Demacker PNM, Bartelink AKM, van Dalen R, Sauerwein RW, Gallati H, Vannice JL, van der Meer JWM (1994) Differential expression of proinflammatory cytokines and their inhibitors during the course of meningococcal infections. J Infect Dis 169: 157–161

Van Deventer SJH, von der Mohlen MAM, Kimmings AN, Wedel NI, Lorenz TL, Nelson B, Hack CE, Friedmann NF (1994) rBPI$_{23}$ inhibits endotoxicity in humans. J Endotoxin Res 1 Suppl 1: 18

Vogel SN, Henricson BE (1990) Role of cytokines as mediators of endotoxin-induced manifestations: a comparison of cytokines induced by doses of LPS and MPL that elicit comparable early endotoxin tolerance. In: Nowotny A, Spitzer JJ, Ziegler EJ (eds) Cellular and molecular aspects of endotoxin reactions. Elsevier, Amsterdam, pp 465–474

Walterspiel JW, Kaplan SL, Mason EO Jr (1986) Protective effect of subinhibitory polymyxin B alone and in combination with ampicillin for overwhelming *Haemophilus influenzae* type B infection in the infant rat: evidence for in vivo and in vitro release of free endotoxin after ampicillin treatment. Pediatr Res 20: 237–241

Wang M-H, Flad H-D, Feist W, Brade H, Kusumoto S, Rietschel ET, Ulmer AJ (1991) Inhibition of endotoxin-induced interleukin-6 production by synthetic lipid A partial structures in human peripheral blood mononuclear cells. Infect Immun 59: 4655–4664

Wang M-H, Flad H-D, Feist W, Musehold J, Kusumoto S, Brade H, Gerdes J, Rietschel ET, Ulmer AJ (1992) Inhibition of endotoxin or lipid A-induced tumor necrosis factor production by synthetic lipid A partial structures in human peripheral blood mononuclear cells. Lymphokine Cytokine Res 11: 23

Warren HS, Danner RL, Munford RS (1992a) Anti-endotoxin monoclonal antibodies. N Engl J Med 326: 1153–1157

Warren HS, Glennon ML, Wainwright N, Amato SF, Black KM, Kirsch SJ, Riveau GR, Whyte RI, Zapol WM, Novitsky TJ (1992b) Binding and neutralization of endotoxin by Limulus antilipopolysaccharide factor. Infect Immun 60: 2506–2513

Warren HS, Amato SF, Fitting C, Black KM, Loiselle PM, Pasternack MS, Cavaillon J-M (1993) Assessment of ability of murine and human anti-lipid A monoclonal antibodies to bind and neutralize lipopolysaccharide. J Exp Med 177: 89–97

Weingarten R, Sklar LA, Mathison JC, Omidi S, Ainsworth T, Simon S, Ulevitch RJ, Tobias PS (1993) Interactions of lipopolysaccharide with neutrophils in blood via CD14. J Leukocyte Biol 53: 518–524

Weinstein SL, June CH, DeFranco AL (1993) Lipopolyssacharide-induced protein tyrosine phosphorylation in human macrophages is mediated by CD14. J Immunol 151: 3829–3838

Weiss J, Elsbach P, Shu C, Castillo J, Grinna L, Horwitz A, Theofan G (1992) Human bactericidal/permeability-increasing protein and a recombinant NH_2-terminal fragment cause killing of serum-resistant gram-negative bacteria in whole blood and inhibit tumor necrosis factor release induced by the bacteria. J Clin Invest 90: 1122–1130

Wenzel RP (1992) Anti-endotoxin monoclonal antibodies – a second look. N Engl J Med 326: 1151–1153

Wenzel RP, Bone R, Fein A, Quenzer R, Schentag J, Gorelick KJ, Wedel NI, Perl T (1991) Results of a second double-blind, randomized, controlled trial of antiendotoxin antibody E5 in gram-negative sepsis. 31st Interscience Conference on Antimicrobial Agents and Chemotherapy, American Society for Microbiology, Washington, DC, p 294

Westendorp RGJ, Brandt A, Thompson J, Dik H, Meinders AE (1990) Experiences with plasma- and leucapheresis in meningococcal septicaemia. Intensive Care Med 16: S102

Wilde CG, Seilhamer JJ, McGrogan M, Ashton N, Snable JL, Lane JC, Leong SR, Thornton MB, Miller KL, Scott RW, Marra MN (1994) Bactericidal/permeability-increasing protein and lipopolysaccharide (LPS)-binding protein. J Biol Chem 269: 17411–17416

Wood DM, Parent JB, Gazzano-Santoro H, Lim E, Pruyne PT, Watkins JM, Spoor ES, Reardan DT, Trown PW, Conlon PJ (1992) Reactivity of monoclonal antibody E5 with endotoxin. I. Binding to lipid A and rough lipopolysaccharides. Circ Shock 38: 55–62

Wortel CH, von der Möhlen MAM, van Deventer SJH, Sprung CL, Jastremski M, Lubbers MJ, Smith CR, Allen IE, ten Cate JW (1992) Effectiveness of human monoclonal anti-endotoxin antibody (HA-1A) in gram-negative sepsis: relationship to endotoxin and cytokine levels. J Infect Dis 166: 1367–1374

Worthen GS, Avdi N, Vukajlovich S, Tobias PS (1992) Neutrophil adherence induced by lipopolysaccharide in vitro. Role of plasma component interaction with lipopolysaccharide. J Clin Invest 90: 2526–2535

Wright SD (1991) CD14 and immune response to lipopolysaccharide. Science 252: 1321–1322

Wright SD, Ramos RA, Tobias PS, Ulevitch RJ, Mathison JC (1990) CD14, a receptor for complexes of lipopolysaccharide (LPS) and LPS binding protein. Science 249: 1431–1433

Wurfel MM, Kunitake ST, Lichenstein H, Kane JP, Wright SD (1994) Lipopolysaccharide (LPS)-binding protein is carried on lipoproteins and acts as a cofactor in the neutralization of LPS. J Exp Med 180: 1025–1035

Yang Z, Khemlani LS, Dean DF, Carter CD, Slauson DO, Bochsler PN (1994) Serum components enhance bacterial lipopolysaccharide-induced tissue factor expression and tumor necrosis factor-α secretion by bovine alveolar macrophages in vitro. J Leukocyte Biol 55: 483–488

Yasui K, Komiyama A, Molski TFP, Sha'afa RI (1994) Pentoxifylline and CD14 antibody additively inhibit priming of polymorphonuclear leukocytes for enhanced release of superoxide by lipopolysaccharide: possible mechanism of these actions. Infect Immum 62: 922–927

Ziegler EJ, Fisher CJ, Sprung CL, Straube RC, Sadoff JC, Foulke JC, Wortel CH, Fink MP, Dellinger RP, Teng NNH, Allen IE, Berger HJ, Knatterud GL, Lobuglio AF, Smith CR, the HA-1A Study Group (1991) Treatment of gram-negative bacteremia and septic shock with HA-1A human monoclonal antibody against endotoxin. N Engl J Med 324: 429–436

Ziegler EJ, Smith CR (1992) Anti-endotoxin monoclonal antibodies. N Engl J Med 326: 1165

Zuckerman SH, Qureshi N (1992) In vivo inhibition of lipopolysaccharide-induced lethality and tumor necrosis factor synthesis by *Rhodobacter sphaeroides* diphosphoryl lipid A is dependent on corticosterone induction. Infect Immun 60: 2582–2587

Anticytokine Strategies for the Treatment of Septic Shock: Relevance of Animal Models

D. HEUMANN and M.P. GLAUSER

1 Introduction . . . 299
2 Experimental Models . . . 301
2.1 TNF Blockade in Animal Models . . . 301
2.1.1 Anti-TNF Antibodies and Parenteral (or Intraperitoneal) LPS Challenge . . . 301
2.1.2 Anti-TNF Antibodies and Parenteral Bacterial Challenge . . . 302
2.1.3 TNF Blockade by Means of Soluble TNF Receptors . . . 302
2.2 Blockade of IL-1 by Means of IL-1 Receptor Antagonist . . . 303
2.3 Cytokine Blockade in Models of Focal Tissue Infections . . . 304
3 Comparison of Parenteral and Tissue Models of Infection . . . 305
4 Clinical Studies . . . 306
4.1 Anti-TNF Antibodies . . . 306
4.1.1 The CB0006 Trial . . . 306
4.1.2 The Bay X1351 Trials (NORASEPT Trial) . . . 307
4.1.3 The MAK 195F Antibody . . . 307
4.2 Soluble TNF Receptor Constructs . . . 307
4.3 IL-1 Receptor Antagonist . . . 308
5 Differences Between Animal Models and Clinical Trials . . . 308
6 Conclusion . . . 309
References . . . 309

1 Introduction

For many years septic shock has been described as a complication of gram-negative bacteremia, and initial efforts at devising therapeutic approaches focused on developing antibodies directed against endotoxin (LPS), the major toxic component of gram-negative bacteria. Because recent advances in immunology have greatly contributed to our understanding of the pathophysiology of septic shock and acute inflammatory response initiated not only by LPS but also by other microbial products, many novel approaches for the treatment of septic shock have been developed in addition to anti-LPS antibodies.

Invading bacteria and bacterial products trigger within the host the release of a complex array of cytokines. Normally cytokines are believed to be autocrine

Division of Infectious Diseases, Department of Internal Medicine, 1040 Lausanne, Switzerland

and paracrine (cell to cell) molecules that act locally at their site of production to control the host response to invading organisms. In fact, by influencing coagulation and leukocyte transmigration and by activating professional phagocytes cytokines assist the host to contain a local infection. It is generally recognized that in septic shock an overproduction of cytokines generates a systemic activation which affects vascular permeability and resistance, cardiac function, and induces many metabolic derangements that may result in tissue necrosis, leading eventually to multiple-organ failure and death.

The concept that multiple-organ failure is related to an uncontrolled systemic inflammatory state originated mainly from animal studies. Central to the process was the discovery that macrophages activated by bacterial products release proinflammatory molecules, especially tumor necrosis factor (TNF) and interleukin (IL)-1. In fact the sole macrophage activation with its subsequent cytokine release or the injection of recombinant cytokines can produce experimentally in animals a syndrome indistinguishable from the response to severe infections. In animal models of infection the cascade of events starting from the initial infectious stimulus and usually ending with the death of the animal follows a very predictible time course. In many animal models the course of septic shock usually lasts hours or a few days. The cytokine production that results from infectious challenges also follow a predictible time course, so that experimental

Table 1. Protection over time by anti-TNF antibodies in various models of endotoxemia and intravenous injections of live bacteria

Reference	Model	Time of administration of anti-TNF antibodies (minutes) relative to challenge				
		<120	–60 to –30	0	+30 to +60	>120
Endotoxemia						
Beutler et al. (1985)	Mouse, i.p. LPS, 16 mg/kg	+		±		–
Mathison (1988)	Rabbit, i.v. LPS, 10 mg/kg		+			
Fiedler et al. (1991)	Rhesus, i.v. LPS/GAL				+	
Bagby et al. (1991)	Rat, i.v. LPS, 1 mg/kg	+	+			
Eskandari et al. (1992)	Mouse, i.p. LPS, 100 mg/kg	–		–		
Zanetti et al. (1992)	Mouse, i.p. LPS/GAL			+		
Emerson et al. (1992)	Baboon, i.v. LPS, 2mg/kg		+			
Jin et al. (1994)	Rat, i.v. LPS, 7.5 mg/kg					
IV injection of live bacteria						
Tracey et al. (1987)	Baboon, i.v. *E. coli*	+	±			
Hinshaw et al. (1990)	Baboon, i.v. *E. coli*				+	
Silva et al (1990)	Mouse, i.v. *E. coli*	+			+	–
Jesmok et al. (1992)	Pig, i.v. *E. coli*			+		
Schlag et al. (1994)	Baboon, i.v. *E. coli*					+
Silva et al. (1990)	Mouse, i.v. *Klebsiella*	–				
Opal et al. (1991)	Rat, oral *P. aeruginosa*			±		
Hinshaw et al. (1992)	Baboon, i.v. *S. aureus*				+	
Martin et al. (1993)	Mouse, i.v. *S. pyogenes*		–			

GAL, Mice sensitized with D-galactosammine; +, efficacy; ±, relative efficacy; –, lack of efficacy.

approaches at blocking one cytokine or another are relatively straightforward. In contrast, the sequence of events leading to septic shock in humans is more complex. The micro-organisms and the sites of infections are very diverse. The course of the illness in patients generally takes days rather than hours, as seen in most animal models. Moreover, most patients with septic shock have a large variety of underlying diseases, which are not present in animal studies. For all these reasons it is difficult to extrapolate from observations in straightforward animal models to the very complex situations in patients.

The current strategy to attenuate the detrimental effect of cytokines in human sepsis is based essentially on observations from animal studies. These studies have led to hypotheses that should now be validated in clinical studies planned or currently under way. The present review considers animal studies that have led to the concept of immunotherapy to limit overproduction of proinflammatory cytokines (TNF, IL-1) and focuses particularly on the experimental conditions. We suggest that conclusions drawn from observations in animal models may have somewhat limited applicability to the human sepsis/septic syndrome, thus stressing the need for the careful performance and analysis of the results of clinical trials.

2 Experimental Models

2.1 TNF Blockade in Animal Models

TNF blockade has been investigated in numerous animal models. Table 1 summarizes the studies that have evaluated the protective efficacy of the anti-TNF antibodies in various animals challenged parenterally with either LPS or gram-negative bacteria. In a few studies gram-positive bacteria were also evaluated and are included in this review.

2.1.1 Anti-TNF Antibodies and Parenteral (or Intraperitoneal) LPS Challenge

When LPS was the challenging agent, administered either intravenously (i.v.) or intraperitoneally (i.p.) (Beutler et al. 1985; Mathison et al. 1988; Fiedler et al. 1991; Bagby et al. 1991; Zanetti et al. 1992; Emerson et al. 1992), protection occurred in most experiments, provided the anti-TNF antibodies were given before or during infusion of LPS. Failure to protect the animals by simultaneous injection of LPS and anti-TNF antibodies was reported only once (Eskandari et al. 1992). However, this failure of the anti-TNF antibody may be related to the very high concentration of LPS used in that experiment, i.e., 100 mg/kg (Eskandari et al. 1992). A poor increase of survival was observed in another report (Jin et al. 1994). With regard to delaying TNF blockade with anti-TNF antibodies after LPS challenge, it is inefficient in endotoxic models. Only one observation of relative success is reported (Beutler et al. 1985), while most failures are probably not reported.

2.1.2 Anti-TNF Antibodies and Parenteral Bacterial Challenge

Several studies reported in Table 1 have investigated TNF blockade after i.v. injections of bacteria. Following i.v. challenge of *Escherichia coli* (Tracey et al. 1987; Hinshaw et al. 1990; Silva et al. 1990; Jesmok et al. 1992), anti-TNF antibody administration was shown to be protective up to 30 min after the infusion of bacteria in various animal species. A delayed administration given 2 h after challenge did not increase survival in mice (Silva et al. 1990), whereas in a recent study conducted in baboons a modest protective effect of the anti-TNF antibody was observed when given up to 4 h after bacterial challenge (50% protection over 100% death in controls) (Schlag et al. 1994). Upon challenge with gram-negative species other than *E. coli* treatment with anti-TNF antibodies partially protected neutropenic rats from an oral challenge with *Pseudomonas aeruginosa* (Opal et al. 1991) but did not protect mice from *Klebsiella* infections (Silva et al. 1990). Finally, after i.v. challenge with gram-positive organisms one study reported protection (Hinshaw et al. 1992) and the other not (Martin et al. 1993).

Taken together, the studies of anti-TNF antibody administration after parenteral live bacterial challenge have shown that protection was not uniformly present, and that early treatment was generally associated with a better protection. Information on delayed therapy for many models is lacking, but it is generally recognized that delayed treatment, as after i.v. LPS challenge, is less efficient than early treatment or treatment given before bacterial challenge.

2.1.3 TNF Blockade by Means of Soluble TNF Receptors

A more recent approach for TNF blockade is based on soluble TNF receptors. Two distinct forms of TNF receptors (a p55 form, and a p75 form) are shed from cell surfaces during inflammatory reactions. The TNF soluble receptor shedding is believed to be a regulatory mechanism that prevents circulating TNF to interact with cellular TNF receptors, thus attenuating the effects of excessive TNF production. Fusion proteins have been developed in which each of the two receptors was linked with hinge regions of the human IgG. This dimeric construction extends the half-life of the molecule in vivo and increases the affinity for TNF. Four variants have been constructed using IgG1 of IgG3 as the partner: sTNFR-IgG1 p75, sTNFR-IgG3 p75, sTNFR-IgG1 p55, and sTNFR-IgG3 p55.

These variants have been investigated in several animals (Table 2). All published experiments but one were conducted in mice, most models being parenteral LPS infections. With respect to the sTNFR p55 construct, protection was very efficient when the construct was given prophylactically, and a variable effect was observed depending on the Ig fusion partner (Lesslauer et al. 1991; Ashkenazi et al. 1991; Loetscher et al. 1993; Evans et al. 1994; Jin et al. 1994).

Table 2. Protection over time by soluble TNF receptors in various models of endotoxemia and of iv injections of bacteria

Reference	Model	Time of administration of constructs (minutes) relative to challenge				
		<120	–60 to –30	0	+30 to +60	>120
p55						
LESSLAUER et al. (1991)	Mouse i.p. LPS, IgG3, p55	+			+	±
ASHKENAZI et al. (1991)	Mouse, i.v. LPS, IGI, p55		+		–	
LOETSCHER et al.(1993)	Mouse LPS, IgG3, p55	+			+	±
JIN et al. (1994)	Rat, i.v. LPS, IgG1, p55		+			
EVANS et al. (1994)	Mouse, i.v. *E. coli*, IgG1, p55		+			–
p75						
MOHLER et al. (1993)	Mouse, i.v. LPS, IgG1, p75			+	+	±
LOETSCHER et al. (1993)	Mouse, i.p. LPS, IgG3, p75	–				
EVANS et al. (1994)	Mouse, i.v. *E. coli*, IgG1, p75		(±)			

+, Efficacy; ±, relative efficacy; –, lack of efficacy.

Similarly, the sTNFR p55 construct appeared effective in a model of i.v. bacterial challenge (EVANS et al. 1994). With respect to the sTNFR75 the IgG1 construct was found protective in a model of bolus LPS injection (MOHLER et al. 1993). After a lethal challenge of live *E. coli* it delayed lethality without influencing overall death (EVANS et al. 1994). The sTNFR-IgG3 p75 construct was not protective in parenteral LPS injection model (LOETSCHER et al. 1993) and was not tested after live bacterial challenge.

Taken together, these experiments showed a promising effect of the sTNFR 55 constructs in limiting TNF toxicity, with a superior efficacy of the construct over conventional anti-TNF antibody in one study (JIN et al. 1994). In some models treatment with the sTNFR 55 could be delayed several hours, but this was not true for all models. In contrast, the potential use of the sTNFR 75 constructs appeared more limited. This difference in efficacy may be related to different functions and TNF-binding affinities of the two receptors that are not yet fully understood.

2.2 Blockade of IL-1 by Means of IL-1 Receptor Antagonist

An alternative strategy aimed at attenuating the effects of overproduction of cytokine release investigated blocking of IL-1 by means of the IL-1 receptor antagonist (IL-1ra). Indeed, the IL-1ra inhibits IL-1 by competing with IL-1 for cell receptor sites. Experimentally, IL-1ra has been shown to improve various hemodynamic parameters in models of parenteral LPS or bacterial challenges and to improve survival (FISCHER et al. 1992; WAKABAYASHI et al. 1991; ALEXANDER et al. 1991; OHLSSON et al. 1990; AIURA et al. 1993; Table 3). Due to the short half-life of

Table 3. Effect of interleukin 1 receptor antagonist in various parenteral models of endotoxemia and of i.v. or s.c. injection of bacteria

Reference	Model	IL-1ra (injections)	Effect
OHLSSON et al. (1990)	Rabbit, i.v. LPS	T_0, then every 2 h for 24 h	Increased survival
ALEXANDER et al. (1991)	Mouse, i.p. LPS	$T_{+20\ min}$, then every 4 h for 24 h	Increased survival
FISCHER et al. (1992)	Baboon, i.v. LPS	T_0, then continuous	No effect
WAKABAYASHI et al. (1991)	Rabbit, i.v. *E. coli*	$T_{-15\ min}$, then continuous	Increased survival and hemodynamics
FISCHER et al. (1992)	Baboon, i.v. *E. coli*	T_0, then continuous	Increased survival and hemodynamics
MANCILLA et al. (1993)	Rat, s.c. *K. pneumoniae*	at T_0 at T_0 and T_{24}	increased survival Decreased survival
AIURA et al. (1993)	Rabbit, i.v. *S. epidermidis*	$T_{-5\ min}$, then continuous	Improved hemodynamics

IL-1ra it must be given in multiple repeated injections or in continuous infusion to be efficacious. Importantly, one report suggested a potential deleterious effect of IL-1 blockade. While a single injection of IL-1ra reduced the lethality of *Klebsiella pneumoniae* sepsis in newborn rabbits, and a second repeated injection 24 h after challenge enhanced lethality (MANCILLA et al. 1993).

2.3 Cytokine Blockade in Models of Focal Tissue Infections

While TNF and IL1 blockade have demonstrated efficacy in increasing survival following parenteral LPS or bacterial challenges, several studies have shown no effect or even a deleterious effect in models of local infections with gram-negative bacteria. In murine or rat models of peritonitis, including cecal/ligature puncture (CLP; ECHTENACHER et al. 1990; ESKANDARI et al. 1992; EVANS et al. 1989; ZANETTI et al. 1992; BAGBY et al. 1991), anti-TNF antibodies failed to afford protection, even when given prophylactically (Table 4). Similarly, soluble TNFR p55 construct was ineffective in murine of peritonitis (LOETSCHER et al. 1993). Only one study reported a protective effect in a rat model of peritonitis caused by *Neisseria meningitidis* (NASSIF et al. 1992). One study reported a transient early survival in rat pups pretreated with anti-TNF antibodies and infected i.p. with group B streptococci (TETI et al. 1993). However, by 96 h this protection was no longer apparent. No study reported the effect of delayed treatment with anti-TNF antibodies in these models. The effect of IL-1 blockade has not been reported in peritonitis models or in the CLP model.

Thus studies aimed at investigating the effect of TNF blockade in focal infections have suggested the blocking TNF is not or is only marginally effective in preventing death and is even deleterious in some experiments.

Table 4. Effect of anti-TNF antibodies and of soluble TNF receptor constructs in various models of focal infections

Reference	Model	Observations
Anti-TNF antibodies		
Evans et al. (1989)	Mouse, CLP	No effect
Echtenacher et al. (1990)	Mouse, CLP	Deleterious
Bagby et al. (1991)	Rat, peritonitis *E. coli*	No effect
Eskandari et al. (1992)	Mouse, CLP	No effect (deleterious)
Zanetti et al. (1992)	Mouse, peritonitis *E. coli*	No effect
Nassif et al. (1992)	Rat, peritonitis *N. meningitidis*	Protective
Teti et al. (1993)	Rat, peritonitis Group B *streptococcus*	delay in lethality
TNFR-IgG3, p55		
Loetscher et al. (1993)	Mouse, i.p. *E. coli*	No effect

3 Comparison of Parenteral and Tissue Models of Infection

When evaluating the potential efficacy of anticytokine strategies, a clear distinction emerges between models of parenteral LPS or live bacterial challenge and models of focal infection.

On the one hand, parenteral models of LPS challenge are to be considered as models of intoxication rather than true models of infections. With regard to live bacterial parenteral challenge model inocula used in most studies have ranged from between 10^9 and 10^{12} CFU/kg. Both the parenteral LPS challenge and the high bacterial numbers induce extremely high concentrations of cytokine (usually 20–150 ng/ml TNF). Such high concentrations of TNF are probably toxic for the host, as suggested by the observations after parenteral challenge which have shown that prophylactic TNF blockade protects against death. TNF response peaks 1 and 2 h following LPS or bacterial challenge, and is followed by a rapid decline despite the persistance of circulating LPS or of live bacteria (Fig. 1a). This rapid decline may explain the failure of delayed TNF blockade after bacterial challenge when TNF is no longer present in the circulation.

On the other hand, models of tissue infection in which bacterial inocula are lower (usually less than 10^6 CFU) are perhaps better resemble true infectious models because bacteria must multiply to invade tissues and eventually bloodstream. The time course of cytokine production is different than that after parenteral challenge. Figure 1b illustrates this in mice challenged i.p. with *E. coli*. TNF levels are at least ten times lower than after parenteral challenge (usually less than 1 ng/ml) and in strong contrast to parenteral challenge models are sustained during the whole observation period. In fact, cytokine profiles in these models of tissue infection are very reminiscent of profiles observed in patients with shock. Indeed, TNF levels in these patients are of low magnitude (usually less than 500 pg/ml) and sustained (up to 10 days after onset of shock (Calandra et al. 1990; Damas et al. 1989).

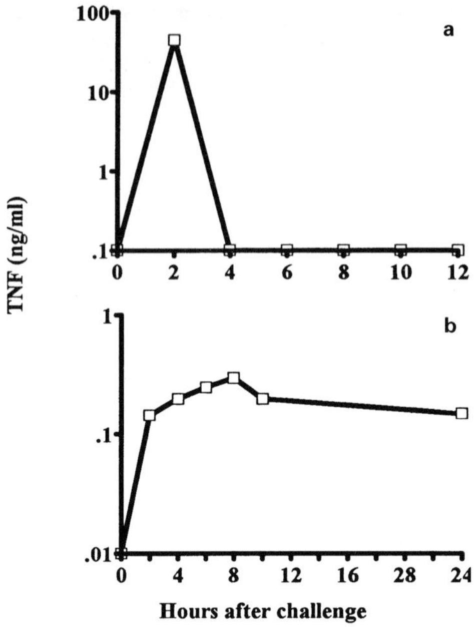

Fig. 1. TNF measured in plasma at various times following challenge. **a** Mice were injected i.v. with 20 mg/kg LPS. **b** Mice were injected i.p. with 10^4 CFU of *E. coli* O111 mixed with mucin/hemoglobin (this inoculum represents the LD_{100} for this strain). (Adapted from ZANETTI et al. 1992)

Given the fact that TNF blockade failed in the models of focal infection or even worsened the prognosis, this might be interpreted as a suggestion that TNF does not play a toxic role in these conditions but rather helps fight infection. Alternatively, one could also postulate that TNF blockade is not efficacious in tissues due to the low penetration of blocking antibodies, allowing TNF to play a detrimental role locally. Indeed, while in parenteral models it has generally been well documented that the anti-TNF antibodies used were administered in sufficiently large does to neutralize circulating TNF levels, similar observations have usually not been made in models of focal infections.

4 Clinical Studies

4.1 Anti-TNF Antibodies

4.1.1 The CB0006 Trial

Results of an open-label, prospective, phase II multicenter trial with escalating doses of an anti-TNF murine mAb (CB0006) have been published (FISHER et al. 1993). A total of 80 patients with severe sepsis or septic shock were enrolled to receive four dosing regimens of CB0006: 0.1, 1, 10 mg/kg, and 1 mg/kg initially and a second dose of 1 mg/kg 24 h later. No placebo group was investigated.

While survival estimates of all patients failed to demonstrate a survival advantage with increasing doses of CB0006, subgroup analysis of those patients with high TNF levels (>50 pg/ml) did suggest a benefit of the intervention.

4.1.2 The Bay X1351 Trials (NORASEPT Trial)

Patients with sepsis syndrome were enrolled, prospectively stratified into shock and nonshock groups, and randomly assigned to receive either 15 or 7.5 mg/kg or a murine anti-TNF mAb, or placebo. The final enrollment was 971 patients, with 49% in shock and 51% in nonshock at randomization (Abraham et al. 1995). Among all infused patients there was no difference in all-cause mortality at 28 days in patients receiving placebo or antibody. Among septic shock patients there was no reduction of mortality in patients receiving anti-TNF antibody. However, a trend toward reduction of mortality at 3 days was observed (15 mg/kg: 44% reduction vs. placebo, p=0.01; 7.5 mg/kg: 49% reduction *vs.* placebo, p=0.004), which was no longer present at 28 days. In contrast, there was a nonstatistically trend toward increased mortality in nonshock patients who received the 15 mg/kg anti-TNF mAb dose regimen.

The Bay X1351 monoclonal antibody has been studied in a European–south African study, the INTERSEPT trial. This study, which randomized only half as many patients as was previously planned, demonstrated no overall effect on mortality but a significant recovery from shock in those patients who survived. The Bay X 1351 monoclonal antibody will be tested in a further trial, NORASEPT II.

4.1.3 The MAK 195F Antibody

The MAK 195F antibody, developed by Knoll, is a Fab fragment of a monoclonal antibody directed against TNF. The results of a phase II study involving fewer than 100 patients showed no effect on overall mortality but apparently a dose-dependent effect on patients with IL-6 levels at the time of randomization of more than 1000 ng/ml. This post hoc finding will be prospectively studied in a phase III trial now underway.

4.2 Soluble TNF Receptor Constructs

A randomized, double-blind trial in patients with sepsis syndrome and hypotension was conducted with escalating doses of sTNFR p75 construct (Agosti et al. 1994). A total of 141 patients were enrolled to receive either placebo or one of three doses of the fusion protein. Inhibition of TNF with the construct was not effective in preventing death in patients at 28 days and increasing doses of the construct were apparently associated with increased mortality. A second trial with the sTNFR p55 construct is under investigation.

4.3 IL-1 Receptor Antagonist

Three trials aimed at blocking IL-1 by means of IL-1ra have been completed. In a preliminary phase II trial 99 patients with sepsis syndrome or septic shock were treated with placebo or escalating doses of IL-1ra (FISHER et al. 1994b). A dose-dependent, 28 day survival benefit was suggested with increasing doses of the drug, but the size of each tratment group was relatively small. A double-blind, phase III confirmatory trial of Il-1ra was undertaken in 893 patients with sepsis or septic shock. While the prospectively defined endpoints did not confirm the preliminary conclusions (FISHER et al. 1994a), since the overall mortality rate in patients receiving placebo was 34% and 29% in patients receiving high-dose IL-1ra, a post hoc analysis suggested a possible survival advantage in patients with increased severity of illness (FISHER et al. 1994ab). A second large phase III study aimed at confirming prospectively this observation was stopped after interim analysis because this effect was not confirmed.

In conclusion, retrospective analyses of the studies aimed at blocking TNF or IL-1 suggest that the interventions benefit only those patients more severely ill (as suggested by the IL-1ra study) or in shock (Intersept study) and/or that the benefit is present early after intervention but vanishes over time and is no longer apparent after 28 days (Norasept study).

5 Differences Between Animal Models and Clinical Trials

Despite great hopes, clinical trials conducted so far in human sepsis have not been as encouraging as animal studies. Several causes may explain this relative and hopefully temporary disappointment.

1. In animal models, the cascade of events starting from the initial stimulus and usually ending with the death of the animal generally follows a predictable time course. The resulting cytokine production also follows a predictible time course, and experimental protocols to block one cytokine cascade or another are relatively straightforward. In many animal models the course of septic shock is usually hours or a few days. In contrast, the sequence of events leading to septic shock in humans is much more complex, and the course of the illness generally lasts days rather than hours, as seen in most animal models.
2. The rationale of clinical trials was based essentially on observations performed in animals receiving parenteral injections of LPS/bacteria, in which cytokine blockade was beneficial, not on observations performed in animal with focal infections, in which such therapy was usually not effcient.
3. The organisms and site of infections are very diverse, and importantly the patients have a large variety of underlying diseases, which is not the case in animal studies.

4. Interventions in animal models have been successful only when applied very early. Indeed, cytokine blockade was efficient in parenteral models, provided blockade was performed prophylactically or very early after challenge.

6 Conclusion

There is one major problem in trying to delineate the best potential interventions of cytokine blockade for the treatment of septic shock, which is the uncertainties about which experimental model best mimics the series of events which lead to septic shock in patients. Simply showing that an antibody can protect mice from a LPS-induced shock is only a crude hint, which wrongly leads one to conclude that the same antibody will be useful in the clinical setting. Notwithstanding, clinical trials based on anticytokine strategies have been based essentially on the lessons learned from models of parenteral injections of LPS or bacteria, not from models of focal infections. With the exception of meningococcal purpura fulminans, most cases of septic shock in human are due to focal infections, a type of infection in which TNF appears to play a protective role in experimental models.

Given the uncertainties of the relevance of animal models in mimicking and predicting outcome in the clinical setting, our understanding of the sole cytokine blockade as a mean of reducing mortality in septic shock relies only on well-planned, well-performed, and well-interpreted clinical studies.

References

Abraham E, Wunderick R, Sliverman H, Perl TM, Nasraway S, Levy H, Bone R, Wenzel RP, Balk R, Allred R, Pennington JE, Wherry JC (1995) Efficacy and safety of monoclonal antibody to human tumor necrosis factor alpha in patients with sepsis syndrome. A randomized, controlled, double-blind, multicenter clinical trial. JAMA 273: 934–941

Agosti JM, Fisher CJ, Opal SM, Lowry SF, Balk RA, Sadoff JC, the STNFR Sepsis Study Group (1994) Treatment of patients with sepsis syndrome with soluble TNF receptors (sTNFR) (Abstr). 34th Interscience Conference on Antimicrobial and Chemotherapy, p 65

Aiura K, Gelfand JA, Burke JF, Thompson RC, Dinarello CA (1993) Interleukin-1 (IL-1) receptor antagonist prevents Staphylococcus epidermidis-induced hypotension and reduces circulating levels of tumor necrosis factor and IL-1 beta in rabbits. Infect Immun 61: 3342–3350

Alexander HR, Doherty GM, Buresh CM, Venzon DJ, Norton JA (1991) A recombinant human receptor antagonist to interleukin 1 improves survival after lethal endotoxemia in mice. J Exp Med 173: 1029–1032

Ashkenazi A, Marsters SA, Capon DJ, Chamow SM, Figari IS, Pennica D, Goeddel DV, Palladino MA Jr, Smith DH (1991) Protection against endotoxin shock by a tumor necrosis factor receptor immunoadhesin. Proc Natl Acad Sci USA 88: 10535–19539

Bagby GJ, Plessala KJ, Wilson LA, Thompson KJ, Nelson S (1991) Divergent efficacy of antibody to tumor necrosis factor-α in intravascular and peritonitis models of sepsis. J Infect Dis 163: 83–88

Beutler B, Milsark IW, Cerami A (1985) Passive immunization against cachectin/tumor necrosis factor protects mice from lethal effect of endotoxin. Science 229: 869–871

Calandra T, Baumgartner JD, Grau GE, Wu MM, Lambert PH, Schellekens J, Verhoef J, Glauser MP, the Swiss-Dutch J5 Study Group (1990) Prognostic values of tumor necrosis factor/cachectin, interleukin-1, alpha-interferon and gamma-interferon in the serum of patients with septic shock. J Infect Dis 161: 982–987

Damas P, Reuter A, Gysen P, Demonty J, Lamy M, Franchimont P (1989) Tumor necrosis factor and interleukin-1 serum levels during severe sepsis in humans. Crit Care Med 17: 975–978

Echtenacher B, Falk W, Männel DN, Krammer PH (1990) Requirement of endogenous tumor necrosis factor/cachectin for recovery from experimental peritonitis. J Immunol 145: 3762–3766

Emerson TE, Lindsey DC, Jesmok GJ, Duerr ML, Fournel MA (1992) Efficacy of monoclonal antibody against tumor necrosis factor alpha in an endotoxemic baboon model. Circ Shock 38: 75–84

Eskandari MK, Bolgos G, Miller C, Nguyen DT, DeForge LE, Remick DG (1992) Anti-tumor necrosis factor antibody therapy fails to prevent lethality after cecal ligation and puncture or endotoxemia. J Immunol 148: 2724–2730

Evans GF, Synder YM, Butler LD, Zuckerman SH (1989) Differential expression of interleukin-1 and tumor necrosis factor in murine septic shock models, Circ Shock 29: 279–290

Evans TJ, Moyes D, Carpenter A, Martin R, Loetscher H, Lesslauer W, Cohen J (1994) Protective effect of 55- but not 75-kD soluble tumor necrosis factor receptor-immunoglobulin G fusion proteins in an animal model of gram-negative sepsis. J Exp Med 180: 2173–2179

Fiedler VB, Loof I, Sander E, Voehringer V, Galanos C, Fournel MA (1991) Monoclonal antibody to tumor necrosis factor-alpha prevents lethal endotoxin sepsis in adult rhesus monkeys. J Lab Clin Med 120: 574–588

Fischer E, Marano MA, Van Zee KJ, Rock CS, Hawes AS, Thompson WA, DeForge L, Kenney JS, Remick DG, Bloedow DC, Thompson RC, Lowry SF, Moldawer LL (1992) Interleukin-1 receptor blockade improves survival and hemodynamic performance in Escherichia coli septic shock, but fails to alter host responses to sublethal endotoxemia. J Clin Invest 89: 1551–1557

Fisher CJ, Opal SM, Dhainaut JF, Stephens S, Zimmerman JL, Nightingale P, Harris SJ, Schein RMH, Panacek EA, Vincent JL, Foulke GE, Warren EL, Gartrard C, Park G, Bodmer MW, Cohen J, Van der Linden C, Cross AS, Sadoff JC, the CB006 Sepsis Syndrome Study Group (1993) Influence of an anti-tumor necrosis factor monoclonal antibody on cytokine levels in patients with sepsis. Crit Care Med 21: 318327

Fisher CJ, Dhainaut JF, Opal SM, Pribble JP, Balk RA, Slotman GJ, Iberti TJ, Rackow EC, Shapiro MJ, Greenman RL, Reines HD, Shelly MP, Thompson BW, LaBrecque JF, Catalano MA, Knaus WA, Sadoff JC, for the Phase III rhIL-1ra Syndrome Study Group (1994a) Recombinant human interleukin 1 receptor antagonist in the treatment of patients with sepsis syndrome. Results from a randomized, double-blind, placebo-controlled trial. JAMA 271: 1836–1843

Fisher CJ, Slotman GJ, Opal SM, Pribble JP, Bone RC, Emmanuel G, Ng D, Bloedow DC, Catalano MA, the I1-ra Sepsis Syndrome Study Group (1994b) Initial evaluation of human recombinant interleukin-1 receptor anatagonist in the treatment od sepsis syndrome: a randomized, open-label, placebo-controlled multicenter trial. Crit Care Med 22: 12–21

Hinshaw LB, Tekamp-Olson P, Chang ACK, Lee PA, Taylor FB, Murray CK, Peer GT, Emerson TE, Passey RB, Kuo GC (1990) Survival of primates in LD_{100} septic shock following therapy with antibody to tumor necrosis factor (TNF alpha). Circ Shock 30: 279–292

Hinshaw LB, Emerson TE Jr, Taylor FB, Chang ACK, Duerr M, Peer GT, Flournoy DJ, White GL, Kosanke SD, Murray CK, Xu R, Passey RB, Fournel MA (1992) Lethal Staphylococcus aureus-induced shock in primates: prevention of death with anti-TNF antibody. J Trauma 33: 668–573

Jesmok G, Lindsey C, Duerr M, Fournel M, Emerson T (1992) Efficacy of monoclonal antibody against human recombinant tumor necrosis factor in E. coli-challenged swine. Am J Pathol 141: 1197–1207

Jin H, Yang R, Marsters SA, Bunting SA, Wurm FM, Chamow SM, Ashkenazi A (1994) Protection against rat endotoxic shock by p55 tumor necrosis factor (TNF) receptor immunoadhesin: comparison with anti-TNF monoclonal antibody. J Infect Dis 170: 1323–1326

Lesslauer W, Tabuchi H, Gentz R, Brockhaus M, Schlaeger EJ, Grau GE, Piguet PF, Pointaire P, Vassalli P, Loetscher H (1991) Recombinant soluble tumor necrosis factor receptor proteins protect mice from lipopolysaccharide-induced lethality. Eur J Immunol 21: 2883–2886

Loetscher H, Angehrn P, Schlaeger EJ, Gentz R, Lesslauer W (1993) Efficacy of a chimeric TNFR-IgG fusion protein to inhibit TNF activity in animal models of septic shock. In: Levin J, Alving CR,

Munford RS, Stütz PL (eds) Bacterial endotoxin: recognition and effector mechanisms. Elsevier, Amsterdam, pp 455–462
Mancilla J, Garcia P, Dinarello CA (1993) The interleukin-1 receptor antagonist can either reduce or enhance the lethality of Klebsiella pneumoniae sepsis in newborn rats. Infect Immun 61: 926–932
Martin RA, Silva AT, Cohen J (1993) Effect of anti-TNF-alpha treatment in an antibiotic treated murine model of shock due to Streptococcus pyogenes. FEMS Microbiol Lett 110: 175–178
Mathison JC, Wolfson E, Ulevitch RJ (1988) Participation of tumor necrosis factor in the mediation of bacterial lipopolysaccharide-induced injury in rabbits. J Clin Invest 81: 1925–1937
Mohler KM, Torrance DS, Smith CA, Goodwin RG, Sremler KE, Fung VP, Madani H, Widmer MB (1993) Soluble tumor necrosis factor (TNF) receptors are effective therapeutic agents in lethal endotoxemia and function simultaneously as both TNF carriers and TNF antagonists. J Immunol 151: 1548–1561
Nassif X, Mathison JC, Wolfson E, Koziol JA, Ulevitch RJ, So M (1992) Tumour necrosis factor alpha antibody protects against lethal meningococcaemia. Mol Microbiol 6: 591–597
Ohlsson K, Björk P, Bergenfeldt M, Hageman R, Thompson RC (1990) Interleukin-1 receptor antagonist reduces mortality from endotoxin shock. Nature 348: 550–552
Opal SM, Cross AS, Sadoff JC, Collins HH, Kelly NM, Victor GH, Palardy GE, Bodmer MW (1991) Efficacy of antilipopolysaccharide and anti-tumor necrosis factor monoclonal antibodies in a neutropenic rat model of Pseudomonas sepsis. J Clin Invest 88: 885–890
Schalg G, Redl H, Davies J, Haller I (1994) Anti-tumor necrosis factor antibody treatment of recurrent bacteremia in a baboon model. Shock 2: 10–18
Silva AT, Bayston KF, Cohen J (1990) Prophylactic and therapeutic effects of a monoclonal antibody to tumor necrosis factor-α in experimental gram-negative shock. J Infect Dis 162: 421–427
Teti G, Manucuso G, Tomasello F (1993) Cytokine appearance and effects of anti-tumor necrosis factor alpha antibodies in a neonatal rat model of group B streptococcal infection. Infect Immun 61: 227–235
Tracey KJ, Fong Y, Hesse DG, Manogue KR, Lee AT, Kuo GC, Lowry SF, Cerami A (1987) Anti-cachectin/TNF monoclonal antibodies prevent septic shock during lethal bacteraemia. Nature 330: 662–664
Wakabayashi G, Gelfand JA, Burke JF, Thompson RC, Dinarello CA (1991) A specific receptor antagonist for interleukin 1 prevents Escherichi coli-induced shock in rabbits. FASEB J 5: 338–343
Zanetti, G, Heumann D, Gerain J, Kohler J, Abbet P, Barras C, Glauser MP, Baumgartner JD (1992) Cytokine production after intravenous of peritoneal Gram negative bacterial challenge in mice. Comparative protective efficacy of antibodies to TN and to LPS. J Immunol 148: 1890–1897

Subject Index

N-acetylglucosamine 19
N-acetylmuramic acid 19
alpha$_1$-acid glycoprotein (AGP) 176, 179, 180
acute pancreatitis 210
acute phase
– proteins 177, 179, 222, 228
– reactant 18
acute pulmonary failure 113
adhesion molecules 49, 170, 175, 178, 221, 253
– endothelial 21
– leukocyte 21
ADP-ribosylation 55
adrenal thrombi 29
adult respiratory distress syndrome (ARDS) 196, 213–215, 217, 218, 222–226, 228, 284
AGP *see* α_1-acid glycoprotein
AIDS 96
albumin 18, 20
allopurinol 252
alveolar arterial gradient 20
N-aminoarginine 199
anaphylatoxins 223, 224
– C3a 26
– C4a 26
– C5a 26
anergic T cells, exhaustive proliferation 91, 92
anergyl 89
angiotensin II 252
antagonistic
– mechanism 58
– preparations 58
anti-core LPS monoclonal antibody 246
anti-endotoxin antibodies 61
anti-inflammatory approach 269
anti-lipid A antibodies 62
anti-LPS 253
anti-O-chain antibodies 62
anti-TNF antibodies 253, 300–302, 305, 306
antibiotics 18, 20
antibodies
– anti-core LPS monoclonal 246
– anti-endotoxin 61
– anti-lipid A 62
– anti-O-chain 62
– anti-TNF 253, 300–302, 305, 306
– cross-protective 42
– lipopolysaccharide 276
– – antibacterial 277
– – antiendotoxic functions 277
– monoclonal (MAbs) 26, 247, 278, 279, 284
– – anti-core antibodies 66–68
– – anti-lipid A antibodies 63
– – lipid A-specific 278
– polyclonal anti-core antibodies 64
– protective function 277
– TNF-α neutralizing 24
antimicrobial treatment 22
antioxidants 228
antithrombin III 28
α_1-antitrypsin 226
APC 91
apolipoprotein A-1 285
apoptosis 89
arachidonic acid metabolites 30
ARDS *see* adult respiratory distress syndrome
L-arginine 252
assay, *Limulus* amebocyte lysate 21
autodestructive inflammation 270
autolysis, bacterial 18
avidity 94

B cells (lymphocytes) 40, 48, 263
– knock-out mice 91
– activated 91
B7 92, 95
bacille calmette-Guérin 174
bacteremia/endotoxemia 241–245
– baboons, measurements in 242
– bacterial counts 242
– endotoxin 243
– experimental, in humans 153
– gram-negative 22, 23
– gram-positive 19
– hemorrhage 243
– histological examinations 241
– hypovolemic-traumatic shock 242
– IL-6 242

bacteremia/endotoxemia
– overwhelming 25, 27
– plasma endotoxin 242
– portal circulation 241
– reticuloendothelial system 243
– rodents 241
– subhuman primate model 241
– systemic circulation 241
– tumor necrosis factor (TNF) 243, 244
bacteria
– gram-negative 17, 20
– gram-positive 17, 19
bacterial
– autolysis 18
– counts 242
– lysis 18
– outer membrane 18
– translocation 228
bactericidal permeability increasing protein (BPI) 18, 47, 246, 247, 253, 276
– recombinant (rBPI) 280, 281
Bacteroides fragilis 45
BAL *see* bronchoalveolar lavage
benzylpenicillin 18
bile salt 251
bioassay 24
blood
– coagulation (*see* coagulation)
– cultures, positive 248
– flow 246
– – maldistribution of microcirculatory 227
– – redistribution 228
– – splanchic 228
– lactate 227
– platelets 120, 123
body temperature 20
BPI (*see* bactericidal permeability increasing protein)
bradykinin 21, 26, 137, 224
bronchoalveolar lavage (BAL) fluid 222, 224
burn injury 210, 245
– bacterial/endotoxin translocation 245
– enteric bacilli 245
– plasma endotoxin 249

C3a 223
C4a 223
C5a 223, 225
Caco-2 251
calpain 54
CAP *see* cationic antimicrobial proteins
capillary
– leakage 61
– permeability 20
Capnocytophaga canimorsus 23
cardiac
– arrest 212, 213
– index 10
– output 20, 26
cardiovascular failure 214, 215
cathepsins 223, 225
cationic antimicrobial proteins (CAP) 276
– CAP7 281
– CAP18 281
– CAP18$^{104-135}$ 281
– CAP37 281
– CAP57 280
CD11b/CD18 225
CD14 18, 49, 51–54, 108
mCD14 282–284
sCD14 283, 284
– recombinant 284
CD28 95
cell
– death, programmed 111
– lysis 277
– wall, bacterial 26
centoxin 63
central nervous system 10
– dysfunction (failure) 215, 216, 218
ceramide 137
cGMP 193, 202
chloramphenicol 18
chloroquine 253
Chromobacterium violaceum 43
chylomicrons 285
circulation
– hyperdynamic 20, 217
– portal 241
– systemic 241
circulatory
– failure 23, 216
– shock 216
clusterin 27
coagulation 10, 124, 125, 170, 221, 224
– disseminated intravascular (DIC) 61, 123, 213, 215, 216, 219, 224, 226
– extrinsic pathway 220
– factor
– – V 28
– – VII 28
– – X 28
– – XI 21
– – XII 21
– system 21, 25, 28
– – extrinsic pathway 28
– – intrinsic pathway 28
coagulopathy 21
collagenase 60, 223, 225
colony-stimulating factor 61
complement 26, 220, 222, 229
– activation factors 27
– alternative pathway 27
– C1-esterase inhibitor 27
– C3 27
– C4 activation products 27
– C5b-9 27
– classical pathway 27
– factors 27
contact system 220, 222, 224

corticosteroids 143, 144, 193
COX-1 138
COX-2 138
creatinine 29
cross-protective antibodies 42
CsA *see* cyclosporin A
CSAIDs *see* cytokine-suppressing anti-inflammatory drugs
cyclic GTP 190
cyclo-oxygenase inhibitors 145
cyclosporin A (CsA) 87
cytokine(s) 2, 6, 7, 9, 11, 21, 23, 133–158, 220, 221, 223, 225, 226, 228, 241–245, 248
– antiinflammatory 25
– bioactive 27
– blockade 304
– proinflammatory 24, 25, 31, 135, 192–194, 198
– receptors, soluble 248
cytotoxic therapy 22
cytokine-suppressing anti-inflammatory drugs (CSAIDs) 148

defensins 18
DIC *see* disseminated intravascular coagulation
dietary ω-3 fatty acids 144
dietary fibers 251
digestive tract 247
disseminated intravascular coagulation (DIC) (*see* also coagulation) 61, 123, 213, 215, 216, 219, 224, 226
dysgonic fermenter 2 23

E5 63, 278, 279
edema, pulmonary 21, 114
EDRF *see* endothelium-derived relaxing factor
eicosanoids 109
ELAM 141
ELAM-1 221, 225
elastase 60, 223, 225, 226
elastase-α_1-antitrypsin complexes 226
encephalopathy, septic 218
β-endorphin 30
endothelial
– cells 26, 28, 29, 48, 49, 106, 120, 220–222, 224–226
– inflammation 220
– leukocyte adhesion molecules (ELAM) 225
endothelin 218, 221
endothelin-1 222
endothelium 220, 221, 223, 225, 226
endothelium-derived relaxing factor (EDRF) 190, 193
endotoxic conformation 46
endotoxin (*see also* lipopolysaccharide) 4, 5, 8, 26, 40, 89, 122, 243, 248, 299
– antagonists 59
– neutralization, intraluminal 250
– translocation 228
– units 20
– plasma 242
– neutralizing protein 50
endotoxinemia 22, 29
enteral nutrition 252
enterobacteriaceae 248
enterotoxins, staphylococcal 83
epidermal growth factor 252
epinephrine 21
erythrocytes 21
Escherichia coli 19, 20, 24, 26, 28, 30, 31, 41, 248, 282, 286, 287
– hemolysin 102
exchange transfusion 288
exotoxins 17, 19, 89
experimental endotoxemia in humans 153
extracorporal lipopolysaccharide adsorption 288
extrinsic pathway of coagulation 220

factor
– XI 224
– XII 224
– XIIa 225
α-Fas 174
fever 5, 139, 141
fibrin
– formation, disseminated intravascular 28
– monomeric 28
– slit products 28
fibrinogen 28
fibrinolysis 28, 124, 221, 224
fibrinolytic system 21, 25, 220
fibrinopeptid A 28

G proteins 19, 106
GalN 174
galactosamine 88, 172
gastric intramucosal pH 228
gastrointestinal failure 215
Vβ-gene segments 85
general malaise 20
generalized autodestructive inflammation 220
generalized inflammatory response 210, 220, 221
glucocorticoids 169, 172, 177, 180
– receptors 87
glutamine 252
glycan 19
gp130 signal transducer 150
gram-negative
– bacteria 51
– sepsis 61
granulocyte-macrophage colony-stimulating factors 222, 225
granulocytes 46, 106
– neutrophilic 120, 121
guanylate cyclase 179, 193
gut
– barrier failure 239, 240, 245
– ischemia 228

gut
– mucosal barrier 250
– permeability 249
– wall 251
gut-associated macrophages 247

HA-1A 278, 279
Haemophilus influenzae 16, 23, 43
Hageman factor (FXII) 28
HDL *see* lipoprotein, high-density
headache 20
heart rate 20
hematological
– failure 218
– malignancies 22
hematopoietic growth factors 222
hemodynamic alterations 6
hemoglobin 47
hemorrhage 61, 243
hepatic failure 21, 213, 215, 219
hepatitis 174
hepatobiliary dysfunction 215
hepatocytes 194
hepatotoxicity 168
high molecular weight kininogen 21, 28, 224
high-density lipoprotein (HDL) *see* lipoprotein
histamine 223
horse shoe crab 29
host defense dysfunction 259–271
host inflammatory response 211
humoral mediator cascades 220
hydrogen peroxide 191, 228
(R)-3-hydroxyfatty acids 43
hydroxyl radical 48, 191, 196
13-hydroxyoctadecadienoic acid 56
hyperdynamic circulation 217
hyperlactatemia 8
hypermetabolism 217
hypotension 21, 61, 141, 168, 171, 178
hypovolemic
– insults 240
– shock 210, 212, 242
hypoxanthine 228
hypoxia 23, 28, 29, 245
– tissue 253

ICAM-1 141, 221, 225
ICAM-1$^{0/0}$ 175
ICE 146
IFN *see* interferon
IgA 251
IL *see* interleukin
immunointervention 269
infection models 305
inflammation
– autodestructive 270
– endothelial 220
inflammatory
– host response 220
– reaction 220
– response 220, 221, 223
iNOS 88
inositol 1,4,5-triphosphate 54, 106
insulin-like growth factor 1 252
intensive care unit 22
interferon(s) (IFN) 151
– γ 86, 134, 174, 175, 177, 178, 180, 192, 197, 198
interferons 151
interleukin(s) 220
– converting enzyme 112, 176, 180
– IL-1 22, 60, 135–158, 172, 175, 176, 180, 192, 221–223, 225, 228, 288
– – α 86
– – β 112, 198
– – blockade 303
– – effects on the vascular wall 142
– – hypotensive effect 141
– – receptor antagonist 21, 152, 153, 176, 303, 304
– – – clinical trials 154
– – receptors, soluble 155
– – signal transduction 137
– IL-2 86, 134, 197
– – receptors 89
– IL-4 25, 86, 149, 169, 176, 193, 197
– IL-5 197
– IL-6 21, 24, 27, 60, 86, 142, 143, 150, 175–177, 221–223, 242, 244, 286, 287
– – receptor 108
– IL-8 21, 24, 27, 60, 176, 221, 222, 223, 225, 228
– IL-10 25, 86, 92, 149, 169, 176, 222
– IL-12 174, 177, 179, 180
– IL-13 25, 149
intraluminal endotoxin neutralization 250
intraperitoneal lipopolysaccharide challenge 301
iron-sulfur centres 196
ischemia 226–229
splanchic 2, 3, 228

jaundice 219

kallikrein 225
– inhibitor 26
kallikrein-kinin system 25
Kawasaki syndrome 96
kidney 11
– thrombi 29
kininogen, high molecular weight 21, 28
Klebsiella pneumoniae 42
Kupffer cell 253
lactate 4, 6, 8, 227
– blood 227
– levels 4
lactic
– acid 29, 227
– acidosis 171
lactoferrin 18, 47, 226, 251

lactulose 251
lavage, bronchioalveolar 20
LBP *see* lipopolysaccharide binding protein
LDL *see* lipoprotein
Legionella pneumophilia 41
lethal septicemia 89
leukemia 22
– inhibiting factor 24
leukocytes 120, 221, 223
– adhesion 253
leukocytosis 21, 61, 218, 219
leukopenia 21, 40, 218, 219
leukotriene(s) 48, 220, 223, 225, 226
limiting-dilution analysis 93
Limulus amebocyte lysate assay 21
Limulus polyphemus (LALF) 29, 50, 281
linoleic acid 57
lipid
– mediators 109
– peroxidation 252
– peroxides 228
lipid A 18, 19, 26, 40, 41, 43, 44, 57, 59
– physical structure 45
lipopolysaccharide(s) (*see also* endotoxin) 17, 40, 41, 87, 134, 246
– adsorption, extracorporal 288
– antibodies 276
– – antibacterial 277
– – antiendotoxic functions 277
– challenge, human 26, 29, 30, 31
– *Escherichia coli* 19, 24, 26, 28, 30, 31
– *Neisseria meningitidis* 24, 25, 27
– physical structure 45
– physicochemical state in plasma 18
– *Pseudomonas fluorescence* 21
– *Pseudomonas pseudomallei* 26
– receptors 49
– releasing effect 18
– *Salmonella abortus equi* 30
– *Salmonella minnesota* 21
– schematic structure 41
lipopolysaccharide-binding protein (LBP) 18, 47, 275
lipoprotein
– high-density (HDL) 18, 276, 285
– – reconstituted 285
– low-density (LDL) 18, 47, 276, 285
– – very low-density (VLDL) 18, 285
– serum 285
lipoteichoic acid 193
lipotropin 30
lipoxygenase(s) 55
– inhibition 144
Listeria monocytogenes 92
liver 10
– dysfunction 215, 219
– failure 213
– thrombi 30
low-density lipoprotein *see* lipoprotein
lpr/lpr mice 90
LPS *see* lipopolysaccharides
lung 9
– thrombi 29
lymphocytes 46
lymphokines (*see also* IL) 86
lysis, bacterial 18
lysozyme 47

MAbs *see* antibodies, monoclonal
α_2-macroglobulin 226
macrophages 48, 194, 195, 220, 221, 222, 228, 245
malaise, general 20
maldistribution of microcirculatory blood flow 227
malignancies, hematological 22
MCP-1 225
Mel-14 86
membrane attack complex 18
meningococcal
– infections 18
– patients 27
meningococcemia, fulminant 23, 25, 29
metabolic acidosis 61
metalloproteinase 169
methylene blue 178–180, 202
methylisothiourea sulfate 195
MHC class II
– deficient mice fail 87
– molecules 83
microorganisms 40
microvascular
– injury 223
– permeability 227
– protein leakage 226
mitochondria 170
mitogen-activated protein kinases 54
MODS *see* multiple organ dysfunction syndrome
MOF *see* multiple organ failure
monoclonal antibodies *see* antibodies
monocyte chemoatractant protein 1 (MCP-1) 222
monocytes 19, 25, 28, 48, 105, 263, 264
monocytes/macrophages 46, 277
monocytic inflammatory potential, down-regulation 264–267
monokines 48
monomeric fibrin 28
mononuclear
– cells 122
– – peripheral 24
– leukocytes 261
monophosphoryl lipid A (MPL) 287
mortality 21, 22
multiple organ dysfunction syndrome (MODS) 16, 210, 211, 216
multiple organ failure (MOF) 209–229
– clinical manifestations 216–219

multiple organ failure (MOF)
- definitions 211
- evolution 215
- incidence 213–215
- mortality 212, 213
- pathogeneses 219–229
- pathophysiology 219–229
- prevalence 213–215
- risk factors 212, 213
multiple organ system failure 210
multiple system(s) organ failure 210
myalgia 20
myocardial
- depressant factor 31, 198
- depression 7, 198
- performance 23
myocardium 24
- ATP-metabolism 31
- blood flow 31
- circulation 31
- contractility 31
- ejection fraction 31

naloxone 30
L-NAME 194, 201
nausea 20
Neisseria meningitidis 16, 18, 22, 24, 25, 27, 43
neurological dysfunction 215
neutropenic patients 22
neutrophil
- adherence molecules 225
- adhesion 128
- rolling 127
neutrophil-endothelial cell interactions 126
neutrophilic granulocytes 120, 121
neutrophils 21, 220, 222–225
NF-κB 283
- inhibited by antibodies to LBP or CD14 283
NH-LBP 284
Nippostrongylus brasiliensis 93
nitric oxide (NO) 26, 31, 48, 140, 170, 175, 178, 189–203, 221, 222, 252
- donors 252
- production 129
- synthases 191
N^G-nitro-L-arginine methylester 88
N^ω-nitro-L-arginine 201, 202
nitroprusside 252
nitrotyrosine 196
L-NMMA 194, 201, 202
NO *see* nitric oxide
non-Hodgkin lymphoma 22
nonspecific immune functions 267, 268
nosocomial infections 250
nuclear factor kB 19
nutrition, parenteral 252

O-specific
- chain 41
- immunoglobulins 62

oliguria 217
opiate receptor antagonist 30
opsonins 19
organ dysfunction (failure) 9–11, 211, 213–215
- criteria 212
- descriptors 211, 212
- hematological 214
- hepatic 214
- indicators 211
- neurological 214
- parameters for quantifying 10
- renal 214
- respiratory 214
- sequence 215
organ perfusion, altered 7
outer membrane vesicles (OMV) 18
oxygen
- consumption 2
- delivery 2, 9
- extraction 226, 227
- free radicals 221, 228, 252
- radicals 23, 222, 225, 226, 228
- uptake, pathological supply dependency 227

PAF 222, 226
PAI *see* plasminogen activator inhibitor
pancreatitis, acute 210
PAR 223
parenteral
- bacterial challenge 302
- lipopolysaccharide challenge 301
- nutrition 252
pathological supply dependency of oxygen uptake 227
pentoxifylline 169
peptidoglycan 17, 19, 26
- receptor 19
permeability 221
- capillary 20
- transcapillary 26
peroxynitrite 196
pertussis toxin 19, 106
PGE2 138
pH, lowered 29
phagocytosis 47
phosphoinositide hydrolysis 111
phospholipase(s) 109
- A_2 30
plasma 21
- contact system 21, 26
- endotoxin 242
- platelet-rich 22
plasmapheresis 288
plasmin 29
plasmin-antiplasmin complexes 224
plasminogen 29
activator inhibitor (PAI) 21, 29, 221, 224
platelet(s) 28
- activating factor 30, 48, 106, 220

pneumolysin 102
polyclonal anti-core antibodies 64
polymorphonuclear leukocytes 47
polymyxin B 247, 249, 253, 282
polysaccharide 41
polythylene glycol 249
polytrauma 215
polyuria 217
pore-forming bacterial toxins 101–115
pores, transmembrane 102
portal
– blood 248
– circulation 241
– vein, TNF concentration 244
prekallikrein 21, 28, 224
pro-inflammatory cytokines 192–194, 198
processing enzymes 146
procoagulant 170
– state 28
programmed cell death 111
prostacyclin 222
prostaglandins 48, 220, 223, 229
– E_2 169, 222, 225
– I_2 222
prostanoids 138
proteases 60, 223
protein C 28
protein kinases 19
proteinases 221
proteinuria 217
Proteus mirabilis 42
prothrombin 28
activation peptice 28
prothrombinase complex 28
Pseudomonas
– *fluorescence* 21
– *pseudomallei* 26
pulmonary
– edema 21, 114, 217
– failure, acute 113
– function 20
– microvascular injury 217
– permeability 226
– vasculature 107
pyrogenicity 40
pyruvate dehydrogenase 227

radioprotection 136
rBPI (*see* bactericidal permealbility increasing protein)
reactive oxygen species 223, 228
receptor(s)
– antagonist 25
– solulable 21, 25
renal
– dysfunction 217, 218
– failure 213
– – acute 215, 216, 218
repair process 113
reperfusion 226–229
respiratory
– burst 106
– failure 215, 216
reticuloendothelial system 243
rheumatoid arthritis 96
Rhodobacter sphaeriodes diphosphoryl lipid A (rsDPLA) 286–288
Rhodopseudomonas viridis 45
rodents 241
rods, gram-negative 20
RTX family 102
RU 486 87
RU38486 172, 175, 176

SAg-induced lethal shock 93
Salmonella
– *abortus equi* 30
– *enterica* 41
– *minnesota* 21
scavanger receptor 49
SD_{50} 94
SDD *see* selective decontamination, digestive tract
SDZ880.431 286
SEA *see Staphylococcus aureus* enterotoxin A
SEB *see* staphylococcal enterotoxin B
SEB-specific cytolytic activity 91
SEC *see* staphylococcal enterotoxin C
selectin 175
selective decontamination, digestive tract
(SDD) 247, 250
sensitization 174
sepsis 212, 213, 218, 220, 222, 225–227
– cellular response 119–130
– clinical trials 214
– gram-negative 22, 61
– more severe forms 213
– signs 5, 6
– syndrome 22, 213, 214, 221
septic
– encephalopathy 218
shock *see* shock, septic
septicemia, lethal 89
sertonin 223
serum lipoproteins 285
shock 213, 215, 228
– baboon model 26, 28, 29
– canine 30
– circulatory 216
– hypovolemic 212
– hypovolemic-traumatic 242
– irreversible 40
– plasma 25
– porcine 30
– septic 1–11, 16, 17, 22, 23–27, 30, 213, 214, 216
– – anticytokine strategies for treatment 299–309
– – definition 2

shock
– – epidemiology 2
– – incidence 3
– – microbiology 3
– – organ dysfunction (failure) 9–11
– – pathophysiological cascade 60
– – prognosis 11
– signs 6–9
Shwartzman reaction 136
signal transduction 54
SIN-1 252
SIRS *see* systemic inflammatory response syndrome
skeletal muscle pO_2 227
sleepiness 141
smooth muscle(s) 23
– cells 49
soluble
– cytokine receptors 248
– guanylate cyclase 190
splanchic
– area, hypoperfusion 30
– blood flow 228
– ischemia 2, 3, 228
– oxygen demand 228
– tissues 228
spleen 19
– thrombi 19
staphylococcal enterotoxin(s) 83
– B (SEB) 84
– C (SEC) 85
Staphylococcus aureus 19, 102
– enterotoxin A (SEA) 84
streptococcal pyrogenic exotoxins 20
Streptococcus
– *pneumoniae* 16, 19
– *pyogenes* 16
streptolysin-O 102, 106
subhuman primate model 241
superantigens 20, 83–96, 197
superoxide 191, 196, 222
– anions 48, 196
supply dependency 227
surgery 210
synergism 114
systemic circulation 241
systemic inflammatory
– host response 219
– response 222
– – syndrome 213, 249

T-cell 48, 261, 262
– activation 96
– anergy 85
– dependent toxic lymphokine syndrome 85
– receptor (TCR) 85
– – antigen specifity 85
– – β-chain 85
– – threshold (SD_{50}) 95
– shock 85
T-lymphocytes *see* T-cells
tachycardia 5, 61
Tachypleus tridentatus (TALF) 281
tachypnea 5
TCR *see* T cell receptor
teichoid acid 17, 19
terminal complement complex 27, 224
TGFβ (*see* transforming growth factor)
thalidomide 169
thermal injury 252
thrombin 28
antithrombin complexes 28, 224
thrombocytopenia 114, 216, 219
thrombomodulin 221
thromboxane 48
A_2 218, 226, 246
– – receptor blocker 251
tissue
– factor 221, 224
– – monocytes 21, 29
– hypoxia 253
– plasminogen activator 21, 29
TNF *see* tumor necrosis factor
tolerance 175
tone, vascular 28
toxic shock syndrome toxin 19
– type 1 (TSST-1) 87, 193
α-toxin 102
– staphylococcal 104, 105
transcapillary
– flux 21
– permeability 26
transferrin 18, 47
transforming growth factor β 25, 149, 176, 193
translocation 248–250
– of bacteria 3
– intervention strategies 250
transmembrane pores 102
trauma 210, 226
traumtic insults 240
TSST-1 *see* toxic shock syndrome toxin type 1
tumor necrosis factor (TNF) 60, 86, 135–158, 167–181, 220–223, 225, 228, 243, 244, 246–249, 253, 282, 284, 286–288
– α 21, 24, 25, 27, 86, 192, 197, 198, 283, 284
– – inhibited by antibodies to LBP or CD14 283
– – receptors, soluble 24
– blockade 301, 302
– receptors 248
– – soluble 156, 244, 302, 303
– neutralizing antibodies to 157
– systemic circulation 244
tyrosine phosphorylation 54

uric acid 29

vascular
– cells 46
– endothelium 124

– permeability 225
– resistance, peripheral 20
– smooth muscle 193, 194
– tone 28
– – reduced 5
vasculature, altered 7
vasoactive intestinal peptide 27, 30
vasodilator 30
VCAM-1 85, 221
ventricles 31
very low-density lipoprotein (VLDL) 18
– /chylomicrons 276
vessels, resistance 23
virulence factor 42
vitronectin 27
vitronectin-thrombin-antithrombin complex 27
VLDL *see* lipoprotein, very low-density

white blood cell count 5

xanthine dehydrogenase 228
xanthine oxidase 228, 252

Current Topics in Microbiology and Immunology

Volumes published since 1989 (and still available)

Vol. 176: **Holland, John J. (Ed.):** Genetic Diversity of RNA Viruses. 1992. 34 figs. IX, 226 pp. ISBN 3-540-54652-9

Vol. 177: **Müller-Sieburg, Christa; Torok-Storb, Beverly; Visser, Jan; Storb, Rainer (Eds.):** Hematopoietic Stem Cells. 1992. 18 figs. XIII, 143 pp. ISBN 3-540-54531-X

Vol. 178: **Parker, Charles J. (Ed.):** Membrane Defenses Against Attack by Complement and Perforins. 1992. 26 figs. VIII, 188 pp. ISBN 3-540-54653-7

Vol. 179: **Rouse, Barry T. (Ed.):** Herpes Simplex Virus. 1992. 9 figs. X, 180 pp. ISBN 3-540-55066-6

Vol. 180: **Sansonetti, P. J. (Ed.):** Pathogenesis of Shigellosis. 1992. 15 figs. X, 143 pp. ISBN 3-540-55058-5

Vol. 181: **Russell, Stephen W.; Gordon, Siamon (Eds.):** Macrophage Biology and Activation. 1992. 42 figs. IX, 299 pp. ISBN 3-540-55293-6

Vol. 182: **Potter, Michael; Melchers, Fritz (Eds.):** Mechanisms in B-Cell Neoplasia. 1992. 188 figs. XX, 499 pp. ISBN 3-540-55658-3

Vol. 183: **Dimmock, Nigel J.:** Neutralization of Animal Viruses. 1993. 10 figs. VII, 149 pp. ISBN 3-540-56030-0

Vol. 184: **Dunon, Dominique; Mackay, Charles R.; Imhof, Beat A. (Eds.):** Adhesion in Leukocyte Homing and Differentiation. 1993. 37 figs. IX, 260 pp. ISBN 3-540-56756-9

Vol. 185: **Ramig, Robert F. (Ed.):** Rotaviruses. 1994. 37 figs. X, 380 pp. ISBN 3-540-56761-5

Vol. 186: **zur Hausen, Harald (Ed.):** Human Pathogenic Papillomaviruses. 1994. 37 figs. XIII, 274 pp. ISBN 3-540-57193-0

Vol. 187: **Rupprecht, Charles E.; Dietzschold, Bernhard; Koprowski, Hilary (Eds.):** Lyssaviruses. 1994. 50 figs. IX, 352 pp. ISBN 3-540-57194-9

Vol. 188: **Letvin, Norman L.; Desrosiers, Ronald C. (Eds.):** Simian Immunodeficiency Virus. 1994. 37 figs. X, 240 pp. ISBN 3-540-57274-0

Vol. 189: **Oldstone, Michael B. A. (Ed.):** Cytotoxic T-Lymphocytes in Human Viral and Malaria Infections. 1994. 37 figs. IX, 210 pp. ISBN 3-540-57259-7

Vol. 190: **Koprowski, Hilary; Lipkin, W. Ian (Eds.):** Borna Disease. 1995. 33 figs. IX, 134 pp. ISBN 3-540-57388-7

Vol. 191: **ter Meulen, Volker; Billeter, Martin A. (Eds.):** Measles Virus. 1995. 23 figs. IX, 196 pp. ISBN 3-540-57389-5

Vol. 192: **Dangl, Jeffrey L. (Ed.):** Bacterial Pathogenesis of Plants and Animals. 1994. 41 figs. IX, 343 pp. ISBN 3-540-57391-7

Vol. 193: **Chen, Irvin S. Y.; Koprowski, Hilary; Srinivasan, Alagarsamy; Vogt, Peter K. (Eds.):** Transacting Functions of Human Retroviruses. 1995. 49 figs. IX, 240 pp. ISBN 3-540-57901-X

Vol. 194: **Potter, Michael; Melchers, Fritz (Eds.):** Mechanisms in B-cell Neoplasia. 1995. 152 figs. XXV, 458 pp. ISBN 3-540-58447-1

Vol. 195: **Montecucco, Cesare (Ed.):** Clostridial Neurotoxins. 1995. 28 figs. XI., 278 pp. ISBN 3-540-58452-8

Vol. 196: **Koprowski, Hilary; Maeda, Hiroshi (Eds.):** The Role of Nitric Oxide in Physiology and Pathophysiology. 1995. 21 figs. IX, 90 pp. ISBN 3-540-58214-2

Vol. 197: **Meyer, Peter (Ed.):** Gene Silencing in Higher Plants and Related Phenomena in Other Eukaryotes. 1995. 17 figs. IX, 232 pp. ISBN 3-540-58236-3

Vol. 198: **Griffiths, Gillian M.; Tschopp, Jürg (Eds.):** Pathways for Cytolysis. 1995. 45 figs. IX, 224 pp. ISBN 3-540-58725-X

Vol. 199/I: **Doerfler, Walter; Böhm, Petra (Eds.):** The Molecular Repertoire of Adenoviruses I. 1995. 51 figs. XIII, 280 pp. ISBN 3-540-58828-0

Vol. 199/II: **Doerfler, Walter; Böhm, Petra (Eds.):** The Molecular Repertoire of Adenoviruses II. 1995. 36 figs. XIII, 278 pp. ISBN 3-540-58829-9

Vol. 199/III: **Doerfler, Walter; Böhm, Petra (Eds.):** The Molecular Repertoire of Adenoviruses III. 1995. 51 figs. XIII, 310 pp. ISBN 3-540-58987-2

Vol. 200: **Kroemer, Guido; Martinez-A., Carlos (Eds.):** Apoptosis in Immunology. 1995. 14 figs. XI, 242 pp. ISBN 3-540-58756-X

Vol. 201: **Kosco-Vilbois, Marie H. (Ed.):** An Antigen Depository of the Immune System: Follicular Dendritic Cells. 1995. 39 figs. IX, 209 pp. ISBN 3-540-59013-7

Vol. 202: **Oldstone, Michael B. A.; Vitković, Ljubiša (Eds.):** HIV and Dementia. 1995. 40 figs. XIII, 279 pp. ISBN 3-540-59117-6

Vol. 203: **Sarnow, Peter (Ed.):** Cap-Independent Translation. 1995. 31 figs. XI, 183 pp. ISBN 3-540-59121-4

Vol. 204: **Saedler, Heinz; Gierl, Alfons (Eds.):** Transposable Elements. 1995. 42 figs. IX, 234 pp. ISBN 3-540-59342-X

Vol. 205: **Littman, Dan R. (Ed.):** The CD4 Molecule. 1995. 29 figs. XIII, 182 pp. ISBN 3-540-59344-6

Vol. 206: **Chisari, Francis V.; Oldstone, Michael B. A. (Eds.):** Transgenic Models of Human Viral and Immunological Disease. 1995. 53 figs. XI, 345 pp. ISBN 3-540-59341-1

Vol. 207: **Prusiner, Stanley B. (Ed.):** Prions Prions Prions. 1995. 42 figs. VII, 163 pp. ISBN 3-540-59343-8

Vol. 208: **Farnham, Peggy J. (Ed.):** Transcriptional Control of Cell Growth. 1995. 17 figs. IX, 141 pp. ISBN 3-540-60113-9

Vol. 209: **Miller, Virginia L. (Ed.):** Bacterial Invasiveness. 1996. 16 figs. IX, 115 pp. ISBN 3-540-60065-5

Vol. 210: **Potter, Michael; Rose, Noel R. (Eds.):** Immunology of Silicones. 1996. 136 figs. XX, 430 pp. ISBN 3-540-60272-0

Vol. 211: **Wolff, Linda; Perkins, Archibald S. (Eds.):** Molecular Aspects of Myeloid Stem Cell Development. 1996. 98 figs. XIV, 298 pp. ISBN 3-540-60414-6

Vol. 212: **Vainio, Olli; Imhof, Beat A. (Eds.):** Immunology and Developmental Biology of the Chicken. 1996. 43 figs. IX, 281 pp. ISBN 3-540-60585-1

Vol. 213/I: **Günthert, Ursula; Birchmeier, Walter (Eds.):** Attempts to Understand Metastasis Formation I. 1996. 35 figs. XV, 293 pp. ISBN 3-540-60680-7

Vol. 213/II: **Günthert, Ursula; Birchmeier, Walter (Eds.):** Attempts to Understand Metastasis Formation II. 1996. 33 figs. XV, 288 pp. ISBN 3-540-60681-5

Vol. 213/III: **Günthert, Ursula; Schlag, Peter M.; Birchmeier, Walter (Eds.):** Attempts to Understand Metastasis Formation III. 1996. 14 figs. XV, 262 pp. ISBN 3-540-60682-3

Vol. 214: **Kräusslich, Hans-Georg (Ed.):** Morphogenesis and Maturation of Retroviruses. 1996. 34 figs. XI, 344 pp. ISBN 3-540-60928-8

Vol. 215: **Shinnick, Thomas M. (Ed.):** Tuberculosis. 1996. 44 figs. XII, 307 pp. ISBN 3-540-60985-7

Springer-Verlag and the Environment

Printing: Saladruck, Berlin
Binding: Buchbinderei Lüderitz & Bauer, Berlin